X-Ray Fluorescence Spectrometry

CHEMICAL ANALYSIS

A SERIES OF MONOGRAPHS ON ANALYTICAL CHEMISTRY
AND ITS APPLICATIONS

Editor
J. D. WINEFORDNER

VOLUME 152

A WILEY-INTERSCIENCE PUBLICATION

JOHN WILEY & SONS, INC.

New York / Chichester / Weinheim / Brisbane / Singapore / Toronto

X-Ray Fluorescence Spectrometry

Second Edition

RON JENKINS

International Centre for Diffraction Data,
Newtown Square, PA

A WILEY-INTERSCIENCE PUBLICATION

JOHN WILEY & SONS, INC.

New York / Chichester / Weinheim / Brisbane / Singapore / Toronto

This book is printed on acid-free paper.

Published simultaneously in Canada.

Library of Congress Cataloging-in-Publication Data:

Jenkins, Ron, 1932–
X-ray fluorescence spectrometry / Ron Jenkins. — 2nd ed.
p. cm. — (Chemical analysis ; v. 152)
"A Wiley-Interscience publication."
Includes bibliographical references and index.
ISBN 0-471-29942-1 (cloth : alk. paper)
1. X-ray spectroscopy. 2. Fluorescence spectroscopy. I. Title.
II. series.
QD96.X2J47 1999
543′.08586—dc21 98-39008

10 9 8 7 6 5 4 3 2 1

CONTENTS

PREFACE TO THE FIRST EDITION

It is now nearly 30 years since the publication, in 1959, of the Wiley/Interscience monograph *X-ray Spectrochemical Analysis* by Verne Birks. In the intervening years the X-ray fluorescence method has come through the birth pains of innovation, has survived the early frustrations of application, and has achieved the status of a reliable, fast, accurate and versatile analytical method. The analytical chemist of today has a vast array of different techniques available for the analysis and characterization of materials, and most would agree that among the more powerful and flexible of these methods are those based on the use of X-ray fluorescence spectrometry. The X-ray fluorescence method is a means of qualitatively and quantitatively determining elements by measurement of the wavelengths and intensities of characteristic emissions. The technique is applicable to all but the very low atomic number elements, with sensitivities down to the low part per million level. In the late 1950s the elements covered by the X-ray fluorescence method ranged from the higher atomic numbers down to titanium ($Z = 22$). By the mid 1960s the advent of first the ethylene diamine d-tartrate (EDDT) crystal and then the penta-erythritol (PE) crystal, along with the chromium and rhodium anode X-ray tubes, increased the coverable atomic number range to include all elements down to and including aluminum ($Z = 13$). Under certain circumstances even magnesium and sodium were measurable albeit with rather poor sensitivity. As we entered the mid 1980s the advent of layered synthetic microstructures (LSM's) has allowed measurements down to carbon ($Z = 6$) with fair sensitivity, and even boron at concentration levels of several percent. The sensitivity of the X-ray fluorescence method for the determination of small quantities of material has also improved significantly. A "small" sample in the late 1950s and early 1960s was typically of the order of milligrams. Today, use of synchrotron or proton source excitation, along with total reflectance geometry, allows measurements at the picogram level. For some, it is difficult to imagine development at the same exciting level over the next two decades. Many believe that X-ray fluorescence has come as far as it will. I personally do not subscribe to this view. I believe that the problems of rapid and efficient sample homogenization will soon be solved. The development of room temperature solid state detectors has much still to yield. Use of the synchrotron is beginning to reveal areas of application of X-ray spectrometry hitherto not even considered. The use of the personal computer has yet to find its full exploitation in automating both quantitative and qualitative analysis. The development of combination X-ray diffractometer/spectrometers is at last beginning to show fruit. Present indications are that X-ray fluorescence spectrometry will continue to be an exciting and dynamic discipline.

PREFACE TO THE SECOND EDITION

I was gratified to learn that the first edition of this book found a place in the teaching of X-Ray Fluorescence Spectrometry. Both the American Chemical Society, and the International Centre for Diffraction Data, have, for a number of years, used the book as a course text in their X-ray fluorescence schools.

In preparing a second edition, I have taken the advantage in expanding the text to give more extensive coverage. In addition to a complete review and update of each chapter, new chapters have been added on "X-Ray Spectra" and "History and Development." The text is now about 30% larger than the first edition. I am grateful to those who have contributed to this work and am especially indebted to Dr. Sue Quick and Don Desrosiers for their painstaking work in proofing the manuscript.

Newtown Square, PA RON JENKINS

CHEMICAL ANALYSIS

A SERIES OF MONOGRAPHS ON ANALYTICAL CHEMISTRY AND ITS APPLICATIONS

J. D. Winefordner, *Series Editor*

Vol. 1. **The Analytical Chemistry of Industrial Poisons, Hazards, and Solvents**. *Second Edition*. By the late Morris B. Jacobs

Vol. 2. **Chromatographic Adsorption Analysis**. By Harold H. Strain (*out of print*)

Vol. 3. **Photometric Determination of Traces of Metals**. *Fourth Edition*
Part I: General Aspects. By E. B. Sandell and Hiroshi Onishi
Part IIA: Individual Metals, Aluminum to Lithium. By Hiroshi Onishi
Part IIB: Individual Metals, Magnesium to Zirconium. By Hiroshi Onishi

Vol. 4. **Organic Reagents Used in Gravimetric and Volumetric Analysis**. By John F. Flagg (*out of print*)

Vol. 5. **Aquametry: A Treatise on Methods for the Determination of Water**. *Second Edition (in three parts)*. By John Mitchell, Jr. and Donald Milton Smith

Vol. 6. **Analysis of Insecticides and Acaricides**. By Francis A. Gunther and Roger C. Blinn (*out of print*)

Vol. 7. **Chemical Analysis of Industrial Solvents**. By the late Morris B. Jacobs and Leopold Schetlan

Vol. 8. **Colorimetric Determination of Nonmetals**. *Second Edition*. Edited by the late David F. Boltz and James A. Howell

Vol. 9. **Analytical Chemistry of Titanium Metals and Compounds**. By Maurice Codell

Vol. 10. **The Chemical Analysis of Air Pollutants**. By the late Morris B. Jacobs

Vol. 11. **X-Ray Spectrochemical Analysis**. *Second Edition*. By L. S. Birks

Vol. 12. **Systematic Analysis of Surface-Active Agents**. *Second Edition*. By Milton J. Rosen and Henry A. Goldsmith

Vol. 13. **Alternating Current Polarography and Tensammetry**. By B. Breyer and H. H. Bauer

Vol. 14. **Flame Photometry**. By R. Herrmann and J. Alkemade

Vol. 15. **The Titration of Organic Compounds** (*in two parts*). By M. R. F. Ashworth

Vol. 16. **Complexation in Analytical Chemistry: A Guide for the Critical Selection of Analytical Methods Based on Complexation Reactions**. By the late Anders Ringbom

Vol. 17. **Electron Probe Microanalysis**. *Second Edition*. By L. S. Birks

Vol. 18. **Organic Complexing Reagents: Structure, Behavior, and Application to Inorganic Analysis**. By D. D. Perrin

Vol. 19. **Thermal Analysis.** *Third Edition*. By Wesley Wm. Wendlandt

Vol. 20. **Amperometric Titrations**. By John T. Stock

Vol. 21. **Reflectance Spectroscopy**. By Wesley Wm. Wendlandt and Harry G. Hecht

Vol. 22. **The Analytical Toxicology of Industrial Inorganic Poisons**. By the late Morris B. Jacobs

Vol. 23. **The Formation and Properties of Precipitates**. By Alan G. Walton

Vol. 24. **Kinetics in Analytical Chemistry**. By Harry B. Mark, Jr. and Garry A. Rechnitz

Vol. 25. **Atomic Absorption Spectroscopy**. *Second Edition*. By Morris Slavin

Vol. 26. **Characterization of Organometallic Compounds** (*in two parts*). Edited by Minoru Tsutsui

Vol. 27. **Rock and Mineral Analysis**. *Second Edition*. By Wesley M. Johnson and John A. Maxwell

Vol. 28. **The Analytical Chemistry of Nitrogen and Its Compounds** (*in two parts*). Edited by C. A. Streuli and Philip R. Averell

Vol. 29. **The Analytical Chemistry of Sulfur and Its Compounds** (*in three parts*). By J. H. Karchmer

Vol. 30. **Ultramicro Elemental Analysis**. By Günther Tölg

Vol. 31. **Photometric Organic Analysis** (*in two parts*). By Eugene Sawicki

Vol. 32. **Determination of Organic Compounds: Methods and Procedures**. By Frederick T. Weiss

Vol. 33. **Masking and Demasking of Chemical Reactions**. By D. D. Perrin

Vol. 34. **Neutron Activation Analysis**. By D. De Soete, R. Gijbels, and J. Hoste

Vol. 35. **Laser Raman Spectroscopy**. By Marvins C. Tobin

Vol. 36. **Emission Spectrochemical Analysis**. By Morris Slavin

Vol. 37. **Analytical Chemistry of Phosphorus Compounds**. Edited by M. Halmann

Vol. 38. **Luminescence Spectroscopy in Analytical Chemistry**. By J. D. Winefordner, S. G. Schulman, and T. C. O'Haver

Vol. 39. **Activation Analysis with Neutron Generators**. By Sam S. Nargolwalla and Edwin P. Przybylowicz

Vol. 40. **Determination of Gaseous Elements in Metals**. Edited by Lynn L. Lewis, Laben M. Melnick, and Ben D. Holt

Vol. 41. **Analysis of Silicones**. Edited by A. Lee Smith

Vol. 42. **Foundations of Ultracentrifugal Analysis**. By H. Fujita

Vol. 43. **Chemical Infrared Fourier Transform Spectroscopy**. By Peter R. Griffiths

Vol. 44. **Microscale Manipulations in Chemistry**. By T. S. Ma and V. Horak

Vol. 45. **Thermometric Titrations**. By J. Barthel

Vol. 46. **Trace Analysis: Spectroscopic Methods for Elements**. Edited by J. D. Winefordner

Vol. 47. **Contamination Control in Trace Element Analysis**. By Morris Zief and James W. Mitchell

Vol. 48. **Analytical Applications of NMR**. By D. E. Leyden and R. H. Cox

Vol. 79. **Analytical Solution Calorimetry**. Edited by J. K. Grime
Vol. 80. **Selected Methods of Trace Metal Analysis: Biological and Environmental Samples**. By Jon C. VanLoon
Vol. 81. **The Analysis of Extraterrestrial Materials**. By Isidore Adler
Vol. 82. **Chemometrics.** By Muhammad A. Sharaf. Deborah L. Illman, and Bruce R. Kowalski
Vol. 83. **Fourier Transform Infrared Spectrometry**. By Peter R. Griffiths and James A. de Haseth
Vol. 84. **Trace Analysis: Spectroscopic Methods for Molecules**. Edited by Gary Christian and James B. Callis
Vol. 85. **Ultratrace Analysis of Pharmaceuticals and Other Compounds of Interest**. Edited by S. Ahuja
Vol. 86. **Secondary Ion Mass Spectrometry: Basic Concepts, Instrumental Aspects, Applications and Trends.** By A. Benninghoven, F. G. Rüdenauer, and H. W. Werner
Vol. 87. **Analytical Applications of Lasers**. Edited by Edward H. Piepmeier
Vol. 88. **Applied Geochemical Analysis** By C. O. Ingamells and F. F. Pitard
Vol. 89. **Detectors for Liquid Chromatography**. Edited by Edward S. Yeung
Vol. 90. **Inductively Coupled Plasma Emission Spectroscopy: Part I: Methodology, Instrumentation, and Performance; Part II: Applications and Fundamentals**. Edited by J. M. Boumans
Vol. 91. **Applications of New Mass Spectrometry Techniques in Pesticide Chemistry**. Edited by Joseph Rosen
Vol. 92. **X-Ray Absorption: Principles, Applications, Techniques of EXAFS, SEXAFS, and XANES**. Edited by D. C. Konnigsberger
Vol. 93. **Quantitative Structure-Chromatographic Retention Relationships**. By Roman Kaliszan
Vol. 94. **Laser Remote Chemical Analysis**. Edited by Raymond M. Measures
Vol. 95. **Inorganic Mass Spectrometry**. Edited by F. Adams, R. Gijbels, and R. Van Grieken
Vol. 96. **Kinetic Aspects of Analytical Chemistry**. By Horacio A. Mottola
Vol. 97. **Two-Dimensional NMR Spectroscopy**. By Jan Schraml and Jon M. Bellama
Vol. 98. **High Performance Liquid Chromatography**. Edited by Phyllis R. Brown and Richard A. Hartwick
Vol. 99. **X-Ray Fluorescence Spectrometry**. By Ron Jenkins
Vol. 100. **Analytical Aspects of Drug Testing**. Edited by Dale G. Deustch
Vol. 101. **Chemical Analysis of Polycyclic Aromatic Compounds**. Edited by Tuan Vo-Dinh
Vol. 102. **Quadrupole Storage Mass Spectrometry**. By Raymond E. March and Richard J. Hughes
Vol. 103. **Determination of Molecular Weight**. Edited by Anthony R. Cooper
Vol. 104. **Selectivity and Detectability Optimization in HPLC**. By Satinder Ahuja
Vol. 105. **Laser Microanalysis**. By Lieselotte Moenke-Blankenburg
Vol. 106. **Clinical Chemistry**. Edited by E. Howard Taylor

Vol. 107. **Multielement Detection Systems for Spectrochemical Analysis**. By Kenneth W. Busch and Marianna A. Busch

Vol. 108. **Planar Chromatography in the Life Sciences**. Edited by Joseph C. Touchstone

Vol. 109. **Fluorometric Analysis in Biomedical Chemistry: Trends and Techniques Including HPLC Applications**. By Norio Ichinose, George Schwedt, Frank Michael Schnepel, and Kyoko Adochi

Vol. 110. **An Introduction to Laboratory Automation**. By Victor Cerdá and Guillermo Ramis

Vol. 111. **Gas Chromatography: Biochemical, Biomedical, and Clinical Applications**. Edited by Ray E. Clement

Vol. 112. **The Analytical Chemistry of Silicones**. Edited by A. Lee Smith

Vol. 113. **Modern Methods of Polymer Characterization**. Edited by Howard G. Barth and Jimmy W. Mays

Vol. 114. **Analytical Raman Spectroscopy**. Edited by Jeanette Graselli and Bernard J. Bulkin

Vol. 115. **Trace and Ultratrace Analysis by HPLC**. By Satinder Ahuja

Vol. 116. **Radiochemistry and Nuclear Methods of Analysis**. By William D. Ehmann and Diane E. Vance

Vol. 117. **Applications of Fluorescence in Immunoassays**. By Ilkka Hemmila

Vol. 118. **Principles and Practice of Spectroscopic Calibration**. By Howard Mark

Vol. 119. **Activation Spectrometry in Chemical Analysis**. By S. J. Parry

Vol. 120. **Remote Sensing by Fourier Transform Spectrometry**. By Reinhard Beer

Vol. 121. **Detectors for Capillary Chromatography**. Edited by Herbert H. Hill and Dennis McMinn

Vol. 122. **Photochemical Vapor Deposition**. By J. G. Eden

Vol. 123. **Statistical Methods in Analytical Chemistry**. By Peter C. Meier and Richard Zund

Vol. 124. **Laser Ionization Mass Analysis**. Edited by Akos Vertes, Renaat Gijbels, and Fred Adams

Vol. 125. **Physics and Chemistry of Solid State Sensor Devices**. By Andreas Mandelis and Constantinos Christofides

Vol. 126. **Electroanalytical Stripping Methods**. By Khjena Z. Brainina and E. Neyman

Vol. 127. **Air Monitoring by Spectroscopic Techniques**. Edited by Markus W. Sigrist

Vol. 128. **Information Theory in Analytical Chemistry**. By Karel Eckschlager and Klaus Danzer

Vol. 129. **Flame Chemiluminescence Analysis by Molecular Emission Cavity Detection**. Edited by David Stiles, Anthony Calokerinos, and Alan Townshend

Vol. 130. **Hydride Generation Atomic Absorption Spectrometry**. Edited by Jiri Dedina and Dimiter L. Tsalev

Vol. 131. **Selective Detectors: Environmental, Industrial, and Biomedical Applications**. Edited by Robert E. Sievers

Vol. 132. **High-Speed Countercurrent Chromatography**. Edited by Yoichiro Ito and Walter D. Conway

Vol. 133. **Particle-Induced X-Ray Emission Spectrometry**. By Sven A. E. Johansson, John L. Campbell, and Klas G. Malmqvist

Vol. 134. **Photothermal Spectroscopy Methods for Chemical Analysis**. By Stephen E. Bialkowski

Vol. 135. **Element Speciation in Bioinorganic Chemistry**. Edited by Sergio Caroli

Vol. 136. **Laser-Enhanced Ionization Spectrometry**. Edited by John C. Travis and Gregory C. Turk

Vol. 137. **Fluorescence Imaging Spectroscopy and Microscopy**. Edited by Xue Feng Wang and Brian Herman

Vol. 138. **Introduction to X-Ray Powder Diffractometry**. By Ron Jenkins and Robert L. Snyder

Vol. 139. **Modern Techniques in Electroanalysis**. Edited by Petr Vanýsek

Vol. 140. **Total-Reflection X-Ray Fluorescence Analysis**. By Reinhold Klockenkamper

Vol. 141. **Spot Test Analysis: Clinical, Environmental, Forensic, and Geochemical Applications**, *Second Edition*. By Ervin Jungreis

Vol. 142. **The Impact of Stereochemistry on Drug Development and Use**. Edited by Hassan Y. Aboul-Enein and Irving W. Wainer

Vol. 143. **Macrocyclic Compounds in Analytical Chemistry**. Edited by Yury A. Zolotov

Vol. 144. **Surface-Launched Acoustic Wave Sensors: Chemical Sensing and Thin-Film Characterization**. By Michael Thompson and David Stone

Vol. 145. **Modern Isotope Ratio Mass Spectrometry**. Edited by T. J. Platzner

Vol. 146. **High Performance Capillary Electrophoresis: Theory, Techniques, and Applications**. Edited by Morteza G. Khaledi

Vol. 147. **Solid Phase Extraction: Principles and Practice**. By E. M. Thurman

Vol. 148. **Commercial Biosensors: Applications to Clinical, Bioprocess, and Environmental Samples**. Edited by Graham Ramsay

Vol. 149. **A Practical Guide to Graphite Furnace Atomic Absorption Spectrometry**. By David J. Butcher and Joseph Sneddon

Vol. 150. **Principles of Chemical and Biological Sensors**. Edited by Dermot Diamond

X-Ray Fluorescence Spectrometry

CHAPTER

1

PRODUCTION AND PROPERTIES X-RAYS

1.1. INTRODUCTION

X-rays are a short wavelength form of electromagnetic radiation discovered by Wilhelm Röntgen in 1895 [1]. X-ray based techniques provided important tools for theoretical physicists in the first half of this century and, since the early 1950s, they have found an increasing use in the field of materials characterization. Today, methods based on absoptiometry play a vital role in industrial and medical radiography. The simple X-ray field units employed in World War I were responsible for saving literally tens of thousands of lives, [2] and today the technology has advanced to a high degree of sophistication. Modern X-ray tomographic methods give an almost complete three-dimensional cross section of the human body, offering an incredibly powerful tool for the medical field. In addition, the analytical techniques based on X-ray diffraction and X-ray spectrometry, both of which were first conceived almost 70 years ago, have become indispensable in the analysis and study of inorganic and organic solids. Today, data obtained from X-ray spectrometers are being used to control steel mills, ore flotation processes, cement kilns, and a whole host of other vital industrial processes (see e.g., [3]). X-ray diffractometers are used for the study of ore and mineral deposits, in the production of drugs and pharmaceuticals, in the study of thin films, stressed and oriented materials, phase transformations, plus myriad other applications in pure and applied research.

X-ray photons are produced following the ejection of an inner orbital electron from an irradiated atom, and subsequent transition of atomic orbital electrons from states of high to low energy. When a monochromatic beam of X-ray photons falls onto a given specimen, three basic phenomena may result, namely, scatter, absorption or fluorescence. The coherently scattered photons may undergo subsequent interference leading in turn to the generation of diffraction maxima. The angles at which the diffraction maxima occur can be related to the spacings between planes of atoms in the crystal lattice and hence, X-ray generated diffraction patterns can be used to study the structure of solid materials. Following the discovery of the diffraction of X-rays by Max Von Laue in 1913 [4], the use of this method for materials analysis has become very important both in industry and research.

Today, it is one of the most useful techniques available for the study of structure dependant properties of materials.

X-ray powder diffractometry involves characterization of materials by use of data dependant on the atomic arrangement in the crystal lattice (see e.g., [5]). The technique uses single or multiphase (i.e., multicomponent) specimens comprising a random orientation of small crystallites, each of the order of 1 to 50 μm in diameter. Each crystallite in turn is made up of a regular, ordered array of atoms. An ordered arrangement of atoms (the crystal lattice) contains planes of high atomic density that in turn means planes of high electron density. A monochromatic beam of X-ray photons will be scattered by these atomic electrons and, if the scattered photons interfere with each other, diffraction maxima may occur. In general, one diffracted line will occur for each unique set of planes in the lattice. A diffraction pattern is typically in the form of a graph of diffraction angle (or interplanar spacing) versus diffracted line intensity. The pattern is thus made up of a series of superimposed diffractograms, one for each unique phase in the specimen. Each of these unique patterns can act as an empirical "fingerprint" for the identification of the various phases, using pattern recognition techniques based on a file of standard single-phase patterns. Quantitative phase analysis is also possible, albeit with some difficulty, because of various experimental and other problems, not the least of which is the large number of diffraction lines occurring from multiphase materials.

A beam of X-rays passing through matter will be attenuated by two processes, scatter and photoelectric absorption. In the majority of cases the greater of these two effects is absorption and the magnitude of the absorption process, that is, the fraction of incident X-ray photons lost in passing through the absorber increases significantly with the average atomic number of the absorbing medium. To a first approximation, the mass attenuation coefficient varies as the third power of the atomic number of the absorber. Thus, when a polychromatic beam of X-rays is passed through a heterogeneous material, areas of higher average atomic number will attenuate the beam to a greater extent than areas of lower atomic number. Thus the beam of radiation emerging from the absorber has an intensity distribution across the irradiation area of the specimen, which is related to the average atomic number distribution across the same area. It is upon this principle that all methods of X-ray radiography are based. Study of materials by use of the X-ray absorption process is the oldest of all of the X-ray methods in use. Röntgen himself included a radiograph of his wife's hand in his first published X-ray paper. Today, there are many different forms of X-ray absorptiometry in use, including industrial radiography, diagnostic medical and dental radiography, and security screening.

Secondary radiation produced from the specimen is characteristic of the elements making up the specimen. The technique used to isolate and measure individual characteristic wavelengths is called X-ray spectrometry. X-ray spectrometry also has its roots in the early part of this century, stemming from Moseley's work in 1913 [6]. This technique uses either the diffracting power of a single crystal to isolate narrow wavelength bands, or a proportional detector to isolate narrow energy bands, from the polychromatic beam characteristic radiation excited in the sample. The first of these methods is called *Wavelength Dispersive Spectrometry* and the second, *Energy Dispersive Spectrometry.* Because the relationship between emission wavelength and atomic number is known, isolation of individual characteristic lines allows the unique identification of an element to be made and elemental concentrations can be estimated from characteristic line intensities. Thus this technique is a means of materials characterization in terms of chemical composition.

Although the major thrust of this monograph is to review X-ray spectroscopic techniques, these are by no means the only X-ray-based methods that are used for materials analysis and characterization. In addition to the many industrial and medical applications of diagnostic X-ray absorption methods already mentioned, X-rays are also used in areas such as structure determination based on single crystal techniques, space exploration and research, lithography for the production of microelectronic circuits, and so on. One of the major limitations leading to the further development of new X-ray methods is the inability to *focus* X-rays, as can be done with visible light rays. Although it is possible to partially reflect X- rays at low glancing angles, or diffract an X-ray beam with a single crystal, these methods cause significant intensity loss, and fall far short of providing the high intensity, monochromatic beam that would be ideal for such uses as an X-ray microscope. Use of synchrotron radiation offers the potential of an intense, highly focused, coherent X-ray beam, but has practical limitations due to size and cost. The X-ray laser could, in principle, provide an attractive alternative, and since the discovery of the laser in 1960, the possibilities of such an X-ray laser have been discussed. Although major research efforts have been, and are still being, made to produce laser action in the far ultraviolet and soft X-ray regions, production of conditions to stimulate laser action in the X-ray region with a net positive gain is difficult. This is due mainly to the rapid decay rates and high absorption cross sections that are experienced in practice.

1.2. CONTINUOUS RADIATION

When a high energy electron beam is incident upon a specimen, one of the products of the interaction is an emission of a broad wavelength band of radiation called *continuum.* This continuum, which is also referred to as

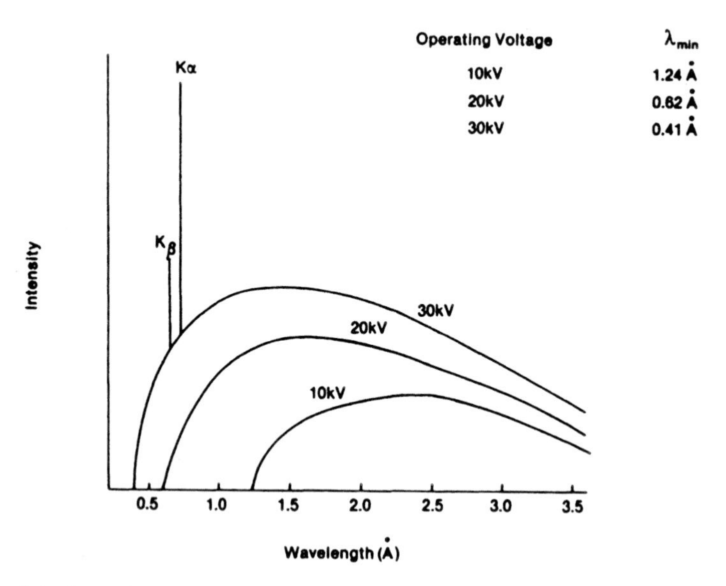

Figure 1.1. Intensity output from a molybdenum target X-ray at 10, 20, and 30 kV.

white radiation or *bremsstrahlung*, is produced as the impinging high energy electrons are decelerated by the atomic electrons of the elements making up the specimen. A typical intensity/wavelength distribution of this radiation is illustrated in Figure 1.1 and is typified by a minimum wavelength λ_{min}, which is roughly proportional to the maximum accelerating potential V of the electrons, that is, $12.4/V$ keV. However, at higher potentials, an experimentally measured value of the minimum wavelength of the continuum may yield a somewhat longer wavelength than would be predicted due to the effect of Compton scatter. The intensity distribution of the continuum reaches a maximum, I_{max}, at a wavelength 1.5 to 2 times greater than λ_{min}. Increasing the accelerating potential causes the intensity distribution of the continuum to shift towards shorter wavelengths. The curves given in Figure 1.1 are for the element molybdenum ($Z=42$). Note the appearance of characteristic lines of Mo $K\alpha(\lambda=0.71$ Å) and Mo $K\beta(\lambda=0.63$ Å), once the K shell excitation potential of 20 keV has been exceeded. Most commercially available spectrometers utilize a sealed X-ray tube as an excitation source, and these tubes typically employ a heated tungsten filament as a source of electrons, and a layer of pure metal such as chromium, rhodium, or tungsten, as the anode. The broad band of white radiation produced by this type of tube is ideal for the excitation of the characteristic lines from a wide range of atomic numbers. In general, the higher the atomic number of the anode material, the more intense the beam of radiation produced by the tube. Conversely, however, because the higher atomic number anode elements generally require thicker exit windows in the tube, the longer wavelength output from such a tube is rather poor and so these high atomic number anode tubes are less satisfactory for the excitation of longer wavelengths from low atomic number samples (see Section 6.2).

1.3. CHARACTERISTIC RADIATION

If a high energy particle, such as an electron, strikes a bound atomic electron, and the energy E of the particle is greater than the binding energy ϕ of the atomic electron, it is possible that the atomic electron will be ejected from its atomic position, departing from the atom with a kinetic energy (E-ϕ), equivalent to the difference between that of the initial particle and the binding energy of the atomic electron. The ejected electron is called a *photoelectron* and the interaction is referred to as the *photoelectric effect*. While the fate of the ejected photoelectron has little consequence as far as the production and use of characteristic X-radiation from an atom is concerned, it should be mentioned that study of the energy distribution of the emitted photoelectrons gives valuable information about bonding and atomic structure [7]. Study of such information forms the basis of the technique of *photoelectron spectroscopy* (see e.g., [8]). As long as the vacancy in the shell exists, the atom is in an unstable state and there are two processes by which the atom can revert back to its original state. The first of these involves a rearrangement that does not result in the emission of X-ray photons, but in the emission of other photoelectrons from the atom. The effect is known as the *Auger effect*, and the emitted photoelectrons are called Auger electrons. The second process by which the excited atom can regain stability is by transference of an electron from one of the outer orbitals to fill the vacancy. The energy difference between the initial and final states of the transferred electron may be given off in the form of an X-ray photon. Since all emitted X-ray photons have energies proportional to the differences in the energy states of atomic electrons, the lines from a given element will be characteristic of that element. The relationship between the wavelength λ of a characteristic X-ray photon and the atomic number Z of the excited element was first established by Moseley (see Chapter 4). Moseley's law is written

$$\frac{1}{\lambda} = K[Z - \sigma]^2, \tag{1.1}$$

in which K is a constant that takes on different values for each spectral series. σ is the shielding constant that has a value of just less than unity. The wavelength of the X-ray photon is inversely related to the energy E of the photon according to the relationship

$$\lambda = \frac{12.4}{E}. \tag{1.2}$$

There are several different combinations of quantum numbers held by the electron in the initial and final states, hence several different X-ray wavelengths will be emitted from a given atom. For those vacancies giving rise to characteristic X-ray photons, a series of very simple selection rules can be used to define which electrons can be transferred. For a detailed explanation of these rules, refer to Section 4.4. Briefly, the principal quantum number n must change by 1, the angular quantum number ℓ must change by 1, and the vector sum of $(\ell + s)$ must be a number changing by 1 or 0. In effect this means that for the K series only $p \rightarrow s$ transitions are allowed,

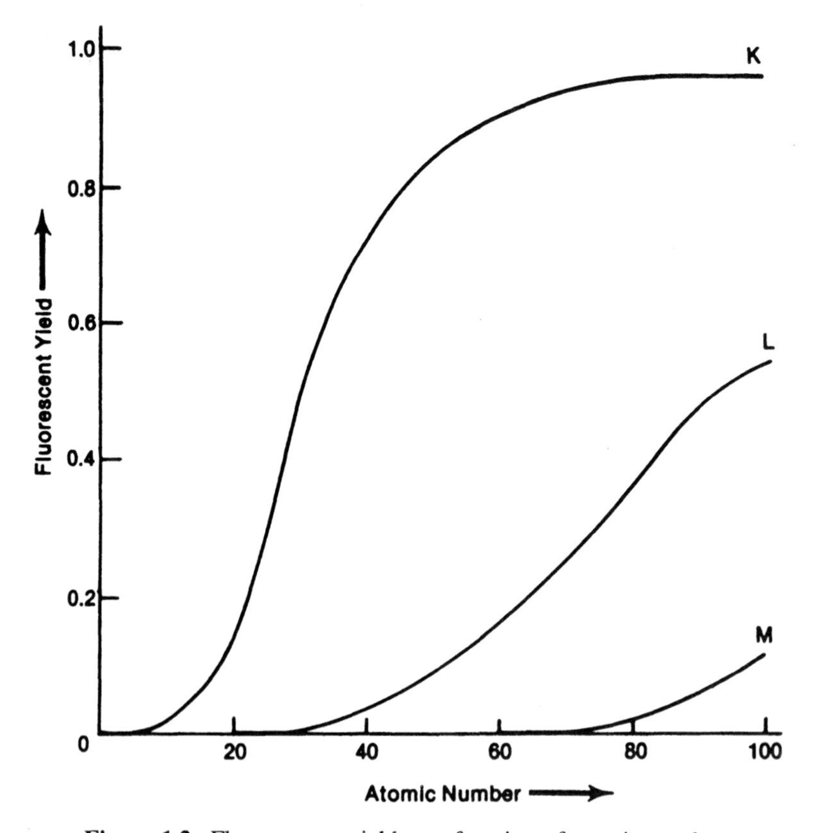

Figure 1.2. Fluorescence yield as a function of atomic number.

yielding two lines for each principle level change. Vacancies in the L level follow similar rules and give rise to L series lines. There are more of the L lines since $p \rightarrow s$, $s \rightarrow p$, and $d \rightarrow p$ transitions are all allowed within the selection rules. In general, electron transitions to the K shell give between two and six K lines, and transitions to the L shell give about twelve strong to moderately strong L lines.

Because there are two competing effects by which an atom may return to its initial state, and because only one of these processes will give rise to the production of a characteristic X-ray photon, the intensity of an emitted characteristic X-ray beam will be dependant upon the relative effectiveness of the two processes within a given atom. As an example, the number of quanta of K series radiation emitted per ionized atom, is a fixed ratio for a given atomic number. This ratio is called the fluorescent yield.

Figure 1.2 shows fluorescent yield curves as a function of atomic number. It will be seen that, whereas the K fluorescent yield is close to unity for higher atomic numbers, it drops by several orders of magnitude for the very low atomic numbers. In practice, this means that if the intensities obtained from pure barium ($Z = 56$) are compared to those of pure aluminum ($Z = 13$), all other things being equal, pure barium would give about 50 times more counts than pure aluminum. Also note from the curve that the L fluorescent yield for a given atomic number is always less than the corresponding K fluorescent yield.

1.4. ABSORPTION OF X-RAYS

All matter is made up of molecules, each consisting of a group of atoms. Each atom is in turn made up of a nucleus surrounded by electrons in discrete energy levels. The number of electrons is equal to the atomic number of the atom, when the atom is in the ground state. When a beam of X-ray photons is incident upon matter, the photons may interact with the individual atomic electrons.

The processes involved in the excitation of a characteristic wavelength λ_f by an incident X-ray beam λ_o are illustrated in Figure 1.3. Here a beam of X-ray photons of intensity $I_o(\lambda_o)$ falls onto a specimen at an incident angle ψ_1. A portion of the beam will pass through the absorber, the fraction being given by the expression

$$I(\lambda) = I_o(\lambda) \times exp(\mu \times \rho \times x), \tag{1.3}$$

where μ is the mass attenuation coefficient of absorber for the wavelength and ρ the density of the specimen. x is the distance traveled through the

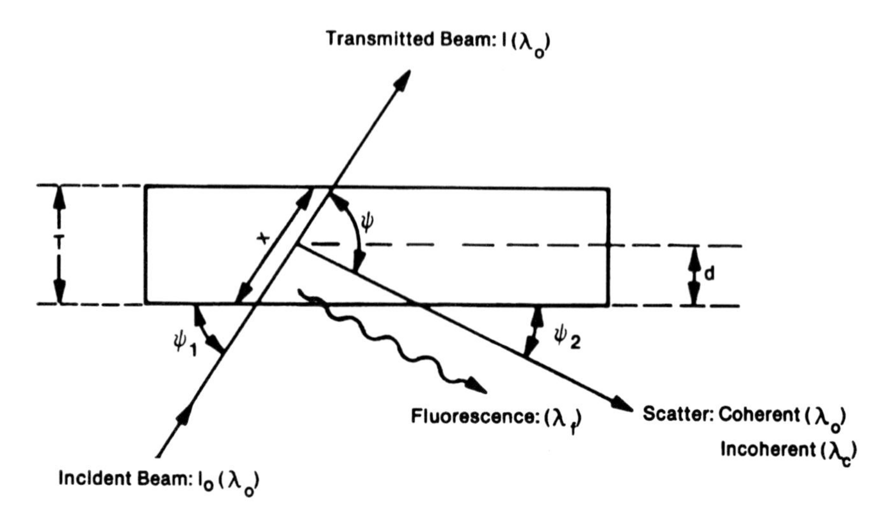

T = Sample Thickness
x = Path Length
d = Penetration Depth
ψ_1 = Incident Angle
ψ_2 = Take-Off Angle
ψ = Scattering Angle
μ = Mass Absorption Coefficient
ρ = Density

- Transmitted Beam Intensity: $I(\lambda_o) = I_o(\lambda_o) \exp(-\mu\rho x)$
- Incoherent Scatter Wavelength: $\lambda_c - \lambda_o = 0.0243\,[1 - \cos\psi]$
- Fluorescence Penetration Depth: $d = x \sin\psi_2$

Figure 1.3. Interaction of the primary X-ray beam with the sample.

specimen and this is related to the thickness T of the specimen by the sine of the incident angle ψ_1. Note that as far as the primary wavelengths are concerned, three processes occur. The first of these is the attenuation of the transmitted beam as described above. Secondly, the primary beam may be scattered over an angle ψ and emerge as coherently and incoherently scattered wavelengths. Thirdly, fluorescence radiation may also arise from the sample. The depth d of the specimen contributing to the fluorescence intensity is related to the attenuation coefficient of the sample for the fluorescence wavelength, and the angle of emergence ψ_2 at the fluorescence beam is observed. It can be seen from Equation 1.3 that a number $(I_o - I)$ of photons has been lost in the absorption process. Although a significant fraction of this loss may be due to scatter, by far the greater loss is due to the photoelectric effect. It is important to differentiate between *mass absorption coefficient* and *mass attenuation coefficient*. The mass absorption coefficient is dependant only on photoelectric interaction between the incident beam and atomic photons produced within the specimen. The mass attenuation coefficient also includes scattering of the primary beam. Thus, for most quantitative X-ray spectrometry, the mass attenuation coefficient should be employed. Photoelectric absorption occurs at each of the energy levels of the atom, thus the total photoelectric absorption is determined by the sum of each individual absorption within a specific shell. Where the absorber is made up of a number of different elements, as is usually the case, the total attenuation is made up of the sum of the products of the individual elemental mass absorption coefficients and the weight fractions of the respective elements. This product is referred to as the total matrix absorption. The value of the mass attenuation referred to in Equation 1.3 is a function of both the photoelectric absorption and the scatter. However, the photoelectric absorption influence is usually large compared to the photoelectric absorption.

The photoelectric absorption is made up of absorption in the various atomic levels and it is an atomic number dependant function. A plot of the mass absorption coefficient as a function of wavelength contains a number of discontinuities called absorption edges, at wavelengths corresponding to the binding energies of the electrons in the various subshells. As an example, Figure 1.4 shows a plot of the mass absorption coefficient as a function of wavelength for the element barium. It can be seen that as the wavelength of the incident X-ray photons becomes longer, the absorption increases. A number of discontinuities are also seen in the absorption curve, one *K* discontinuity, three *L* discontinuities and five *M* discontinuities. Since it can be assumed that the mass attenuation coefficient is proportional to the total photoelectric absorption, as indicated at the bottom of the figure, an equation can be written that shows that the total photoelectric absorption is made up of absorption in the *K* level, plus absorption in the three *L* levels, and so on. The figure also shows the magnitudes of the various contributions of the levels to the total value of the absorption coefficient of $26^2/g$ for barium at a wavelength of 0.3 Å. It can be seen that the contributions from *K*, *L*, and *M* levels, respectively, are 18, 7.5, and 0.45, with 0.05 coming from other outer levels. If a slightly longer wavelength were chosen, say perhaps 0.6 Å, the wavelength would fall on the long wavelength side of the *K* absorption edge. This being the case, these photons are insufficiently energetic to eject *K* electrons from the barium, and since there can be no photoelectric absorption in the level, the term for the *K* level drops out of the equation. Since this *K* term is large in comparison to the others, there is a sudden drop in the

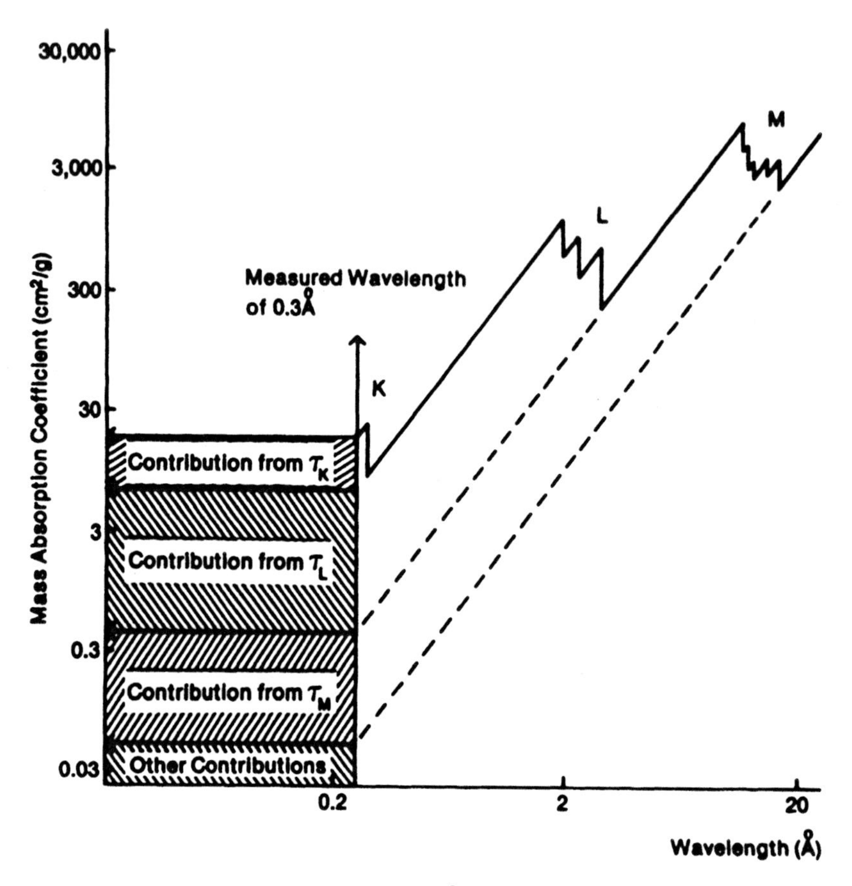

Total Mass Absorption Coefficient for 0.3Å Radiation by Barium = 26 cm²/g

$$\begin{aligned}\tau_{Total} &= \tau_{K_I} + [\tau_{L_I} + \tau_{L_{II}} + \tau_{L_{III}}] + [\tau_{M_I} + \tau_{M_{II}} + \tau_{M_{III}} + \tau_{M_{IV}} + \tau_{M_V}] + \text{others} \\ &= 18 + 7.5 + 0.45 + 0.05 \\ &= 26\end{aligned}$$

Figure 1.4. Mass absorption coefficient for barium, showing individual contributions from *K*, *L* and *M* subshells.

absorption curve corresponding exactly to the wavelength at which photoelectric absorption in the *K* level can no longer occur.

Clearly, a similar situation occurs for all other levels giving the discontinuities in the absorption curve indicated in the figure. Since each unique atom has a unique set of excitation potentials for the various subshells, each atom exhibits a characteristic absorption curve. This particular effect is very important in quantitative X-ray spectrometry because the intensity of a beam of characteristic photons leaving a specimen is dependent upon the relative absorption effects of the different atoms making up the specimen. This effect is called a matrix effect and is one of the reasons why a curve of characteristic line intensity, as a function of element concentration, may not be a straight line. In earlier years, it was common practice to use tables of mass attenuation coefficient versus wavelength to generate absorption data. With the advent of rapid computer methods for applying matrix correction methods, the shortcomings of the use of lookup tables soon became apparent, and algorithms were developed allowing the simple calculation of absorption data by use of polynomials or least squares analysis [9].

The distances traveled by X-ray photons through solid matter are not very great for the wavelengths and energies of the characteristic lines used in X-ray fluorescence analysis. The x term in Equation 1.3 represents the distance traveled, in centimeters, by a monochromatic beam of X-ray photons, in a matrix of average mass absorption coefficient expressed in cm^2/g. The equation can be rearranged in the logarithmic form

$$2.3 \times \log_{10} \times \left(\frac{I}{I_o}\right) = \mu\rho x. \tag{1.4}$$

For 99% absorption, that is, $I_o = 100$ and $I = 1$, Equation 1.4 reduces to

$$x(\mu m) = \frac{46\,000}{\mu\rho}, \tag{1.5}$$

where x is now expressed in microns. Since densities of most solid materials are in the range of 2 to 7 g/cm^3 and values of mass absorption coefficients are typically in the range 50 to 5000 cm^2/g, values for x range from a few microns to several hundred microns (see Chapter 9). It can be seen from Figure 1.3 that, in the optical arrangement typically employed in commercial X-ray spectrometers, the path length is related to the depth of penetration d of a given analyte wavelength

$$d = x \times \sin\psi_2, \tag{1.6}$$

where ψ_2 is the takeoff angle of the spectrometer. Although depth of penetration is a rather arbitrary measure, it is nevertheless useful, since it does give an indication as to the thickness of the sample contributing to the measured fluorescence radiation from the specimen for a given analyte wavelength or energy. In most spectrometers, the value of ψ_2 is between 30° and 45°, so the value of d is generally about one half the value of the path length x.

1.5. COHERENT AND INCOHERENT SCATTERING

Scattering occurs when an X-ray photon interacts with the electrons of the target element. When this interaction is elastic, that is, no energy is lost in the collision process, the scatter is referred to as coherent (Rayleigh) scatter. Since no energy change is involved, the coherently scattered radiation will retain exactly the same wavelength as that of the incident beam. The origin of the coherently scattered wave is best described by thinking of the primary photon as an electromagnetic wave. When such a wave interacts with an electron, the electron is oscillated by the electric field of the wave, and, in turn, radiates wavelengths of the same frequency as the incident wave. All atoms scatter X-ray photons to a lesser or greater extent, the intensity of the scatter being dependent on the energy of the incident ray and the number of loosely bound outer electrons. In other words, the intensity of the scatter is dependent upon the average atomic number. The scattered photon can also give up a small part of its energy during the collision, especially where the electron that collides with the photon is only loosely bound. In this instance, the scatter is referred to as incoherent (Compton scatter). Compton scatter is best presented in terms of the corpuscular nature of the X-ray photon. In this instance, an X-ray photon collides with a loosely bound outer atomic electron. The electron recoils under the impact, removing a small portion of the energy of the primary photon, which is then deflected with the corresponding loss of energy, or increase of wavelength. There is a simple relationship between the incident λ_o and incoherently scattered wavelength λ_c, this being

$$\lambda_c - \lambda_o = 0.0242(1 - \cos\psi). \quad (1.7)$$

ψ is the angle over which the X-ray beam is scattered which, in most commercial spectrometers, is equal to 90°. Since the cosine of 90° is zero, there is generally a fixed wavelength difference between the coherently and incoherently scattered lines, equal to around 0.024 Å. This constant difference gives a very practical means of predicting the angular position of an incoherently scattered line. Also, the coherently scattered line is much broader than a coherently scattered (diffracted) line because the scattering angle is not a single value but a range of values due to the divergence of the primary X-ray beam.

1.6. INTERFERENCE AND DIFFRACTION

X-ray diffraction is a combination of two phenomena, coherent scatter and interference. At any point where two or more waves cross one another they are said to interfere. Interference does not imply the impedance of one wave-train by another, but rather describes the effect of superposition of one wave upon another. The principal of superposition is that the resulting displacement at any point and at any instant may be formed by adding the instantaneous displacements that would be produced at the same point by independent wave-trains, if each were present alone. Figure 1.5 illustrates this effect and shows two waves out of phase with each other, giving a resultant ray of zero; and two waves in phase with each other, giving a resulting wavelength of twice the original amplitude. Thus, under certain geometric conditions, wavelengths that are exactly in phase may add to one another, and those that are exactly out of phase, may cancel each other out. Under such conditions, coherently scattered photons may constructively interfere with each other, giving diffraction maxima.

As illustrated in Figure 1.6(a), a crystal lattice consists of a regular arrangement of atoms, with layers of high atomic density existing throughout the crystal structure. Planes of high atomic density mean, in turn, planes of high electron density. Since scattering occurs between impinging X-ray photons and the loosely bound outer orbital atomic electrons, scattering also occurs when a monochromatic beam of radiation falls onto the high atomic

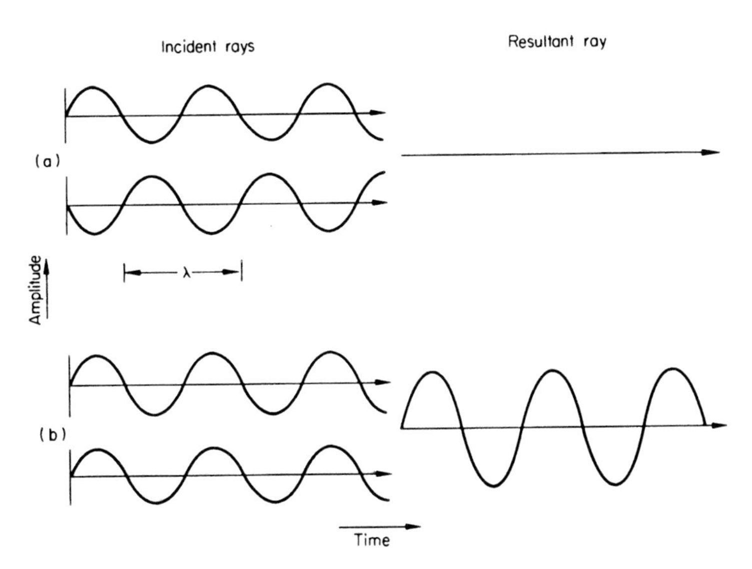

Figure 1.5. Destructive and nondestructive interference.

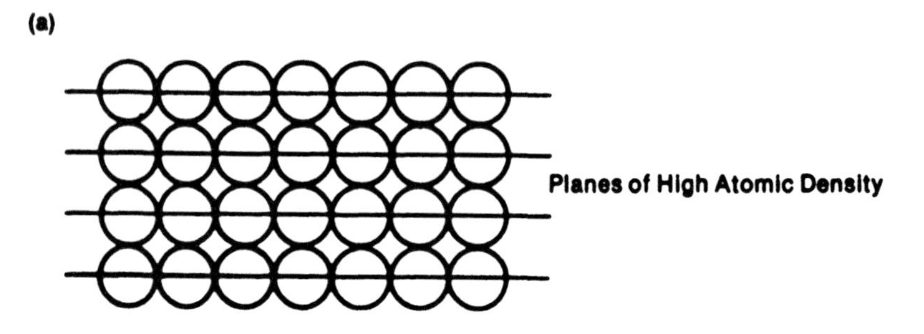

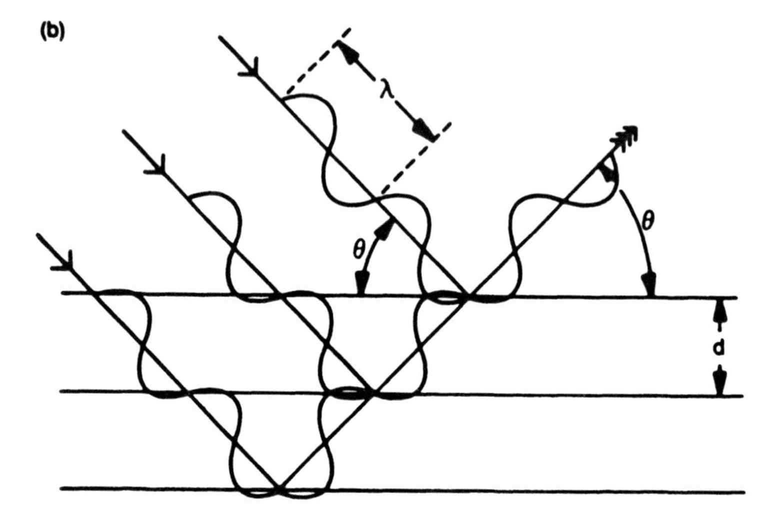

In order to ensure that the scattered waves remain in phase, the path length difference between successive waves ($2d.\sin\theta$) must equal a whole number (n) of wavelengths (λ).

i.e. $n\lambda = 2d.\sin\theta$

Figure 1.6. Diffraction by an ordered array of atoms.

density layers. In order to satisfy the requirement for constructive interference, it is necessary that the scattered waves originating from the individual atoms, that is, the scattering points, be in phase with one another. The geometric conditions required for this to occur are illustrated in Figure 1.6(b). Here, a series of parallel rays strike a set of crystal planes at an angle θ and are scattered as previously described. Reinforcement occurs when the difference in the path-lengths of the two interfering waves is equal to a whole number of wavelengths.

Hence the overall condition for reinforcement is

$$n\lambda = 2d \sin\theta, \tag{1.8}$$

a statement of Bragg's Law. The diffraction effect is important since, by using a crystal of fixed $2d$, each unique wavelength will be diffracted at a unique diffraction (Bragg) angle. Thus, by measuring the diffraction angle, the value of the wavelength can be determined. Since there is a simple relationship between wavelength and atomic number, as given by Moseley's law, Equation 1.1, the atomic number(s) of the element(s) from which the wavelengths were emitted can be established.

BIBLIOGRAPHY

[1] W. C. Röntgen,"On a new kind of rays: second communication," *Ann. Phys. Chem.*, **64**, 1–11 (1898).

[2] W. R. Nitske, *The Life of Wilhelm Conrad Röntgen, Discoverer of the X-ray*, Univ. of Arizona Press, 355 pp. (1971).

[3] H. K. Herglotz and L. S. Birks, *X-ray Spectrometry*, Dekker: New York, 513 pp. (1978).

[4] W. Freidrich, P. Knipping and M. Von Laue, "Interference phenomena with Röntgen rays," *Ann. Physik*, **41**, 971–988 (1913).

[5] H. P. Klug and L. E. Alexander, *X-ray Diffraction Procedures, 2nd ed.*, Wiley: London, 966 pp. (1974).

[6] H. G. J. Moseley, (1913/14), "High frequency spectra of the elements," *Phil. Mag.*, **26**, 1024–1034 (1913), and **27**, 703–713 (1914).

[7] Bonnelle and Mande, *Advances in X-ray Spectroscopy*, Chapter 20, "Potential Characteristics and Applications of X-ray Lasers" by D. J. Nagel, pp. 371–410, Pergamon: Oxford (1982).

[8] K. Siegbahn et al., Eds., *Atomic, Molecular And Solid State Structure Studied By Means of Electron Spectroscopy*, Almquist: Uppsala (1967).

[9] G. R. Lachance and F. Claisse, *Quantitative X-ray Fluorescence Analysis*, Wiley: New York, 402 pp. (1994).

CHAPTER

2

INDUSTRIAL APPLICATIONS OF X-RAYS

2.1. INTRODUCTION

The industrial uses of X-ray-based techniques are many and varied, representing some 100 years of research and application [1]. As was described in Chapter 1, the three basic properties of X-ray photons are those of scatter, photoelectric absorption, and fluorescence. Each of these three properties has been utilized for different industrial applications [2]. Figure 2.1 illustrates eight examples. The first of these is the X-ray fluorescence technique, in which the wavelengths 2.1(a) or energies 2.1(b) of fluorescence emission lines are measured to establish the elemental composition of a sample. Use of the intensities of the wavelengths can also allow quantification of each identified element. Chapter 1 discussed the possibility of using X-ray powder diffraction patterns as an empirical *fingerprint* to establish which compounds (phases) make up a specimen 2.1(c). The powder method is generally considered to be an empirical method, because the calculations required to decide which lines arise from which set(s) of interplanar spacings can be rather complicated, due mainly to the superposition of lines and uncertainties about the line intensities.

A second important application of the X-ray diffraction method is the *single crystal* method, Figure 2.1(d). In the single crystal approach, a single crystal is fixed in a specific orientation relative to the X-ray beam, and diffraction maxima sought. The orientation of the crystal is then changed and again diffraction maxima sought. This process is repeated as many times as necessary to build up a detailed picture of the intensity distribution of diffracted radiation coming from a specimen. The effect of repeating many measurements at different crystal orientations is to reduce the rather complicated three-dimensional lattice calculations to a series of much simpler calculations.

Thus, from single crystal diffraction data, it is possible to establish very detailed information about the structure of a specimen. Even though the calculations are greatly simplified, they are still relatively complicated, and need all of the power of a modern computer for their solution. This limitation, and the fact that the technique is only applicable to pure phases from which small single crystals can be grown, make it the purview of

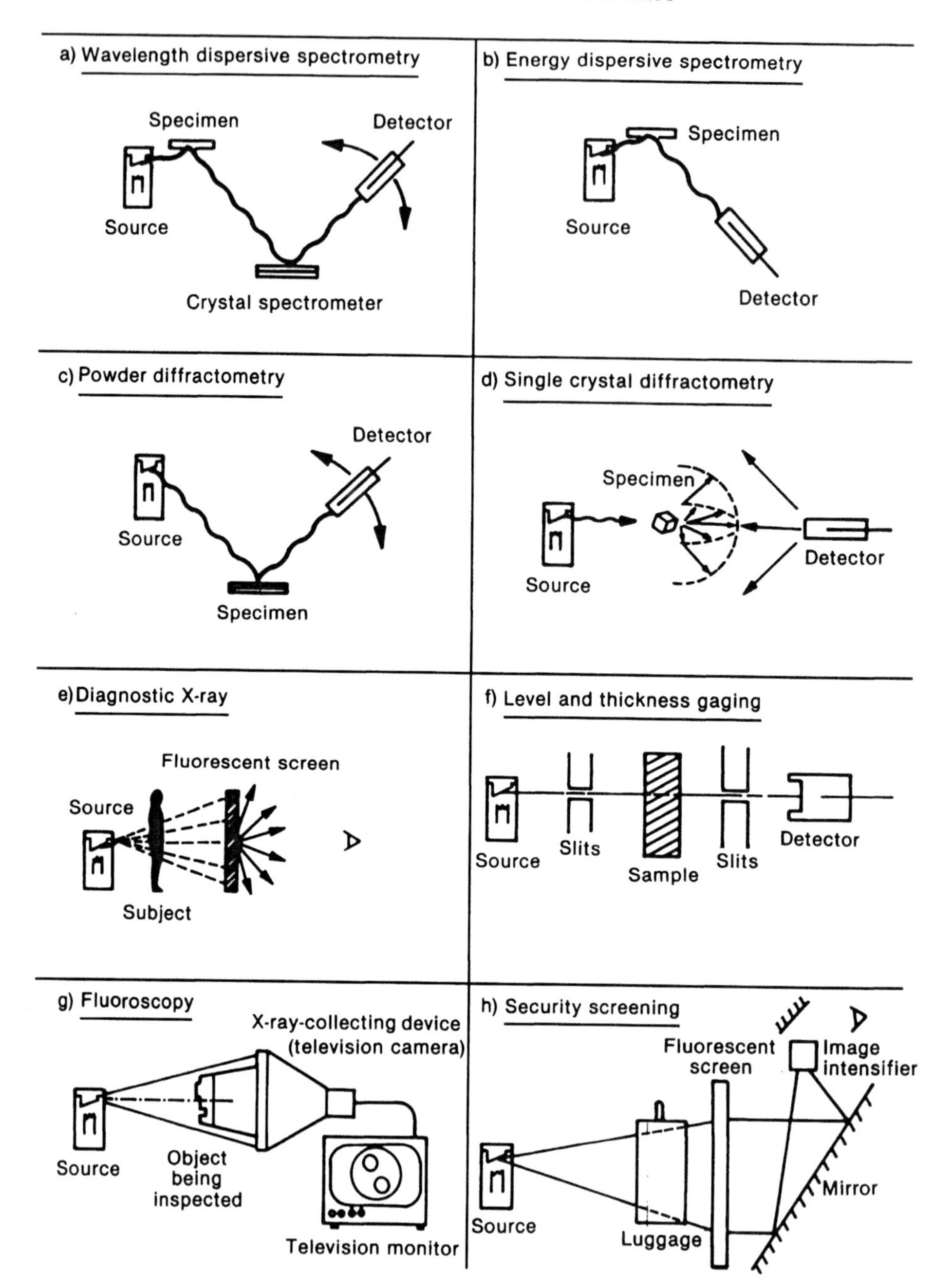

Figure 2.1. Uses of X-ray methods in industry and research.

the research laboratory, rather than finding much direct application in industry.

The other four examples given in Figure 2.1 are all based on the absorption of X-rays. In the following sections, each of these last four methods will be discussed in detail. Briefly, the diagnostic use of X-rays, Figure 2.1(e), is based on the fact that X-ray photons are more strongly absorbed by high atomic numbers than by low atomic numbers. Hence, if a human limb is placed between the X-ray source and a film, a shadow-graph of the bone structure will be observed. The absorption technique can also be used for measuring thickness and depth of absorbing liquids in a opaque container, as in level and thickness gauging, Figure 2.1(f). Where the technique is used for the examination of the gross structure of metal and similar objects, the technique is referred to as X-ray fluoroscopy, Figure 2.1(g). In the last technique, security screening, Figure 2.1(h), X-rays are used to evaluate the contents of luggage.

2.2. DIAGNOSTIC USES OF X-RAYS

One of the first practical applications of the use of X-rays was the radiographic analysis of the human body. Röntgen's original paper, published in 1898, included a photograph of his wife's hand clearly showing the bone structure of the fingers. The potential of this technique was quickly realized and methods became very sophisticated by the early 1900s. It has been estimated that in World War I, as many lives were saved by the use of radiographic techniques as were actually lost in the conflict. Radiographic technique depends on the fact that the absorption of X-rays varies as a third power of atomic number. Hence, in general, higher atomic number elements absorb X-ray photons of a given wavelength more than lower atomic number elements. The bone structure of the human body is high in calcium ($Z = 20$). Bones absorb X-rays to a greater extent than the flesh and soft tissue; thus if a limb is placed between a polychromatic beam of X-rays and a piece of photographic film, transmitted photons falling onto the film will produce a shadowgraph of the bone structure of the limb. In the design of systems for radiographic analysis, one of the variables is the response characteristics of the film. This, in turn, is dependant upon the graininess of the emulsion. Both contrast and sharpness of the film image quality are important to the radiologist, who is invariably seeking to observe minor flaws and discontinuities in the image. Exposure time and image quality can be improved by use of a sandwich arrangement of double-sided emulsion between phosphor screens. A major disadvantage of film is that it represents a passive rather than an active detection device. If a given exposure fails to reveal the detail sought by the radiographer, the exposure may have to be repeated several times.

By direct use of a fluorescence screen in place of the film, it is, in principle, possible to obtain a real-time image. However, in practice, a rather intense primary source would be required in order to give an observable signal. Use of solid state amplifiers as an image intensifier overcomes this problem and today there is a wide variety of these devices available [3]. In its simplest form the solid-state amplifier comprises an X-ray sensitive photocathode plus an electroluminescent screen. The photoconductor changes the local field strength of the electroluminescent screen as a function of X-ray flux, producing light and dark areas. A closed circuit TV system is typically coupled to the output port of the image intensifier for fluoroscopic examination. An important development in real-time imaging was the vidicon tube [4]. The term *vidicon* was first applied to the tubes that replaced the photoelectric charge storage/TV tube combination, the development of which opened up the way towards real-time imaging. The original vidicons used a Sb_2S_3 photoconductor that was scanned by an electron beam at the rear surface. Most modern vidicon systems use PbO, or amorphous selenium, as the photoconductive surface. As the beam scans across the surface of the photoconductor, it deposits a small negative charge, bringing the scanned area to the same potential as the cathode. X-ray photons falling onto the front surface induce conductivity, producing a positive potential proportional to the quantity of radiation. When the scanning beam returns to this particular area, a certain negative charge is required to return the area to cathode potential, and it is the deposition of this negative charge that is used to generate a display signal.

Conventional real-time fluoroscopy uses a vidicon type detector and allows viewing of the image on a TV monitor screen. Use of such a system enables a real-time visualization of dynamically changing situations within the living body. Obtaining sufficient contrast between the volume of interest and the immediate surroundings remains a major experimental problem. Unfortunately, blood and soft tissue have very similar absorption characteristics since their average atomic numbers are almost identical. Contrast can sometimes be enhanced by ingestion or injection of contrast agents immediately before or even during fluoroscopic examination. A simple example would be the ingestion of a *barium meal* by a patient before fluoroscopic examination of the digestive system. In arteriography, injection of image enhancers into the blood stream is employed.

Another method of image improvement is the use of data processing techniques such as image subtraction or image enhancement. In the former method, the significant intrinsic noise associated with the vidicon/TV tube system can be almost completely removed by first storing the experimental image in a suitable digitized form, then subtracting a similar digitized stored image representing the average dark field pixel pattern. Use of lower noise level Reticon systems also helps to reduce the overall background level. These systems have about a six-fold better signal/noise ratio than the vidicon system. As an example, a system is described [5] in which a Reticon camera with a 32×32 photodiode array was used to collect images which were then fed, via an analog to digital converter, to a magnetic disk. This system has been used on an experimental basis to detect pulmonary pulsations in monkeys. Image enhancement techniques are based on taking the signal from an X-ray detector, storing it as a digital image, applying computer-based data processing, and redisplaying the treated image. This can be thought of as a high speed, noise reduction, edge enhancement process, which acts as an interface between the detector system and the final TV output. In many ways it is similar to the sequence employed in analytical X-ray instrumentation where raw data are collected, stored, smoothed, peak maxima sought, and the treated data redisplayed. For real-time applications, this processing of fluoroscopic data allows rates of around 10 MHz pixels. That translates to the processing of $512 \times 512 = 262{,}144$ pixel words in a TV frame in about 1/30th of a second.

2.3. TOMOGRAPHY

One of the newer and extremely valuable techniques of radiography is the method of tomography. In normal radiography, in order to keep the image as sharp as possible during an exposure, relative movement of source, object, and detection medium should be kept at an absolute minimum. In tomography, suitable relative movement is introduced in all planes of the object except for the one of interest. This renders all nondesirable planes blurred except the plane of interest, which is kept sharp. Figure 2.2 compares normal projective imaging with tomographic imaging. In the former case, the two orthogonally placed objects produce an image of similar sharpness. In the latter case, movement of the detector during the course of a contiguous series of exposures focuses just one of the orthogonal members. Tomography can thus be described as sectional röntgenography. By giving the X-ray tube a curvilinear motion, synchronized with the recording medium in the opposite direction, the shadow of the selected plane remains stationary on the film while all other planes have a relative displacement and appear blurred or obliterated. This gives a three–dimensional effect to the viewed image. The tomographic method is invaluable to the radiographer in that it allows revelation of three-dimensional detail that would be completely masked by conventional fluoroscopy.

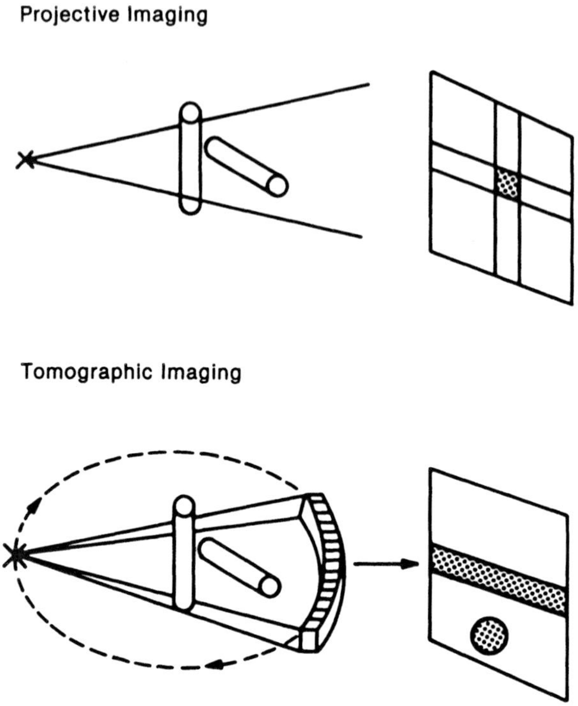

Projective Imaging

Tomographic Imaging

Figure 2.2. X-ray fluoroscopy by projective and tomographic imaging.

The actual hardware for tomographic work typically allows the acquisition of multiple exposures. Traditionally this is done using a single X-ray source and a multiple array detector. As an example, the Philips Tomoscan 310 uses a rotating anode X-ray tube that runs at 280 kW, generating a pulse width of 2 msec and a pulse frequency of 8 msec. The detector is a xenon gas-ionization chamber with 576 individual detector elements. This gives excellent quality pictures in extremely short exposure times, typically of the order of a few seconds. Very sophisticated computer software is also available with these systems, allowing great flexibility in terms of pattern storage, enhancement, redisplay of areas of interest, and so on. Newer developments in tomographic instrumentation utilize multiple X-rays sources, allowing an even higher degree of flexibility. A typical system consists of 28 X-ray sources arranged in a semicircle with 28 corresponding image intensifiers. This entire system permits mathematical reconstruction imaging of a cylindrical three-dimensional volume of 23 cm axially and 30 cm diameter transaxially. This system is applicable to the study of dynamic relationships of anatomic structure and to the function of moving organs, such as the heart and lungs, and circulation. This technique should prove invaluable for the study of patients with cardiopulmonary disabilities or abnormalities of vascular anatomy.

In vascular radiography, the principle of geometric magnification is applied to show the smallest possible anatomical structures. This technique essentially uses a fan-type beam from the X-ray source, which diverges through the object under examination. Details inadequately imaged in a normal radiograph become strikingly visible using this approach, making comprehensive and accurate diagnoses much more probable. The same basic idea can be applied to tomography, but with even greater advantage in that the inefficient use of detectors generally encountered in conventional fan-beam geometry, is avoided by coupling the source and detector together. A direct fan-beam scanner with a rotating X-ray source/detector assembly attached to it is mounted on a rigid frame that can be moved radially with respect to the isocenter. Mechanical movements and adjustments can be made automatically, within the whole gantry, without repositioning the patient. Reconstructive zoom and variable image regeneration capability provide increased visibility of the area of interest without rescanning.

In recent years, the use of X-ray radiography has diminished, with the growth of less invasive imaging techniques based on ultrasound methodology. More recently, the technique known as *optical coherence technology* [OCT] has been developed. It is believed that OCT will soon replace most conventional biopsy methods [6]. OCT is analogous to ultrasound imaging, except that infrared radiation is employed instead of acoustic waves. Early research indicates that OCT may offer resolution at a higher order of magnitude than traditional imaging techniques.

2.4. LEVEL AND THICKNESS GAUGING

The thickness of a solid homogeneous material is readily measurable by X-ray absorptiometry. This method can be applied for both the estimation of bulk thickness and for the measurement of layer thickness. Figure 2.3 shows the three basic techniques that are employed for X-ray gauging and indicates a specimen being irradiated by a collimated primary beam of X-ray photons that may, or may not, have been filtered to change the spectral distribution. In the back-scatter arrangement (Figure 2.3(a)), a collimated detector with an optional filter is placed at an angle from the incident beam, but on the same side of the specimen as that beam. Both scattered and fluorescent radiation

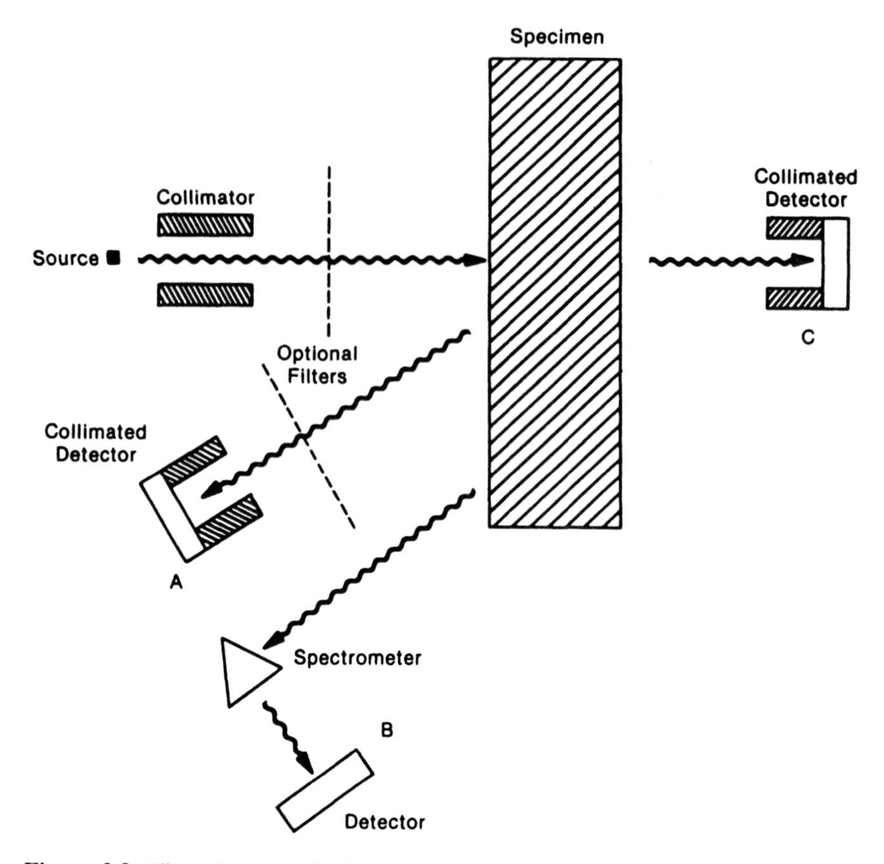

Figure 2.3. Three basic methods for X-ray thickness gauging: (a) back scattering position; (b) fluorescence position; (c) transmission position.

from the specimen can enter the detector. In the fluorescence arrangement (Figure 2.3(b)), a spectrometer is interposed between the specimen and the detector, such that only a selected wavelength(s) can enter the detector. Finally, in the transmission arrangement (Figure 2.3(c)) a collimated detector is placed on the opposite side of the specimen than the source, such that only radiation transmitted by the specimen can enter the detector.

2.5. X-RAY THICKNESS GAUGING

The most common method of measuring the thickness or consistency of a material is using the transmission mode (Figure 2.3(c)). As is seen from

Equation 1.3, the attenuation of the X-ray beam is dependant upon specimen density, thickness, and mass attenuation coefficient. Thus, for a given material, for example, steel, where the density and attenuation coefficient are known, the ratio of incident to transmitted beam intensity can be used to measure the thickness of the steel. As an example, this technique has been employed for many years for the completely automatic measurement of the thickness of sheet steel. Alternatively, where the apparent thickness is known, the same intensity ratio can be used to monitor the variations in absorption coefficient (and, therefore, average atomic number) of the material contained within this layer. As an example, this technique has been used for the determination of the plutonium content and distribution in flat reactor fuel plates [7]. When thickness and mass attenuation coefficient are known, the intensity ratio can be used to measure changes in bulk density due to the presence of voids or holes. This latter technique has been employed with success for porosity measurements in the manufacture of materials such as leather, textiles, storage batteries, and so on. A solution contained within a pipe, for example, has a measurable effect on the total attenuation of an X-ray beam. Thus the absorption method can be used to establish the level of liquid within the closed pipe.

X-ray methods have also been employed with great success for the measurement of layer and coating thickness. Methods based on both back-scatter and the fluorescence arrangement have been used. Figure 2.4 illustrates the different schemes that are typically employed. A polychromatic beam of radiation falling onto a specimen excites characteristic radiation from both layer and substrate elements, in addition to being scattered by the sample. The total back-scatter signal, being atomic number dependant, can provide a simple and inexpensive means of estimating layer thickness or layer average atomic number. This technique is, however, somewhat empirical, and methods based on the selection of the fluorescence signal from layer or substrate wavelength are usually more successful. In Figure 2.4, method (a) is applicable when the attenuation of the radiation intensity from the substrate layer by the surface layer can be used to measure layer thickness. This method is satisfactory provided that the layer thickness is within critical depth, that is, an increase in layer thickness still gives a measurable change in the transmitted intensity. Method (b) can be used when a single wavelength can be selected from the substrate layer radiation. By use of filters or a spectrometer channel, the calculation of layer thickness becomes much easier, and much more precise, because only one value of the mass absorption coefficient has to be considered. Method (c) is used when the layer thickness is within critical depth of the measured wave length. In this instance, the intensity of characteristic radiation from a layer element can be used to estimate the layer thickness.

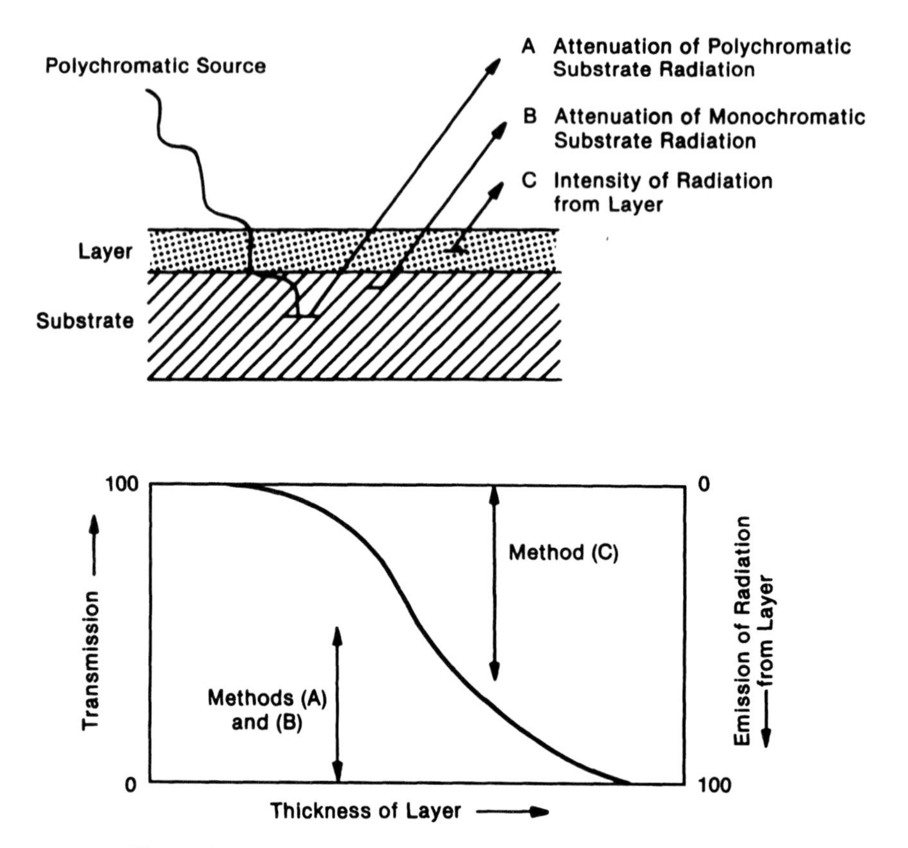

Figure 2.4. Methods for the measurement of coating thickness.

2.6. NONDESTRUCTIVE TESTING

X-ray imaging tests are among the methods most widely used to examine interior regions of metal castings, fusion weldments, composite structures, and brazed components. Radiographic tests are made on pipeline welds, pressure vessels, nuclear fuel rods and other critical materials, and components that may contain three-dimensional voids, inclusions, gaps, or cracks aligned so that the critical areas are parallel to the X-ray beam. Since penetrating radiation tests depend upon the absorption properties of materials on X-ray photons, the tests can reveal changes in thickness and density and the presence of inclusions in the material under examination.

Essentially two basic techniques are employed. In X-ray radiography, the sample being examined is placed between an X-ray source and a film and an exposure produces an absorption shadow-graph of the object being irradiated. In X-ray fluoroscopy, the film is replaced by some form of X-ray transducer that converts the transmitted X-ray signal into a voltage level. Radiography is an effective means of nondestructive testing inspection limited only by the time required for exposure and development of the film. Evaluation must, of course, in this instance take place off-line. This can produce complications, particularly if a large number of objects is being examined and a large number of exposures has to be correlated with these objects. X-ray fluoroscopy avoids this problem and allows direct, on-line examinations to be made. In this case, defects such as porosity or shrink holes show up as contrasting areas on the fluorescent screen. A simple fluoroscopic setup is shown in Figure 2.1(g). The fluoroscopic technique also allows the object under inspection to be manipulated, bringing it into an optimum viewing position. It can thus be observed in motion on a screen in front of the operator providing a facility for virtual three-dimensional observation of the size and shape of a defect or inclusion. In most commercially available systems, the object on the screen can be enlarged, allowing optimization of the image quality. Fluoroscopy can be integrated into a production or quality control situation in such a way that each inspection takes only a few minutes and no intermediate file of paperwork is required. For the inspection of heavy or awkward castings, manipulation of the object itself may be virtually impossible. In such cases, the X-ray tube must be transported to the object and maneuvered to obtain the desired angle for inspection. Castings that can be inspected within an enclosed area, for example a specially designed lead-lined room, may be examined by tubes mounted on mobile or fixed stands, or on overhead suspension systems. Such an arrangement provides an inspection system of maximum flexibility and accessibility that is capable of handling diverse castings of almost any shape and size.

In recent years, technological advances in fluoroscopy have drastically reduced inspection/interpretation times, while providing a level of picture quality and resolution able to meet extremely rigid specifications for light alloy and iron/steel castings, at a thickness used in the majority of cast components. At the same time, the incentive to tighten inspection criteria for castings has increased. As an example, the drive to conserve energy in the automobile industry, both during and after manufacture, has led to the use of more lightweight alloys and introduced newer design concepts. In many areas of modern day research it is necessary to establish what happens inside a given sample. Fluoroscopic techniques allow such examinations to be made. Examples include the examination of nuclear fuel rod containers, or discovering how a projectile behaves inside a gun barrel. In the food processing and packaging industry, X-ray inspection is becoming increasingly important. As the output and use of mass production increases, so do the chances of accidentally introducing foreign bodies, such as glass or metallic particles, into food.

2.7. SECURITY SCREENING SYSTEMS

Within a year of the discovery of X-rays by Röntgen, X-ray imaging was being employed, not just for medical applications, but also by customs and security personnel for examining the contents of packages and parcels. As an example, at the turn of this century, X-ray security screening was used in Paris for detecting the contents of parcels sent to the French Chamber of Deputies. Post office authorities used X-rays to detect coins, mailed hidden in newspaper or sealing wax, contrary to the regulations of the time. Some attempts were also made to apply the technique in the areas of zoology and paleontology. Objects, including ancient mummies and carcasses of sacred Egyptian ibises, have been examined at the Vienna Museum of Natural History. Although the full potential of the penetrating power of X-rays was quickly appreciated, its implementation at that time was rather impractical and hence proved of minimal value for use in security screening, in contrast to its use by the medical profession. The use of X-rays as a basis for security systems was very limited for a period of almost 70 years, even though manufacturers were offering industrial or medical units for security use. These machines were primarily designed for other functions, and the units failed to develop a viable market with no more than a few dozen units ever being used for security screening. However, there was a renewal of interest in the use of X-ray methods for security screening in the early 1960s, mainly for bomb disposal work, and for the screening of mail of prominent people. The need for large dose rates and subsequent safety hazards required the use of rather massive units that were, in consequence, both bulky and slow in operation. Nearly all of this early equipment was still primarily based on designs really intended for medical/industrial use, and they comprised essentially a cabinet type unit fitted with a fluorescent screen viewed through lead glass. X-ray machines finally became viable for security applications in the late 1970s [8]. The main reason was the ability to generate an X-ray image with such a low dose rate of X-rays, that the radiation would not even damage photographic film.

The requirements of security X-ray systems generally demand that they are enclosed, self-contained units (called *cabinet* X-ray systems), that can be operated by relatively unskilled personnel. The units must also be absolutely radiation safe under all operating conditions. A low dose rate unit employs an X-ray dose rate, something like 50 000 times less in terms of radiation flux, compared with high dose systems with direct viewing fluorescent screens. These requirements can be translated into hardware by making the X-ray system in a cabinet, with specifications depending on the degree of inspection detail wanted, inspection rate required, maximum parcel size to be examined, and physical layout of the actual hardware.

Low dose systems are usually designed around a radiation shielded cabinet in which the radiation source itself is contained within a radiation safe enclosure. A conventional system has a control panel, X-ray generator, viewing system, and a door or port through which the luggage to be inspected is loaded. Because low dose systems require much less shielding than high dose systems, items to be inspected can be transported through them at high speed via flexible lead curtains that act as baffles. A typical unit is illustrated in Figure 2.5. The system operates continuously with a low

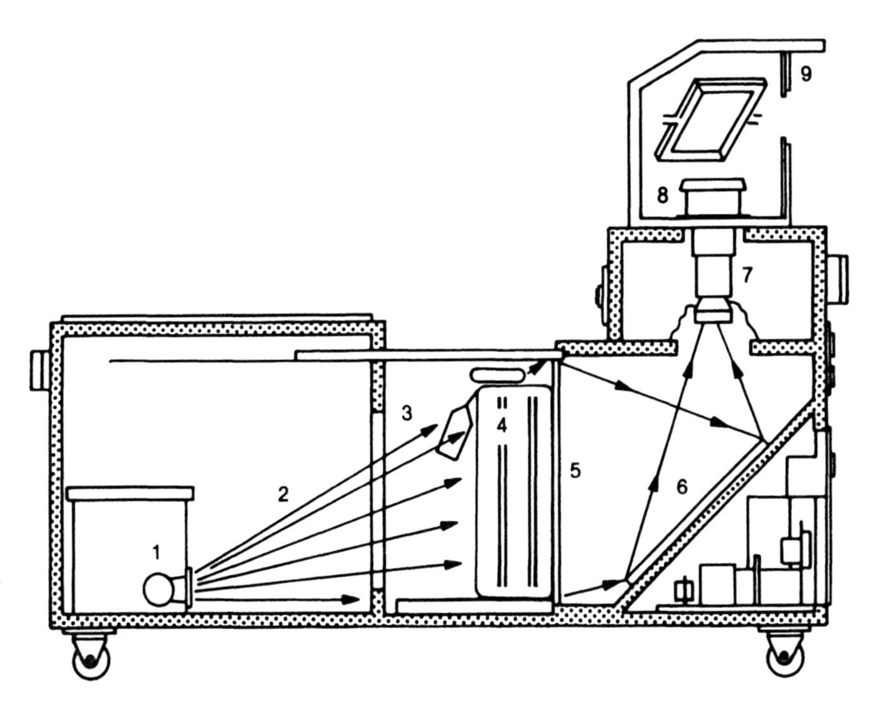

Figure 2.5. Low-dose X-ray security screening system: (1) X-ray tube; (2) X-ray Beam; (3) X-ray chamber; (4) item being inspected; (5) fluorescent screen; (6) mirror; (7) light amplifier or television camera; (8) adjustable viewing mirror; (9) adjustable viewing hood.

level X-ray beam which, when on, causes a fluorescent screen to illuminate. The screen is viewed by a multistage light amplifier that is attached to a closed circuit television camera for viewing on a TV monitor, or an output lens for direct viewing of the phosphor of the last light amplifier module. The pulsed X-ray system produces a single X-ray pulse, which momentarily causes the screen to be illuminated. During this time a closed circuit TV camera views the screen, and the video signal is stored. The image is then repeatedly displayed on a TV monitor. The storage device must keep refreshing the TV monitor picture, and is cleared before the next inspection cycle. The scanning X-ray beam system operates on a two dimensional raster scan in which the beam moves in a vertical line while a conveyor belt moves the parcel under examination in the horizontal plane. By chopping the vertical beam and storing it in an x/y raster that is displayed on the TV monitor, the X-ray image is presented as a series of closely spaced dots of differing brightness. The storage device keeps refreshing the TV monitor picture in the same manner as the pulsed X-ray system. By the early 1980s, there were approximately 1500 low dose systems in the United States, mainly at transportation and public sites. In addition, there are around 200 to 300 high dose systems at governmental and private security locations. All of the low dose systems are of the *film-safe* type, since a high percentage of the traveling and tourist public carry film and cameras on their persons. In the United States, several thousand prohibited weapons and prohibited articles are detected by these screening systems each year. In addition to their actual ability to detect threat items, they also offer a tremendous psychological deterrence to both amateur and professional criminals.

2.8. X-RAY LITHOGRAPHY

The past decade has seen a tremendous growth in the use of microelectronic circuits. Contemporary designs have features that are about two orders of magnitude smaller than can be seen with the naked eye, and the trend to even smaller devices seems likely to continue, giving structural dimensions down to the level of about 100 Å. At this size of component, optical magnifying devices have limited use, and the use of X-ray and ultraviolet based lithographic techniques have successfully provided the fine-line instrumentation required for the manufacture of these devices. Lithography may be simply defined as the art of transferring an image from one surface to another, and is used industrially as a means for making micro-patterns for the high speed reproduction of components. X-ray lithography is essentially proximity printing using soft X-rays [9]. X-ray lithography is used for the

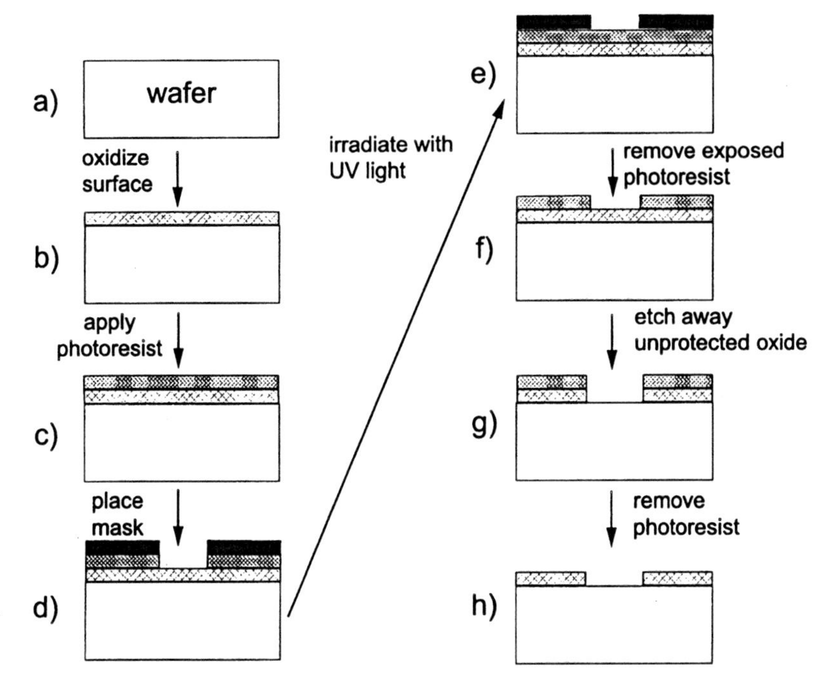

Figure 2.6. Example of X-ray Lithography.

manufacture of the replicate pattern, then chemical etching is used to reproduce this pattern on the required substrate.

The major steps in the lithographic process [10] are illustrated in Figure 2.6. A *p*-type silicon wafer (a) is first oxidized to give a thin surface layer of insulating silicon oxide (b). A layer of photoresist is then applied (c), and after a suitable mask is inserted (d), the wafer is irradiated with ultraviolet (or X-ray) photons (e). During this treatment, the physical properties of the *exposed* photoresist are changed, whereas the *unexposed* photoresist remains unchanged. After removing the mask, solvent is used to dissolve away the exposed photoresist (f). The wafer is then baked to harden the unexposed photoresist and treated with chemical etchants to remove the insulating layer of silicon oxide where it is no longer protected by a photoresist layer (g). Finally, the last photoresist is removed by acid treatment, giving the final product (h).

X-ray lithography is really the application of X-ray microscopic techniques to the fabrication of electronic microcircuits. In X-ray microscopy, an absorption shadow-graph is produced on a photographic plate and then the developed film is viewed using an optical microscope. Although the use of X-ray lithography is a relatively new technique dating back only to the

1970s, there have been rapid advances in the field, and today the technique is widely accepted. An X-ray lithographic exposure station has three essential parts: 1) a highly intense X-ray source; 2) a mask consisting of the device pattern to be transferred, made on some highly absorbing medium, mounted, in turn, on a thin membrane substrate; and 3) an X-ray resist of high resolution suitable for the subsequent device fabrication steps as the X-ray sensitive material. The maximum absorption for most useful resist materials occurs at about 500 Å, but the best replica resolution is obtained at around 75 Å. The sources used to deliver this highly intense X-ray beam might be a rotating anode X-ray tube running at 15 to 25 kW, an electron storage ring which emits synchrotron radiation, or multimillion degree plasmas heated by lasers or electric discharges [11]. Some success has also been achieved with specially designed stationary X-ray sources. The material for the production of the absorbing mask is typically gold supported on a plastic substrate such as mylar, polymethyl methacrylate (PMMA), beryllium, polyimide, and so on. In the actual production of a device using the X-ray lithographic technique, a mixture of ethylene diamine, pyrocatecol, and hydrofluoric acid will etch all types of silicon, except silicon which is heavily doped with boron. The absorber pattern is first transferred to a thin silicon wafer which is doped with boron. The silicon is then etched through the rear side through windows in the protected silicon layer, leaving the boron doped membrane stretched over the frame of silicon which was protected during etching. Membranes produced in this way are transparent to red light and are useful for the wavelength region from 7 to 15 Å. Typical thicknesses are 1 to 2 μm. The major advantages of X-ray lithography relative to other techniques mostly stem from the fact that X-rays are not reflected and are only slightly refracted. In effect, this means that to all intents and purposes X-rays travel in straight lines and are able to accurately reproduce pattern characteristics. The high aspect ratios obtained with X-ray lithography cannot be obtained with any other lithographic technique at this time. It also appears that mask life, using X-rays, is better than using electron beam methods. Against this, however, the disadvantages of the X-ray method is the low overall efficiency of X-ray sources, requiring rather expensive high power generators. Also precise alignment with X-rays is difficult and the first mask is invariably produced with an electron beam. It is hoped that future development of focusing X-ray systems will avoid this costly and inconvenient intermediary step.

2.9. X-RAY ASTRONOMY

One of the oldest sciences is astronomy. For thousands of years man has sought to understand the mysteries of the visible universe. In the 19th century, classical methods of astronomy based on the use of optical telescopes were greatly enhanced by the use of two contemporary inventions, the spectroscope and the camera. In the mid-1940s, a concentrated source of radiowaves was discovered in the constellation Cygnus and the new science of radio astronomy was born. In the early 1960s, another branch of astronomy was initiated, namely X-ray astronomy [12]. X-ray astronomy allows exploration of the universe way beyond the reach of optical and radio astronomy. In 1962, an Aerobee rocket equipped with three Geiger counters was launched from New Mexico and the first measurement of X-rays from a source beyond the solar system was made [13]. In the succeeding 20 years or so, X-ray astronomy has made enormous strides. Today, the latest detectors are able to detect sources many orders of magnitude weaker than were observed with the detectors used in 1962. Cosmic X-rays are generated in regions of extreme violence, either where electrons have been accelerated to very high speeds, or, more commonly, where there is gas at a temperature of millions of degrees. X-ray astronomy thus exposes some of the most exotic and exciting places in the universe. It has, for example, revealed invisible clouds of ultrahot gases on all scales, from the atmosphere of stars to enormous pools that envelop whole clusters of galaxies, as well as streams of gas apparently about to disappear from our universe down the mouths of black holes.

A major experimental problem is that the earth's atmosphere absorbs most cosmic X-ray photons before they reach the earth's surface, the absorption being almost complete at a height of about 30 km. This means that observations can only be made from rockets or orbiting satellites. The first observation of extraterrestrial X-rays was made in 1948 when Freidman's group at the U.S. Naval Research Laboratory in Washington designed experiments in which captured German V2 rockets equipped with Geiger counters were used to detect X-rays from the sun's corona. The NRL group unsuccessfully attempted to find X-ray sources beyond the solar system, but an AS&E / MIT team [14] was able to pinpoint a very intense X-ray source, Sco X-1, in the constellation Scorpius. The strongest X-ray sources in our galaxy are double star systems where a compact neutron star is drawing gas from a normal companion star. As this gas spirals down onto the neutron star's surface, it heats up to about 10 million degrees centigrade and produces copius quantities of X-ray photons. For example, the gas streams in Sco X-1 emit 10 000 million times as much power in X-rays as a star like the sun. Thus the X-ray output from Sco X-1 is a thousand times greater than the sun's power output at all wavelengths. The development of detector arrays allowed directional information to be obtained, and collimators made observations from a single source without influence from others possible. The X-ray telescope was developed in order to obtain more detailed information about specific sources. Since X-ray photons cannot be bent with a lens like visible light, grazing incidence techniques are used. X-rays can be reflected only if they strike a polished metal surface at a very shallow angle. Thus X-ray telescopes have very different configurations from optical and radio reflectors. An X-ray telescope, used simply to concentrate X-rays, uses a slightly tapering cylinder, highly polished on the inside. An imaging X-ray telescope adds a confocal hyperboloid of revolution. X-rays entering parallel to the axis of the telescope and near the edge of the tube strike the inner surfaces at grazing incidence, reflecting to reach the focus further along the axis.

The first major breakthrough in the use of X-ray telescopes above the earth's atmosphere came with the Einstein observatory, which was launched in 1978 (see Figure 2.7). This probe used imaging X-ray mirrors and was able to pinpoint sources with great accuracy. All orbiting observatories have a limited lifetime, which is predicated by the lifetime of batteries or electronics packages, or perhaps until they run out of the gas which is used for driving mechanical movements to sight on one source or another. The longest lived satellite was Copernicus, whose X-ray telescope was still working well when the satellite was turned off after eight years of use. The Einstein observatory ran out of gas in April 1981 and was then closed down after two and one-half years of use, which, despite gyroscope problems in later months, far surpassed its predicted life of one year.

In addition to the use of X-ray techniques for examination and mapping of the universe, they have also provided the means of direct analysis of the

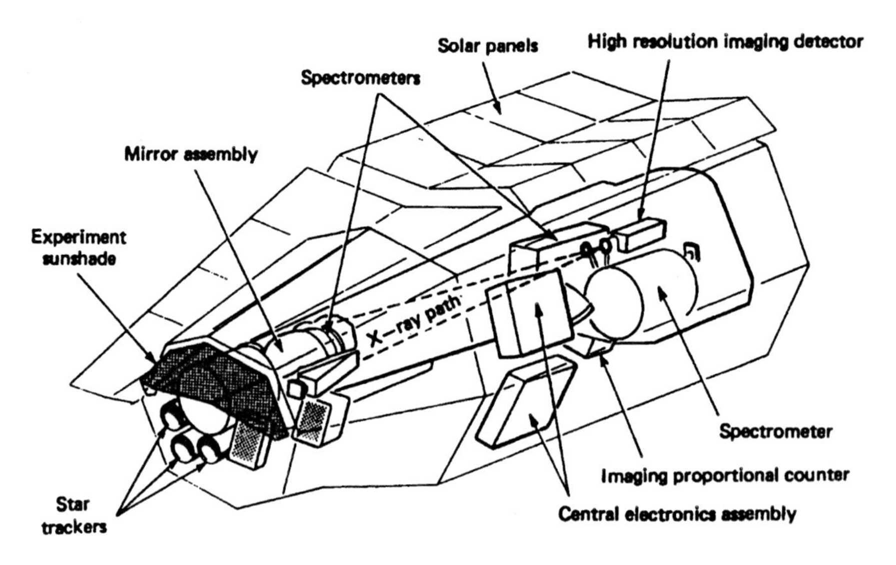

Figure 2.7. The Einstein observatory.

surfaces of some of our neighboring planets. As an example, during the years 1976 and 1977, Surveyors V, VI, and VII provided the first complete and accurate analysis of lunar soils [15]. One of the more recent examples of the use of a transportable instrument is the α-proton, X-ray spectrometer (APXS) for the Mars Pathfinder Mission [16]. The APXS instrument will be carried on the microrover system and has a instrument sensor head which can be remotely placed at soil or rock samples of interest. The principle of the APXS technique is based on three interactions of α particles from a radioisotope source with matter. The first of these interactions is elastic (Rutherford) back scattering. Rutherford scattering is based on the fact that the energy of a back scattered α particle is related to its initial energy and the mass of the target atom. The second process is proton emission. α particles can merge with target atoms, forming a compound nucleus in a highly excited state, followed by the emission of a proton (and, in some instances, γ-radiation). The third interaction is akin to the photoelectric effect resulting in the production of characteristic X-radiation. A pair of silicon detectors is employed to collect emitted and characteristic radiation.

BIBLIOGRAPHY

[1] A. Michette and S. Pfauntsch, *X-rays, the First Hundred Years*, Wiley: New York, 262 pp. (1996).

[2] R. Jenkins, "X-ray Technology," in *Encyclopedia of Chemical Technology*, Vol. 24, Kirk & Othmer, Wiley: New York, pp. 678–708.

[3] W. Kuhl, "Real time radiolographic imaging: medical and industrial applications," *ASTM STP-716*, p. 33 (1980).

[4] E. E. Sheldon, *U.S. Patent #2,817,781*, "Image Storage Device," Dec. (1957).

[5] M. C. Ziskin and C. M. Phillips, "Real time radiolographic imaging: medical and industrial applications," *ASTM STP-716*, p. 294 (1980).

[6] M. Brezinsky, "Optical coherence technology," *Science*, June 17th, p. 2,037 (1997).

[7] M. C. Lambert, *U.S. Atomic Energy Commission OM Rept. HW-57941* (1958).

[8] D. J. Haas, "Now practical 'see through' security," *Security World*, August, 20–24 (1976).

[9] G. A. Garretson, "X-ray lithography primer," *IEEE X-ray Standards Workshop*, October (1981).

[10] S. Muroga, "Very large scale integration design," in *Encyclopedia of Physical Science and Technology*, **14**, 306–327 (1987).

[11] E. Spiller and R. Feder, "X-ray Lithography," in *Topics in Applied Physics*, **22**, Springer-Verlag: New York, p. 35 (1977).

[12] W. Tucker and R. Giacconi, *The X-ray Universe*, Harvard University Press: Cambridge, MA, 201 pp. (1985).

[13] H. Freidman, *Space Research IV*, N. Holland Publishing Co.: Amsterdam (1964) 966 pp.

[14] R. Giacconi, H. Gursky, F. R. Paolini, and B. Rossi, "Evidence for X-rays from sources outside the solar system," *Phys. Rev. Letters*, **9**, 435–439 (1962).

[15] A. Turkevich, et al., "Chemical composition of lunar surface in Mare Tranquillitatis, *Science*, **165**, 277–279 (1969).

[16] R. Rieder, H. Wänke, T. Economou, and A. Turkevich, "Determination of the chemical composition of Martian soil and rocks: the alpha proton X-ray spectrometer," *J. Geophysical Res.*, **102**, E2, 4027–4044 (1997).

CHAPTER

3

X-RAY DIFFRACTION

3.1. USE OF X-RAY DIFFRACTION TO STUDY THE CRYSTALLINE STATE

A crystal consists of atoms or molecules arranged in patterns that are repeated regularly in three dimensions. The smallest repeat unit, the *unit cell*, of this three-dimensional unit may consist of one or many atoms. In a given crystal, all of the individual unit cells must be identical in orientation and composition. While the term *crystallinity* is commonly used to explain diffraction phenomena, it is not an ideal term since it is generally related to physical characteristics, such as shape and luster, observable by eye or with the optical microscope. In practice, there are many cases of materials giving diffraction patterns, where such physical properties are not visibly apparent, because the size of the particles is so small. In fact, the property that gives rise to the interference of coherently scattered X-rays, and hence to the diffraction pattern, is *order*. Almost all solid materials exhibit some degree of regular order and, as shown in Section 1.6, under certain experimental circumstances, this order will give rise to an X-ray diffraction pattern. The X-ray pattern is characteristic of the material from which it was derived, because each unique compound is made up of a similarly unique combination and arrangement of atoms. X-ray diffraction patterns can thus be used to characterize materials. This is the basis of the X-ray powder method.

X-ray patterns are recorded with an almost monochromatic X-ray source and each diffraction peak angle corresponds to one or more d-spacings. Bragg's law, Equation 3.1, is used to convert each observed peak maximum measured in degrees 2θ to d-spacing. When using Bragg's law for powder diffraction, it is customary to write it in a form in which the n does not appear explicitly. That is

$$\lambda = (2d/n)\sin\Theta. \tag{3.1}$$

The d/n term is the d-value that is almost invariably used in powder diffraction. It should be noted that d/n is a submultiple of the separation between adjacent planes that pass through lattice points. The value of λ is known to a few parts per million. Thus measurements of angles at which diffraction occurs make the determination of the d-values possible, with relative ease and with high precision. Bragg's law also shows that since the maximum value of θ is 90°, and hence the maximum value of $\sin\theta$ is unity, the minimum detectable value of d will be equal to $\lambda/2$. The maximum value of d is not limited in any fundamental way but is generally determined by the experimental arrangement.

Useful additional information can be derived by differentiating Bragg's Law

$$\Delta\lambda = (2d \times \cos\theta\Delta\theta) + (2\Delta d \times \sin\theta). \quad (3.2)$$

Since λ is a constant, $\Delta\lambda$ equals zero giving

$$\frac{\Delta d}{d} = -\cot\theta \times \Delta\theta. \quad (3.3)$$

Equation 3.3 is a very useful measure of the relative error in d that results from a given error in θ, that is from $\Delta\theta$. The latter must be expressed in radians. As θ approaches 90°, cot θ approaches zero. Therefore, at 90°, the value of $\Delta d/d$ will always be zero regardless of the error of angular measurement. It is evident that the most precise results will be made at higher angles.

3.2. THE POWDER METHOD

Of all of the methods available to the analytical chemist for materials characterization, only X-ray diffraction is capable of providing general purpose qualitative and quantitative information about the presence of phases (e.g., compounds) in an unknown mixture. While it is true that techniques such as *differential thermal analysis* will provide some information on specific phase systems, such methods cannot be classified as general purpose. As described in the previous section, a diffraction pattern is characteristic of the atomic arrangement within a given phase and, to this extent, it acts as a *fingerprint* of that particular phase. The powder method derives its name from the fact that the specimen is typically in the form of a microcrystalline powder, although, as has been previously indicated, any material which is made up of an ordered array of atoms will give a diffraction pattern. The possibility of using a diffraction pattern as a means of phase identification was recognized many years ago but it was not until the late 1930s that a systematic means of unscrambling, or *search/matching* the superimposed diffraction patterns was proposed [1,2]. The search/matching technique is based on the use of a file, called the *Powder Diffraction File* (PDF),[1] which consists of a collection of single-phase reference patterns, characterized in the first stage by their strongest reflections. The search technique is based on finding potential matches by comparing the strong lines in the unknown pattern with these standard pattern lines. A potential match is then confirmed by a check using the full pattern in question. The identified pattern is then subtracted from the experimental pattern and the procedure repeated on the residue pattern until all lines are identified.

Techniques for the search/matching process have changed little over the years and although, in the hands of experts, manual search/matching is an extremely powerful tool, for the less experienced user it can be rather time consuming. The responsibility for the maintenance of the Powder Diffraction File lies with the International Centre for Diffraction Data, Newtown Square, Pennsylvania, U.S.A. This group is made up of a staff of permanent employees along with a large number of academic and industrial scientists who are active in the field of X-ray powder diffractometry. The PDF is a unique assembly of good quality single-phase patterns and is used by thousands of chemists, geologists, materials scientists, etc., all over the world. In recent years, the automation of the search/match process has made the routine use of the method even more widespread since this has done much to relieve the tedium associated with manual search/matching. The two basic parameters used in the search/match process, are the *d*-values, which have been calculated from the measured 2θ values, and the relative intensities of the lines. Whereas the *d*-value can be precisely measured, perhaps with an accuracy of better than 0.5% in routine analysis, the measured intensities are rather unreliable and can be subject to large errors, due mainly to orientation problems. The ease of any qualitative search procedure is also greatly influenced by the quality of the standard d/I values employed for comparison. The quality of the data in the PDF is quite variable, and is probably of the order of 1 to 20 parts per thousand in *d*, and 5 to 50% in terms of intensity. Because of the great uncertainty in high credence to the *d*-values but much less to the intensities, it is important, that the greatest care be taken in the estimation of experimental *d*-values.

3.3. USE OF X-RAY POWDER CAMERAS

The measured parameters in X-ray powder diffraction are the 2θ maxima and their relative intensities. Each 2θ value is converted to a *d*-spacing, using

[1]PDF is a registered trademark of the JCPDS, International Centre for Diffraction Data Newtown Square, PA.

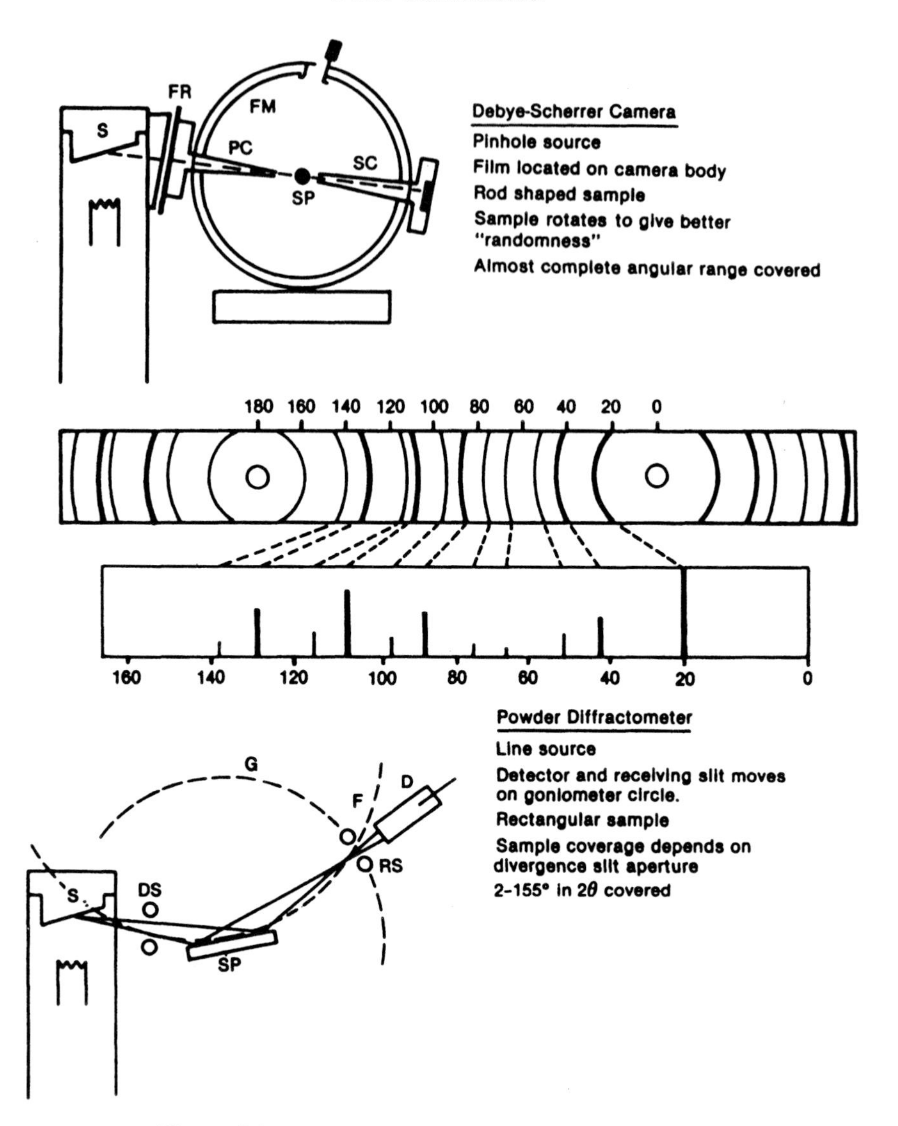

Figure 3.1. Measurement of X-ray diffraction patterns.

Bragg's law to generate a list of *d*-spacings and intensities called a *d/I* list [3]. There are two basic classes of instrument which are used for the measurement of a diffraction pattern, *diffractometers* and film *cameras* (see Figure 3.1). While most modern laboratories employ diffractometers for routine work, the somewhat slower camera methods offer certain advantages, and these still find useful application. Two camera methods are in fairly common use, the Debye-Scherrer method and the Guinier method. Many factors can contribute to the decision to select a camera technique rather than the powder diffractometer for obtaining experimental data, including lack of adequate sample quantity, the need to investigate sample texture, the desire for maximum resolution, and, possibly most important, the lower initial cost.

The upper portion of Figure 3.1 illustrates a typical Debye-Scherrer camera. The component parts include the light-tight camera body and cover that holds the sample mount *SP* and film *FM*, a secondary collimator *SC*, which also acts as the main beam stop, and an incident beam collimator *PC*. A thin metal filter *FR* is placed between the X-ray source *S* and the entrance collimator, to remove the unwanted $K\beta$ component plus some continuum, from the incident beam, while passing most of the $K\alpha$. During operation, the incident beam collimator directs a narrow beam of X-rays onto the sample mounted in the camera center. A second collimator is generally placed behind the specimen to prevent scattered radiation from spoiling the film. The sample is packed in a lithium glass capillary or mounted on the tip of a glass fiber. The sample on its mount is placed at the camera center and aligned to the main beam before film loading, by viewing through the collimator, or an alignment microscope. Since X-ray diffraction is a bulk or mass dependant phenomena, the intensities of the diffracted beams are proportional to the crystalline sample volume. Exposure times are dependant on the size and crystallinity of the sample. An ideal Debye-Scherrer sample is polycrystalline and is comprised of thousands of crystallites of around 5 to 10 μm in size. The sample is rotated during exposure causing the crystallites to assume all possible orientations in the X-ray beam. A crystallite is oriented properly to the beam for diffraction when the incident beam makes an angle θ, with a set of lattice planes. At this instant, a diffracted beam is emitted that also makes an angle θ with the lattice planes and, therefore, an angle of 2θ to the main beam. If the sample is ground fine enough, a smooth and uniform circle of diffracted intensity will appear. If the particle size is too large, however, the circle of diffracted intensity will appear spotty and possibly incomplete. The line intensities from Debye-Scherrer films are normally estimated on a scale of one to ten by visual comparison with a series of lines of varying intensity registered on a standard exposed film. Two factors that contribute to the low resolution of the Debye-Scherrer camera types are first, that incident beam divergence still exists in a collimator designed to yield the narrowest beam with usable intensity, and second, the X-ray filter still passes a broad wavelength band that includes $K\alpha_1$ and $K\alpha_2$ radiation.

Higher resolution can be obtained by going to the Guinier focusing geometry camera. The Guinier geometry makes use of an incident beam monochromator, that is cut and oriented to diffract the $K\alpha_1$ component of the incident radiation. In addition to diffracting a very narrow wavelength band, the monochromator curvature is designed to use the whole surface of the crystal to diffract simultaneously, thus yielding a large diffracted intensity. As a consequence of this curvature and the orientation of the crystal itself, the monchromator not only easily separates $K\alpha_1$from $K\alpha_2$, but also converts the divergent incident beam into an intense convergent diffracted beam focused onto a sharp line. Although the quality of data from a well-aligned Guinier camera is high, the camera does have its limitations, especially in the estimation of line intensities. A well-equipped X-ray diffraction laboratory should probably have access to both Debye-Scherrer and Guinier camera techniques, in addition to a diffractometer.

3.4. THE POWDER DIFFRACTOMETER

The lower portion of Figure 3.1 illustrates the powder diffractometer. The powder diffractometer was developed in the late 1940s and has changed little from the original concept. The major difference that is found in modern instrumentation is the use of the minicomputer for control, data acquisition, and data processing. The geometric arrangement employed in the powder diffractometer is known as the Bragg-Brentano parafocusing system and is typified by a diverging beam from an X-ray line source *S*, falling onto a large flat specimen *SP*, being diffracted and passing through a receiving slit *RS* to the detector *D*. The amount of divergence is determined by the effective focal width of the source and the aperture of the divergence slit. Use of the narrower divergence slit will give a smaller specimen coverage at a given diffraction angle, thus allowing the attainment of lower diffraction angles where the specimen has a larger apparent surface. (Thus larger values of *d* are attainable.) Axial divergence is controlled by two sets of parallel plate collimators (Soller slits) placed between focus and specimen, and specimen and scatter slit, respectively. Two circles are generated by this geometry, the focusing circle *F*, and the goniometer circle *G*. The source, the specimen, and the receiving slit, all lie on the focusing circle, which has a variable radius. The specimen lies at the center of the goniometer circle and this has a fixed radius. The instrumental line width of the diffracted profile will be determined mainly by the angular aperture of the receiving slit, but the intensity of this line will be dependant both on slit aperture and the focal spot characteristics of the X-ray tube.

A photon detector, typically a scintillation detector, is placed behind the receiving slit and this converts the diffracted X-ray photons into voltage pulses. These pulses may be integrated in a ratemeter to give an analog signal on an x/t recorder. In modern instruments, the pulses are processed by a computer and displayed directly on the computer terminal. By synchronizing the scanning speed of the goniometer with the recorder, a plot is obtained of degrees 2θ versus intensity. This plot is called a diffractogram. A timer/scaler is also provided for quantitative work and this is used to obtain a measure of the integrated peak intensity of a selected line(s) from each analyte phase in the specimen. A diffracted beam monochromator may also be used to improve signal to noise characteristics. The output from the diffractometer is a powder diagram, essentially a plot of intensity as a function of diffraction angle, which may be in the form of a strip chart or a hard copy from a computer graphics terminal.

3.5. QUALITATIVE APPLICATIONS OF THE X-RAY POWDER METHOD

As has been previously stated, a polycrystalline phase diffracts a monochromatic beam of X-rays into a spectrum or *pattern* of lines that can be recorded by diffractometer or camera techniques. Each line in the diffraction pattern, in terms of its position and intensity, represents a particular family of crystal planes in the phase. Since the sequence of planes is unique to the phase, the X-ray pattern is, in turn, unique enough to act as a fingerprint for the identification of the phase. The patterns of individual phases differ widely in the sequence and intensities of their lines and these effects are used in the qualitative identification process generally known as search/matching. Figure 3.2 gives an example of a single entry (*card*) from the PDF. A search manual is used for the identification of a phase provided that the pattern of that phase is in the manual. The complete PDF comprises nearly 115 000 phases as of 1998. A simple alphabetical index based upon the name or the chemical formula of a material provides one practical means of very effectively using this large reference library of patterns. Because of the problems of pattern recognition with a data file this large, the PDF is also available as subsets. Currently available are subfiles of inorganics, organics, minerals, metals and alloys, NBS patterns, forensics, zeolites, explosives, superconductors, and common phases; and new potential subfiles are always under consideration. These data are available in card form, microfiche, books, and on computer disks, magnetic tape, or CD-ROM. Concurrently with the buildup of the PDF, there has been considerable effort to develop both manual and computer-based retrieval methods, to match an unknown pattern to one in the PDF. The retrieval of patterns is somewhat analogous to the matching of human fingerprints which makes use of the characteristic loops, whorls, arches, pockets, etc. The problem of pattern matching in X-ray powder diffraction is generally easy for patterns obtained from single phases,

46-1045

SiO2

Silicon Oxide

Quartz, syn

Rad.: CuKa1 λ: 1.540598 Filter: Ge Mono d-sp: Diff.

Cut off: Int.: Diffract. I/Icor.: 3.41

Ref: Kern, A., Eysel, W., Mineralogisch-Petrograph. Inst., Univ. Heidelberg, Germany, ICDD Grant-in-Aid, (1993)

Sys.: Hexagonal S.G.: $P3_221$ (154)

a: 4.91344(4) b: c: 5.40524(8) A: C: 1.1001

α: β: γ: Z: 3 mp:

Ref: Ibid.

Dx: 2.649 Dm: 2.660 SS/FOM: F_{30} = 539(.0018 , 31)

εα: ηωβ: 1.544 εγ: 1.553 Sign: + 2V:

Ref: Swanson, Fuyat, Natl. Bur. Stand. (U.S.), Circ. 539, 3, 24 (1954)

Color: White
Integrated intensities. Pattern taken at 23(1) C. Low temperature quartz. 2θ determination based on profile fit method. 02 Si type. Quartz group. Silicon used as an internal stand. PSC: hP9. To replace 33-1161. Mwt: 60.08. Volume[CD]: 113.01.

Wavelength= 1.54056 *

d(A)	Int	h	k	l	d(A)	Int	h	k
4.2549	16	1	0	0	1.0638	<1	4	0
3.3434	100	1	0	1	1.0477	1	1	0
2.4568	9	1	1	0	1.0438	<1	4	0
2.2814	8	1	0	2	1.0346	1	2	1
2.2361	4	1	1	1	1.0149	1	2	2
2.1277	6	2	0	0	.9896	<1	1	1
1.9798	4	2	0	1	.9872	<1	3	1
1.8179	13	1	1	2	.9783	<1	3	0
1.8017	<1	0	0	3	.9762	<1	3	2
1.6717	4	2	0	2	.9608	<1	3	2
1.6591	2	1	0	3	.9285	<1	4	1
1.6082	<1	2	1	0	.9181	<1	3	2
1.5415	9	2	1	1	.9161	2	4	0
1.4528	2	1	1	3	.9152	2	4	1
1.4184	<1	3	0	0	.9089	<1	2	2
1.3821	6	2	1	2	.9008	<1	0	0
1.3749	7	2	0	3	.8972	<1	2	1
1.3718	5	3	0	1	.8889	1	3	1
1.2879	2	1	0	4	.8813	<1	1	0
1.2559	3	3	0	2	.8782	<1	4	1
1.2283	1	2	2	0	.8598	<1	3	0
1.1998	2	2	1	3	.8458	<1	1	1
1.1977	<1	2	2	1	.8407	<1	5	0
1.1839	2	1	1	4	.8359	<1	4	0
1.1801	2	3	1	0	.8296	1	2	0
1.1529	1	3	1	1	.8254	2	4	1
1.1406	<1	2	0	4	.8189	<1	3	3
1.1145	<1	3	0	3	.8117	3	5	0
1.0815	2	3	1	2	.8097	<1	3	3

Figure 3.2. An example of a PDF card image.

but becomes more complicated for actual unknowns where the pattern may represent a mixture of several phases.

The most commonly employed manual search/match procedure is the Hanawalt system. The Hanawalt search manual is organized into groups and subgroups. Two lines from the table of *d*-values serve to locate the entry of the pattern in the manual. The *d*-value of the strongest line of the pattern determines the entry group, sometimes called the Hanawalt Group, and the second strongest line determines the subgroup, that is, the location within that group. An individual entry also lists the *d*-values of the next six strongest lines in the pattern in order of decreasing intensities. In order to identify an unknown, the first and second strongest lines are chosen as the line-pair with which to look for a matching entry in the search manual. If the remaining six lines of a qualifying entry also match lines in the unknown pattern, then the identification is probably quite certain. The final step in the procedure is to confirm the identification by going to the PDF to check the complete standard reference pattern. If the choice of line-pair does not lead to identification, successively weaker lines are chosen to use with first line. If still unsuccessful, a new start is made with a different choice of first line.

Although the Hanawalt method for searching the powder data file is the most popular search/match method, it has not always proven completely successful for identifying unknown diffraction patterns in which the intensity values are markedly different from those in the file patterns, even though the *d*-values themselves might be very accurate. The problem may arise when fewer than two of the three strong lines in the experimental pattern are recognized for use in the search procedure. This situation often occurs when data are obtained by electron or neutron diffraction. In the X-ray diffraction case, it may also occur when the sample shows a very strong preferred orientation, or where there are insufficient crystallites in the sample to achieve ideal randomness. An alternative search method, which partially compensates for such intensity problems, is the Fink method [4]. The Fink method relies almost exclusively on the largest *d*-spacings of the pattern. The basic assumption is that, although there may be significant differences in relative intensities in, for example, the electron diffraction pattern when compared with the corresponding X-ray pattern, most of the strong lines are observable in both patterns. The Fink search manual lists the first eight *d*-values of each reference pattern. To start the search, the user lists the largest *d*-values from the unknown material and then attempts to find a match based on the *d*-values.

It is apparent that a basic sequence of search/match steps is amenable to modern computer techniques. In the mid-1960s, three main-frame computer-based search programs were described almost simultaneously. In 1965, Frevel et al. [5] described a program to perform an entire search and match process, based on standard files prepared from the PDF. The program was called *ZRD*, because it uses atomic number *Z*, functional group information *R*, and *d*-spacing data *D*. Other computer programs were described by Johnson and Vand [6] and Nichols [7]. In the following years, each of the programs has been modified in some manner, either to improve operation, generalize the application, include many types of data files and computers, or incorporate successful features found in other search/match programs [8,9]. More recently, most major X-ray instrument manufacturers have developed search/match programs capable of performing searches on a personal computer, of the size typically supplied with automated powder diffractometers. All are highly dependent on data quality, both that of the unknown pattern and the reference pattern. In order to quantify the success of computer search/matching and to compare with hand-searching techniques, the ICDD has run several round-robin tests in which standard samples and data sets have been circulated to typical users for analysis. In each case it was found that high success rates are observed when unknown and standard reference patterns are of high quality [10, 11].

The rapid incorporation of minicomputers into diffraction instrumentation has provided a great opportunity to perform rapid data collection and sophisticated data reduction, followed by search/match procedures on an automated or semiautomated basis, completely in-house and unattended (see e.g., [12, 13]). In the United States today more than 90% of all new powder diffractometers sold are automated to some degree, although still less than 10% or so of the 30 000 diffraction users in the world are currently using computers for search/matching. The advent of the CD-ROM in the late 1980s, meant that the whole Powder Data File could be stored easily, along with suitable search/match programs, to be run on a personal computer [14]. This CD-ROM offers 540 megabytes of storage and is available at reasonable cost. Stand-alone PC based search systems are available [15], allowing even the most modestly equipped X-ray laboratory the opportunity to use the very latest automated search/match software.

3.6. QUANTITATIVE METHODS IN X-RAY POWDER DIFFRACTION

Once the presence of a phase has been established in a given specimen, it can, at least in principle, be determined how much of that phase is present, by use of the intensities of one or more diffraction lines from the phase. The intensities of the diffraction peaks are dependant on a number of different factors, falling into three categories:

- *Structure dependant.* A function of atomic size and atomic arrangement, plus some dependance on the scattering angle and temperature.
- *Instrument dependant.* A function of diffractometer conditions, source power, slit widths, detector efficiency, etc.
- *Specimen dependant.* A function of phase composition, specimen absorption, particle size, distribution, and orientation.

For a given phase, or selection of phases, all structure dependant terms are fixed and in this instance have no influence on the quantitative procedure. Provided that the diffractometer terms are constant, these effects can also be ignored. Thus if the diffractometer is calibrated with a sample of the pure phase of interest, and then the same conditions are used for the analysis of the same phase in an unknown mixture, only the random errors associated with a given observation of intensity have to be considered. The biggest problem in the quantitative analysis of multiphase mixtures remains the specimen dependant terms, and specifically those dependant upon particle size and distribution, plus effects of absorption.

The absorption effect has already been referred to in Section 1.4 and clearly, in a multiphase mixture, different phases will absorb the diffracted photons by different amounts. As an example, the mass attenuation coefficient for Cu $K\alpha$ radiation is 308 (cm^2/g) for iron, but only 61 (cm^2/g) for silicon. Thus iron atoms are five times more efficient in absorbing Cu $K\alpha$ photons than are silicon atoms. There are a variety of standard procedures for correcting for the absorption problem. The most common by far is the use of an internal standard. In this method, a standard phase is chosen which has about the same mass attenuation coefficient as the analyte phase. A weighed amount of this material is added to the unknown sample, and the intensities of lines from the analyte phase, and the internal standard phase, are then used to estimate the relative concentrations of internal standard and analyte phases. The relative sensitivity of the diffractometer for these two phases is determined by a separate experiment. Other procedures are available for the analysis of complex mixtures, but these are beyond the scope of this particular work. For further information, the reader is referred to specific texts dealing with the X-ray powder method (see e.g., [16]).

While the treatment of the absorption effect is fairly simple, the handling of particle problems is much more complex. As has been previously stated, ideally the powder method utilizes a specimen that is randomly oriented, since the geometry of the system requires that an equal number of crystallites be in the correct position (i.e., orientation) to diffract at any angle of the goniometer. When particles lie in a preferred orientation, there will be more particles available to diffract at the angle corresponding to this orientation and, what is equally important, fewer particles available to diffract at other diffraction angles. The overall effect may be to enhance some intensities and to diminish others. In other words, the intensities are now dependant upon particle distribution and orientation. Some materials, just by virtue of their crystal habit, may become preferred during sample preparation. As an example, mica, being a rather *platy* material, will prefer to stack one plate on top of another rather than take up a random orientation. The overall effect of preferred orientation can vary from insignificance, to errors of the order of tens of percent. Careful specimen preparation is always called for and preparation may include grinding, sieving, spray-drying, or other suitable techniques. Tests have been made comparing these various methods and, as was suggested previously, in extreme cases the intensities vary by as much as an order of magnitude [17].

The areas of application of quantitative X-ray powder diffraction are many and varied, and hundreds of analysts are using this technique on a daily basis. Some of the more common applications include ore and mineral analysis, quality control of rutile/anatase mixtures, retained austenite in steels, determination of phases in airborne particulates, various thin film applications, study of catalysts, and analysis of cements. The current state of the art in the quantitative analysis of multiphase materials is that accuracies of the order of one percent or so can be obtained in those cases where the particle orientation effect is either nonexistent, or its effects have been adequately compensated.

3.7. OTHER APPLICATIONS OF X-RAY DIFFRACTION

In addition to the many applications of qualitative and quantitative phase analysis already described, there are a variety of other applications of X-ray diffraction that should be mentioned. While the diagrams shown in Figure 3.3 by no means represent an exhaustive list, it should at least give an indication of the versatility of X-ray diffraction methods. Figure 3.3(a) illustrates the use of X-ray powder diffractometry at nonambient temperatures. Most major instrument manufacturers supply different attachments to a standard powder diffractometer which allow measurements at temperatures from ambient up to about 2000 °C, or down to the temperature of liquid nitrogen. Such attachments are invaluable for the study of phase transformations, especially where computerized systems allow the display of a three-dimensional *intensity* 2θ angle/temperature diagram. This technique can be combined with the use of inert or other special atmospheres within the sample chamber.

Other applications of high temperature diffraction include the study of thermal expansion, recrystallization, thermal decomposition, solid state addition and replacement reactions, etc. The second diagram, Figure 3.3(b),

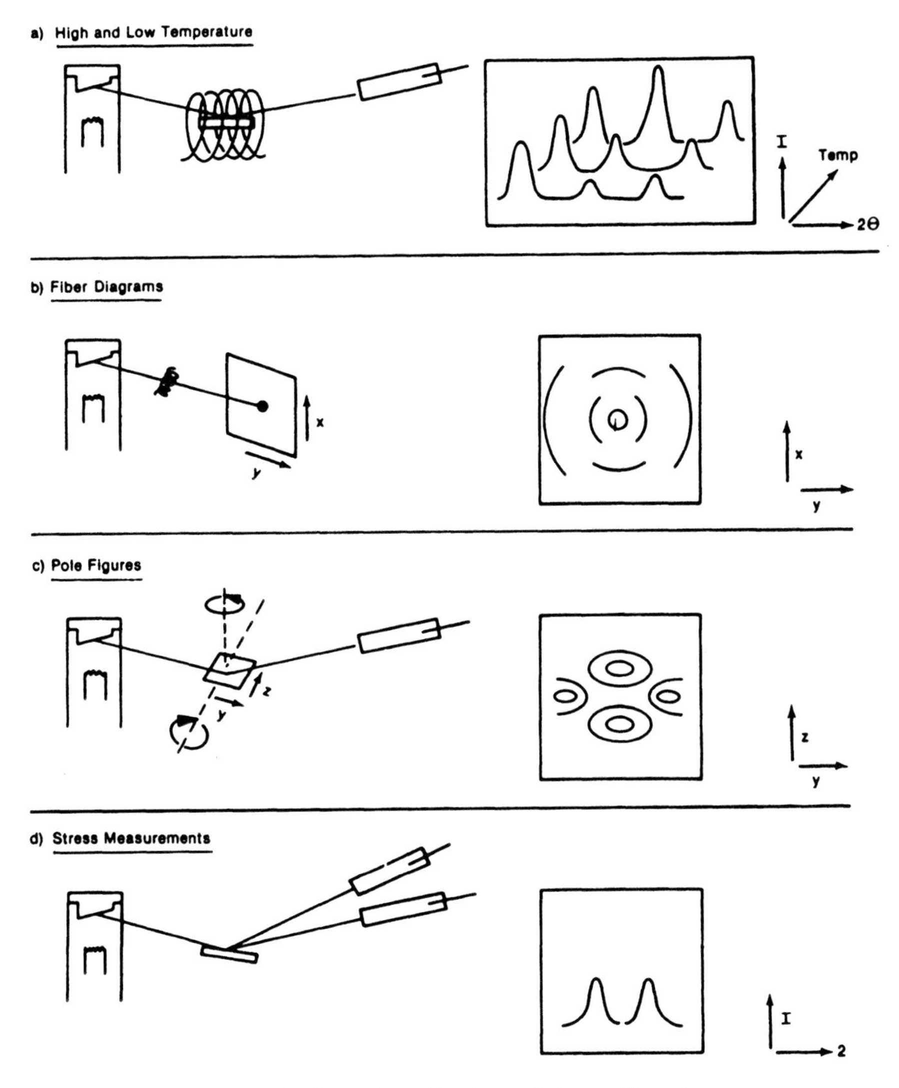

Figure 3.3. Special applications of X-ray diffraction.

illustrates the use of X-ray diffraction for the study of fibers. In this experimental arrangement, the fiber is mounted in a specific direction between the X-ray source and a flat piece of film. When the exposure is complete and the film is developed, a number of noncontinuous diffraction rings will be seen on the film. The lengths, widths, and positions of these rings can be used to establish the preferred orientation and the degree of crystallinity of the specimen (see e.g., [16], Chapter 10). As was stated in the last section, where nonrandom orientation of particles is present in a specimen, the intensities around a given diffraction ring will be nonuniform. Most natural and artificial fibers show some of this preferred orientation because of the orientation, during growth, of the long chain molecules from which they are made. The physical and mechanical properties of the fiber are very dependant upon this orientation. The study of multidimensional intensity distribution can also be used for the examination of flat specimens, by representing a stereographic projection of the intensity as a *Pole Figure* diagram, as illustrated in Figure 3.3(c). In a pole figure (texture) goniometer, two additional movements are used to rotate and tilt the specimen at a fixed $\theta/2\theta$ angle setting. Coupling of the tilt and rotation allows the three-dimensional space above the specimen to be sampled for X-ray intensity. The best way to conceive of this movement, is to imagine a continuous line drawn from a point on a ball, spiraling away from the point until the line reaches the maximum diameter of the ball. A line normal to the surface of the specimen, and at its center point (the pole), tracks this continuous line exactly. Special charts are used to plot the pole figure information. Pole figures are used for the study of preferred orientation, or *texture*, in metals, plastics, thin films, etc., and are invaluable for the study and prediction of flow and plastic deformation due to mechanical treatments such as rolling and pressing.

Another important use of X-ray diffraction in industrial processes is the study of residual stress, illustrated in Figure 3.3(d). When a uniform metal bar is subject to tension, either by pulling, compression, or bending, stress will cause deformation of the individual unit cells of the crystal structure. Such deformation will cause shifting and/or broadening of the diffracted line profile from a given reflection. Use of two measurements, one in the correct Bragg position and one slightly displaced from the Bragg position, avoids the need for an unstressed comparison sample, and allows the individual principle stresses to be calculated. The X-ray method is particularly useful for the determination of residual stress because it is applicable to the specimen as is, and does not involve pretreatment of the sample. Automated machines are available today that allow the rapid measurement of stress both on laboratory test specimens, as well as large finished pieces such as weldments in pipe lines, aircraft wings, and so on.

BIBLIOGRAPHY

[1] J. D. Hanawalt and H. W. Rinn, "Identification of crystalline materials", *Ind. Eng. Chem.*, Anal. Ed., **8**, 244–247 (1936).

[2] J. D. Hanawalt, H. W. Rinn, and L. K. Frevel, "Chemical analysis by X-ray diffraction—classification and use of X-ray diffraction patterns," *Ind. Eng. Chem., Anal. Ed.*, **10**, 457–512 (1938).

[3] R. Jenkins and R. L. Snyder, *X-ray Powder Diffractometry*, Wiley Interscience: New York, 403 pp. (1996).

[4] W. C. Bigelow and J. V. Smith, "Two new indexes to the powder diffraction file," *ASTM Spec. Tech. Publ. STP-372*, 54–89 (1965).

[5] L. K. Frevel, C. E. Adams, and L. E. Ruhberg, "A fast search program for powder diffraction analysis," *J. Appl. Cryst.*, **9**, 300–305 (1976).

[6] G. G. Johnson, Jr. and V.A. Vand, "A computerized powder diffraction identification system," *Ind. Eng. Chem.*, **59**, 19 (1965).

[7] M. C. Nichols, "A FORTRAN II program for the identification of X-ray powder diffraction patterns." Lawrence Livermore Laboratory [Rep.] UCRL-70078 (1966).

[8] R. G. Marquart, et al., "A search-match system for X-ray powder diffraction data," *J. Appl. Cryst.*, **12**, 629–634 (1979).

[9] J. W. Edmonds, "Generalization of the Frevel ZRD search-match program for powder diffraction analysis," *J. App. Cryst.*, **13**, 191–192 (1980).

[10] R. Jenkins, "A round robin test to evaluate computer search/match methods for qualitative powder diffraction," *Adv. X-ray Anal.*, **20**, 125–137 (1977).

[11] R. Jenkins and C. R. Hubbard, "A preliminary report on the design and results of the second round robin to evaluate search/match methods for qualitative powder diffraction," *Adv. X-ray Anal.*, **22**, 133–142 (1978).

[12] R. Jenkins, et al., "Automated powder diffractometry, new dimensions in instrumentation and analytical software," *Norelco Reporter Special Issue*, February (1983).

[13] R. G. Marquart, et al., "A search/match system for X-ray powder diffraction data," *J. App. Cryst.*, **12**, 629–634 (1979).

[14] R. Jenkins and M. A. Holomany, "PC-PDF a search/display system utilizing the CD-ROM and the complete powder diffraction file," *Powder Diff.*, **2**, 215–219, (1987).

[15] R. G. Marquart, "μPDSM: mainframe search/match on an IBM PC," *Powder Diff.*, **1**, 34–39 (1986).

[16] H. P. Klug and L. E. Alexander, *X-Ray Diffraction Procedures, 2nd ed.*, Wiley: New York, 966 pp. (1974).

[17] L. D. Calvert, A. F. Sirianni, G. J. Gainsford and C. R. Hubbard, "A comparison of methods for reducing preferred orientation," *Adv. X-ray Anal.*, **26**, 105–110 (1982).

CHAPTER

4

X-RAY SPECTRA

4.1. INTRODUCTION

Section 1.3 shows that one of the methods by which an excited atom can revert to its original ground state is by transferring an electron from an outer atomic level to fill the vacancy in the inner shell. An X-ray photon is then emitted from the atom with an energy equal to the energy difference between the initial and final states of the transferred electron. In practice, millions of atoms are involved in the excitation of a given specimen and all possible deexcitation routes are taken. It was also indicated that the various deexcitation routes can be defined by a simple set of selection rules that account for the majority of the observed wavelengths. The classically accepted nomenclature system for the observed lines is that proposed by Seigbahn in the 1920s, and is referred to as *the Siegbahn system.* The wavelengths that fit the selection rules are called *normal* or *diagram* lines.

4.2. ELECTRON CONFIGURATION OF THE ELEMENTS

Atoms are made up of nuclei surrounded by electrons, the number of electrons being equal to the atomic number (i.e., the number of nuclear protons) of the atom. Thus magnesium, atomic number 12, has 12 electrons; barium, atomic number 56, has 56 electrons; and so on. The configuration of the electrons within an atom follows a definite pattern and simple rules can be applied to predict their states. Each electron represents a certain amount of energy and this energy can be described by four parameters, the so-called *quantum numbers* of the electron. These quantum numbers are n, the *principal quantum number*, which can take positive integral values 1, 2, 3, 4, ...; ℓ, the *angular quantum number*, which can have values of O through $(n-1)$; m, the *magnetic quantum number*, which can have values of ℓ through O through $+\ell$; and s, the *spin quantum number*, which has values of $\pm 1/2$.

The Pauli exclusion principle states that no two electrons within an atom can have the same set of quantum numbers, thus there can only be two

Table 4.1. Atomic Structures of the First Three Principal Shells.

Shell / Electrons	n	l	m	s	Orbitals	J
K (2)	1	0	0	$\pm\frac{1}{2}$	1s	$\frac{1}{2}$
L (8)	2	0	0	$\pm\frac{1}{2}$	2s	$\frac{1}{2}$
	2	1	1	$\pm\frac{1}{2}$		
	2	1	0	$\pm\frac{1}{2}$	2p	$\frac{1}{2};\frac{3}{2}$
	2	1	−1	$\pm\frac{1}{2}$		
M (18)	3	0	0	$\pm\frac{1}{2}$	3s	$\frac{1}{2}$
	3	1	1	$\pm\frac{1}{2}$		
	3	1	1	$\pm\frac{1}{2}$	3p	$\frac{1}{2},\frac{3}{2}$
	3	1	−1	$\pm\frac{1}{2}$		
	3	2	2	$\pm\frac{1}{2}$		
	3	2	1	$\pm\frac{1}{2}$		
	3	2	0	$\pm\frac{1}{2}$	3d	$\frac{3}{2},\frac{5}{2}$
	3	2	−1	$\pm\frac{1}{2}$		
	3	2	−2	$\pm\frac{1}{2}$		

electrons in the first principal subshell where $n = 1$. Similarly, there are eight combinations for $n = 2$, 18 for $n = 3$, etc. In general, there will be $2n^2$ possible combinations. Table 4.1 shows the orbitals that make up the principal subshells, and it can be seen that where $\ell = 0$ the orbital is an s (sharp) orbital; where $\ell = 1$ the orbital is a p (principle) orbital; where $\ell = 2$ the orbital is a d (diffuse) orbital; and where $\ell = 3$ the orbital is an f (fundamental) orbital. Thus there is a logical order in the actual electron configurations for all of the elements in the periodic classification and it can be seen that, with certain exceptions, elements of increasing atomic number are built up by a regular sequence of electron additions in the order 1s, 2s, 2p, 3s, 3p, 4s, 3d, 4p, etc. A good deal of interaction is possible between the electrons, particularly where one is considering the energy required to displace an electron from its normal or ground state configuration. One of the most important of these synergistic effects is described by the coupling of the ℓ and s quantum numbers, the result being a vector sum of the ℓ and s moments. Thus

$$J = \ell + s, \tag{4.1}$$

where J is the total moment. The combination of these quantum numbers gives rise to the so-called transition levels.

4.3. FLUORESCENT YIELD

The process of the displacement of an electron from its normal or *ground state* is called *excitation.* Figure 4.1(a) illustrates the removal of a K level electron by a primary photon having an energy equal to $h\nu - \phi_K$ where ϕ_K is the binding energy of the K electron, leaving the atom in the K^+ state. The atom can return to the nonexcited or *ground state* by various processes, two of which are dominant. The first of these is where an electron from one of the upper levels (i.e., a level with a value of n larger than that of the ionized level) falls to the excited level. Such an electron transference will always be accompanied by the emission of radiation. The second process is similar in its initial stage to the first, but in this case, the radiation produced following the transference of an electron from an upper level is not emitted, but gives up its energy by further ionization within the atom.

This second process is known as the Auger process [1] and is illustrated in Figure 4.1(b). Here the initial excitation stage produced a K vacancy and this was followed by a deexcitation involving transference of an L electron. Such a transfer results in the production of $K\alpha$ radiation but, in this instance,

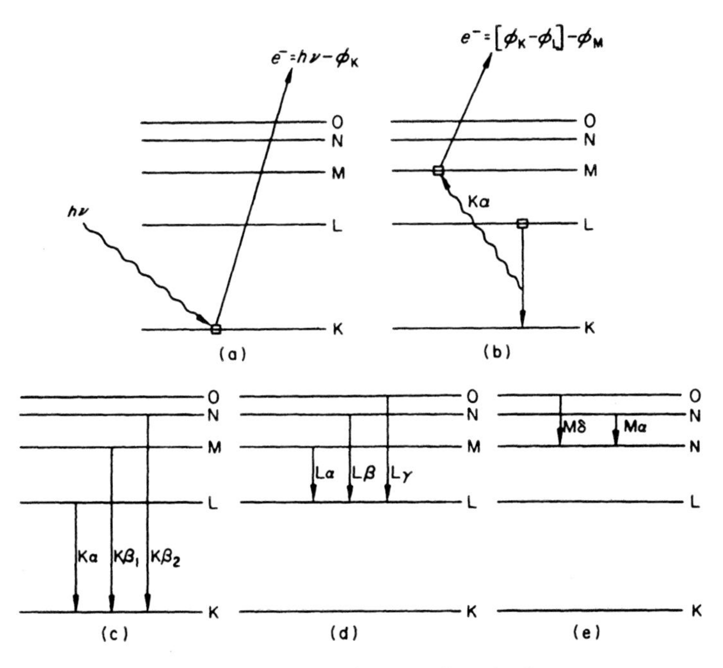

Figure 4.1. Excitation and deexcitation.

the photon of radiation does not leave the atom but itself ejects an M electron. This effect is called *autoionization* and is very much like the phenomena found in the excitation observed in optical spectra. As in the initial excitation, this photoelectron will have an energy equal to that of the incident radiation (in this case $\phi_K - \phi_L$), less that required for the ejection of the electron, that is the binding energy of the ejected electron. Note that the Auger process can lead to the production of two (or more) vacancies in the upper levels, in this instance L^+ and M^+. This double vacancy production is an important factor in the formation of satellite lines.

The quantity of radiation from a certain level will be dependent upon the relative efficiency of the two opposing deexcitation processes involved. This relative efficiency is usually expressed in terms of the *fluorescent yield.* The fluorescent yield ω is defined as the number n of X-ray photons emitted within a given series, divided by the total number N of vacancies formed in the associated level, each with the same time period. A series represents a set of wavelengths all of which arise from transitions to the same atomic subshell. In general terms for the K series

$$\omega_K = \frac{\Sigma(n)_K}{N_K} = \frac{n(K\alpha_1) + n(K\alpha_2) + n(K\beta_1) + n(K\beta_2) + \cdots}{N_K}. \tag{4.2}$$

For example, if within a given period of time, 80 vacancies were formed, and 30, 15, 5, and 1 X-ray photons were emitted corresponding to $K\alpha_1$, $K\alpha_2$, $K\beta_1$ and $K\beta_3$, respectively, the K fluorescent yield would be

$$\omega_K = \frac{30 + 15 + 5 + 1}{80} = \frac{51}{80} = 0.638. \tag{4.3}$$

Note that the fraction of vacancies filled via the Auger process is always one minus the fluorescent yield or, in the given example, 0.362 that is, 29 vacancies.

4.4. RELATIONSHIP BETWEEN WAVELENGTH AND ATOMIC NUMBER

Figure 4.2 shows Moseley diagrams for the K, L, and M series. In order to simplify the figure, only the strongest lines in each series have been shown. These diagrams are plots of the reciprocal of the square root of the wavelength, as a function of atomic number. As indicated by Moseley's law, and as shown in the figure, such plots should be linear. A scale directly in wavelength is also shown, to indicate the range of wavelengths over which a given series occurs.

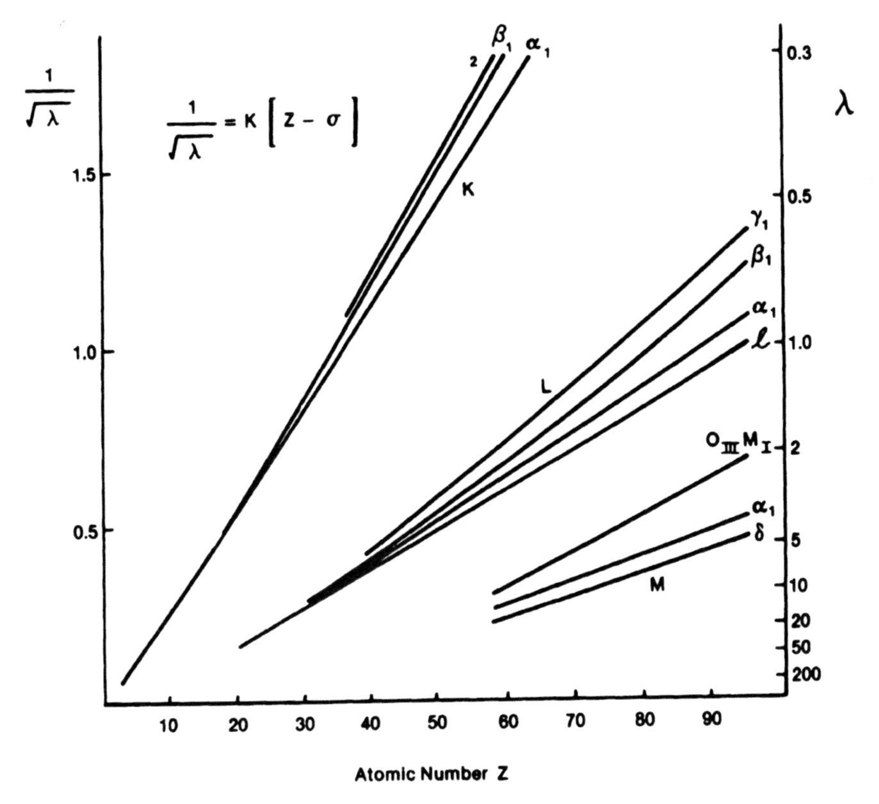

Figure 4.2. Moseley diagrams for the *K*, *L*, and *M* series.

Table 4.1 gives the atomic structures of the first three principle shells. The first shell, the *K* shell, has a maximum of two electrons and these are both in the 1*s* level. Since the value of *J* must be positive in this instance, the only allowed value is 1/2. In the second shell, the *L* shell, there are eight electrons, two in the 2*s* level and six in the 2*p* levels. In this instance, *J* has a value of 1/2 for the 2*s* level and 3/2 or 1/2 for the 2*p* level, thus giving a total of three possible *L* transition levels. These are referred to as L_{I}, L_{II} and L_{III} respectively. In the *M* level, there are a maximum of eighteen electrons, two in the 3*s* level, eight in the 3*p* level and ten in the 3*d* level. Again, with the value of 1/2 for the 3*s* level, values of 3/2 or 1/2 for *J* in the 3*p* level, and 5/2 and 3/2 in the 3*d* level, a total of five *M* transition levels are possible. Similar rules can be used to build up additional levels, *N*, *O*, etc. The selection rules for the production of diagram lines state that the principal quantum number n must change by at least one, the angular quantum number ℓ must change by only one, and the J quantum number must change by zero or one.

Transition groups may now be constructed as illustrated in Table 4.2, based on the appropriate number of transition levels. Application of the selection rules indicates that, in for example, the K series, only $L_{II} \rightarrow K$ and $L_{III} \rightarrow K$ transitions are allowed for a change in the principle quantum number of one. There are equivalent pairs of transitions for $n = 2$, $n = 3$, $n = 4$, etc. Figure 4.3 shows the lines that are observed in the K series. Three groups of lines are indicated. The *normal* lines are shown on the left-hand side of the figure. These consist of three pairs of lines from the L_{II}/L_{III}, M_{II}/M_{III}, and N_{II}/N_{III}, subshells, respectively. While most of the observed fluorescent lines are normal, certain lines may also occur in X-ray spectra that do not fit the basic selection rules at first sight.

These lines are called *forbidden lines* and are shown in the center portion of the figure. Forbidden lines typically arise from outer orbital levels where there is no sharp energy distinction between orbitals. As an example, in the transition elements, where the $3d$ level is only partially filled and is energetically similar to the $3p$ levels, a weak forbidden transition (the β_5) is observed. A third type of line, *satellite lines*, which arise from dual ionizations, may also occur.

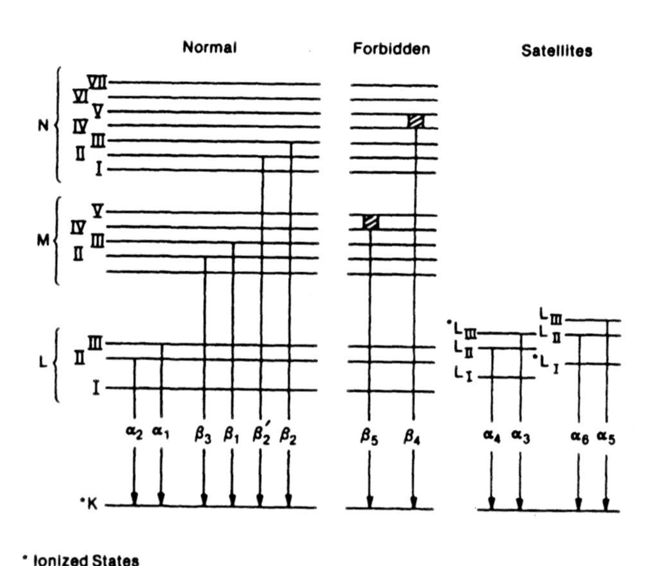

Figure 4.3. Normal, forbidden, and satellite lines in the K series.

Following the ejection of the initial electron in the photoelectric process, a short, but finite, period of time elapses before the vacancy is filled. This time period is called the *lifetime* of the excited state. For the lower atomic number elements, this lifetime increases to such an extent that there is significant probability that a second electron can be ejected from the atom before the first vacancy is filled. The loss of the second electron modifies the energies of the electrons in the surrounding subshells, and other pairs of X-ray emission lines are produced, corresponding to the α_1/α_2. In the K series, the most common of these satellite lines are the α_3/α_4 and α_5/α_6 doublets. These lines are shown at the right-hand side of the figure. Because they are relatively weak, neither forbidden transitions nor satellite lines have great analytical significance. However, they may cause some confusion in qualitative interpretation of spectra and may even be misinterpreted as coming from trace elements. In practice, the number of lines observed from a given element will depend upon the atomic number of the element, the excitation conditions, and the wavelength range of the spectrometer employed. Generally, commercial spectrometers cover the range 0.3 to 20 Å (newer instruments may allow measurements in excess of 100 Å), and three X-ray series are covered by this range. The K series, L series, and M

Table 4.2. Construction of Transition Levels.

Transition Group	n	l	J
K	1	0	$\frac{1}{2}$
L_{I}	2	0	$\frac{1}{2}$
L_{II}	2	1	$\frac{1}{2}$
L_{III}	2	1	$\frac{3}{2}$
M_{I}	3	0	$\frac{1}{2}$
M_{II}	3	1	$\frac{1}{2}$
M_{III}	3	1	$\frac{3}{2}$
M_{IV}	3	2	$\frac{3}{2}$
M_{V}	3	2	$\frac{5}{2}$
N_{I}	4	0	$\frac{1}{2}$
N_{II}	4	1	$\frac{1}{2}$
N_{III}	4	1	$\frac{3}{2}$
N_{IV}	4	2	$\frac{3}{2}$
N_{V}	4	2	$\frac{5}{2}$
N_{VI}	4	3	$\frac{5}{2}$
N_{VII}	4	3	$\frac{7}{2}$

Table 4.3. Correspondence between Siegbahn and IUPAC Notations for Diagram Lines.

Siegbahn	IUPAC	Siegbahn	IUPAC	Siegbahn	IUPAC
$K\alpha_1$	K–L_3	$L\alpha_1$	L_3–M_5	$L\gamma_1$	L_2–N_4
$K\alpha_2$	K–L_2	$L\alpha_2$	L_3–M_4	$L\gamma_2$	L_1–N_2
$K\beta_1$	K–M_3	$L\beta_1$	L_2–M_4	$L\gamma_3$	L_1–N_3
$K\beta_2^I$	K–N_3	$L\beta_2$	L_3–M_5	$L\gamma_4$	L_1–O_3
$K\beta_2^{II}$	K–N_2	$L\beta_3$	L_3–N_5	$L\gamma'_4$	L_1–O_2
$K\beta_3$	K–M_2	$L\beta_4$	L_1–M_3	$L\gamma_5$	L_2–N_1
$K\beta_4^I$	K–N_5	$L\beta_5$	L_1–M_2	$L\gamma_6$	L_2–O_4
$K\beta_4^{II}$	K–N_4	$L\beta_6$	L_3–$O_{4,5}$	$L\gamma_8$	L_2–O_1
$K\beta_{4x}$	K–N_4	$L\beta_7$	L_3–N_1	$L\gamma'_8$	L_2–$N_{6,7}$
$K\beta_5^I$	K–M_5	$L\beta_7$	L_3–$N_{6,7}$	$L\eta$	L_2–M_1
$K\beta_5^{II}$	K–M_4	$L\beta_9$	L_1–M_5	$L\ell$	L_3–M_1
		$L\beta_{10}$	L_1–M_4	L_s	L_3–M_2
		$L\beta_{15}$	L_3–N_4	L_t	L_3–M_3
		$L\beta_{17}$	L_2–M_3	L_u	L_3–$N_{6,7}$
				L_v	L_2–$N_{6,7}$

series correspond to transitions to K, L, and M levels, respectively. Each series consists of a number of groups of lines. The strongest group of lines in the series is denoted α, the next strongest β, and the third γ. There is a much larger number of lines in the higher series, and for a detailed list of all of the reported wavelengths, referred to the work of Bearden [2].

In X-ray spectrometry, most of the analytical work is carried out using either the K or the L series wavelengths. The M series may, however, also be useful especially in the measurement of higher atomic numbers. One practical problem in working with the M series is that many of the lines have not been given names within the Siegbahn system. In recent years the International Union of Pure and Applied Chemistry has addressed the problems of spectroscopic nomenclature and suggested an alternative nomenclature system based on edge designation [3]. This system is referred to as the IUPAC notation. Table 4.3 compares the Seigbahn notation and the IUPAC notation for the strongest lines in the K and L series.

4.5. NORMAL TRANSITIONS (DIAGRAM LINES)

The normal transitions (diagram lines) are defined by three simple atomic selection rules:

- $\Delta n \geq 1$
- $\Delta \ell = \pm 1$
- $\Delta J = \pm 1$, or 0.

Unfortunately, the nomenclature of the associated X-ray wavelengths is archaic and unsystematic, but several broad generalizations can be made. For instance, the final resting place of the transferred electron *always* determines the series of the associated radiation. Thus, ionization in the K shell followed by the filling of the K vacancy leads to the production of K series radiation. The filling of an L vacancy gives rise to L series radiation, and so on. Further, the strongest line in a given series is called the α line and weaker lines are called β, γ, δ, and so on, although the relative intensities of these lines bear little resemblance to the sequence of labeling.

Study of the selection rules will indicate that for an $L \rightarrow K$ transition, $2p \rightarrow 1s$ can give rise to more than one line since the $2p$ electron being transferred has two states, depending upon the sign of the spin quantum number. Where the spin is $-1/2$, the J quantum number will be $+1/2$, since for a p electron $\ell = 1$. The spectroscopic nomenclature in this case is $^2P_{1/3}$. Similarly, if the spin is $+1/2$, J will equal $3/2$ and the spectroscopic nomenclature is $^2P_{3/2}$. Thus two $K\alpha$ lines occur, $K\alpha_1$and the $K\alpha_2$.

Since both ℓ and J are used to define the state of an electron within a given shell, it is possible to construct so-called *levels* between which electrons are transferred. Table 4.2 shows the construction of these levels and that the number of levels is always $(2n - 1)$, thus 1 K-level, 3 L-levels, 5 M-levels, etc. Note that only positive values of J are possible, hence only a single J value arises from the state $\ell = 0$. This is because the s orbital (when $\ell = 0$, the orbital is s) is spherically symmetrical and therefore unpolarizable.

It is often confusing for the newcomer to the field of X-ray spectroscopy to find that there are no less than three different ways of expressing a transition. For example, the $K\alpha_1$ and $K\alpha_2$ lines already mentioned could be explained as follows:

$$K\alpha_1 : L_{III} \rightarrow K \quad \text{or} \quad {}^2P_{3/2} \rightarrow 1s \quad \text{or} \quad {}^2P_{3/2} \rightarrow {}^1S_o$$

$$K\alpha_2 : L_{II} \rightarrow K \quad \text{or} \quad {}^2P_{1/2} \rightarrow 1s \quad \text{or} \quad {}^2P_{1/2} \rightarrow {}^1S_o.$$

The first column, will be found in most textbooks and tabular values of wavelengths. Although it is, at first sight, the most convenient method to use, it is of little use in the discussion of satellite lines and forbidden transitions. The second will be generally familiar to the chemist and has the great merit of indicating exactly which electron states are involved in a given transition. The third is simply the correct spectroscopic nomenclature for the previous method and indicates the electron states. This last method is generally not favored by the analytical spectroscopist, and further description in this text will utilize mainly the second scheme.

4.6. SATELLITE LINES

Satellite lines were observed even by very early workers in the field of X-ray spectroscopy, and much of the theory for their production is well established [4, 5]. Satellite lines are lines which occur following dual ionization of the atom, that is to say, a second ionization occurs within the lifetime of the first excited state. Any wavelength emission following transfer of an electron to a vacancy within an atom, where a second vacancy exists, will differ from that produced where the second vacancy does not exist. This will be apparent, since the second vacancy must cause some perturbation and a general increase in the energy levels of the atom. Thus the wavelengths associated with dual vacancies will be shorter than their single vacancy counterparts. Dual ionization can take place by various means. One of these, the Auger process of autoionization, has already been discussed. The Auger process is indeed the dominant process for the production of double vacancies in all levels above the *L* levels, and it is due to this process that the majority of satellites in the *M*, *L*, *N*, etc., series lines occur. However, this would not be the reason for dual ionization involving *K* and *L* levels, since the Auger process begins following the internal conversion of an X-ray photon which will, in this instance, have come from transference between the two levels in question. It seems certain, therefore, that *K* series satellites exist due to double ionization by single electron impacts.

4.7. CHARACTERISTIC LINE SPECTRA

The types of transitions allowable are best illustrated by means of a series of transition diagrams and their appropriate emission spectra. The following can be taken as typical examples for most of the analytical wavelength region.

4.8. *K* SPECTRA

K spectra arise following the transference of electrons to *K* shell vacancies. *K* Spectra are relatively simple and generally consist of two doublets with an extra line occurring for the higher atomic number elements.

4.8.1. The Tin *K* Spectrum

Tin is atomic number 50 and has, in the ground state, filled *K*, *L*, and *M* levels, filled 4*s*, 4*p*, 4*d*, and 5*s* levels, plus two electrons in the 5*p* level. Figure 4.4 shows the energy level diagram and indicates the approximate binding energies of the various subshells [6]. Following ejection of a $K(1s)$ electron, seven lines may be observed. According to the selection rules, only six lines should be allowable, that is, three doublets corresponding to transitions from the 4*p*, 3*p*, and 2*p* levels. These lines do indeed occur; however, only the α_1 and α_2 appear on the emission diagram as a recognizable doublet. This is because the energy required to polarize the *p* orbitals becomes smaller with the decrease of electron density. For example, in the case of tin, the difference between the states $^2P^{3/2}$ and $^2P^{1/2}$ is 227 eV and between the states $^3P^{3/2}$ and $^3P^{1/2}$ only 42 eV. The visual effect on the emission spectrum is readily seen, since the angular dispersion of a spectrometer is directly proportional to the absolute value of the energy difference between two lines. Thus, whereas the α_1 and α_2 are clearly seen as two sharp and well-separated lines, the β_1/β_3 doublet is completely unresolved. This is even more so in the case of the $4p \rightarrow 1s$ transition, which is generally not referred to as a doublet at all but simply the β_2.

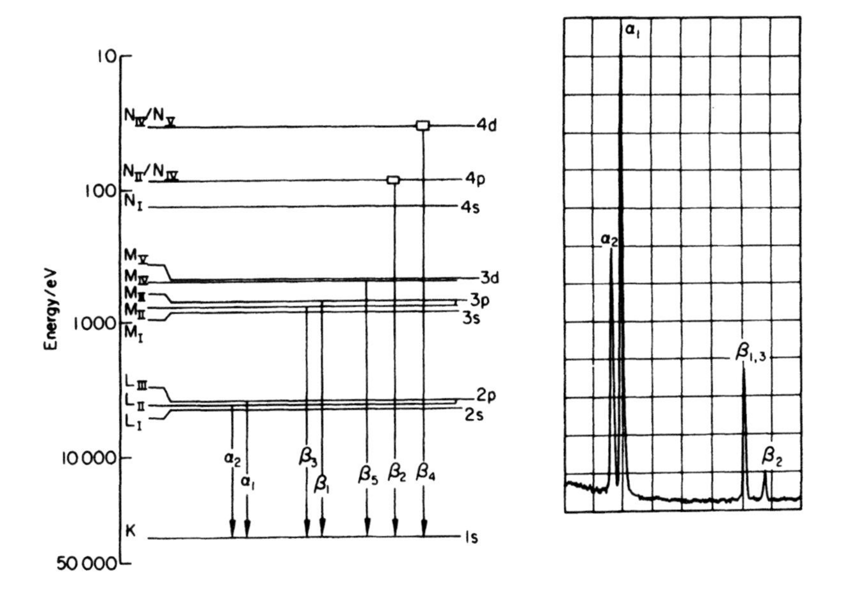

Figure 4.4. The *K* emission spectrum of tin.

Two lines do in fact exist, however, and these are identified by β_2' and β_2''. Two additional transitions occur in the tin K spectrum and these are the forbidden transitions $3d \rightarrow 1s$ and $4d \rightarrow 1s$. These transitions are classified as forbidden since they both represent a $\Delta\ell$ of 2 and, hence, disobey the normal selection rules. Like all forbidden lines, they are very weak and are not generally visible in the measured spectrum. A further point of interest is that the transition (doublet) $5p \rightarrow 1s$ would be allowed, but in the ground state, tin has just two electrons in the $5p$ level. Of course, chemical bonding may increase this number to the full complement of six. A very weak line corresponding to this transition (29.195 keV) is indeed observed in certain tin compounds.

4.8.2. The Copper *K* Spectrum

Copper is atomic number 29 and has filled K and L levels, filled $3s$, $3p$, and $4s$ levels, and an almost full $3d$ level. Thus, unlike tin, it has no electrons in the $4p$ level while in the ground state, although there might be some partial filling of this level during chemical bonding. Figure 4.5 shows the effect of the unfilled $4p$ level since the β_2 line is absent. The absence of the β_2 line does, however, allow the very weak β_5 line, which comes from the forbidden $3d \rightarrow 1s$ transition, to be seen. A further difference to be noted between the

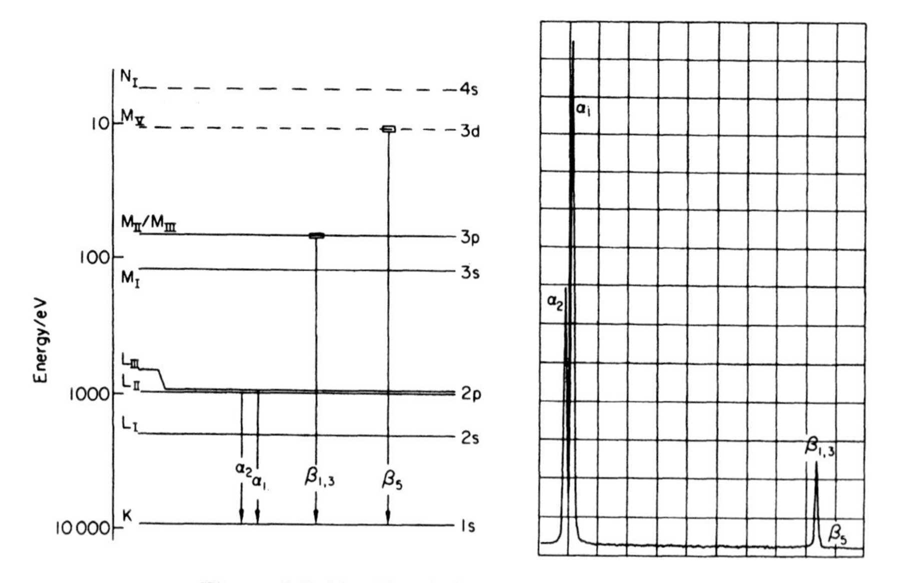

Figure 4.5. The K emission spectrum of copper.

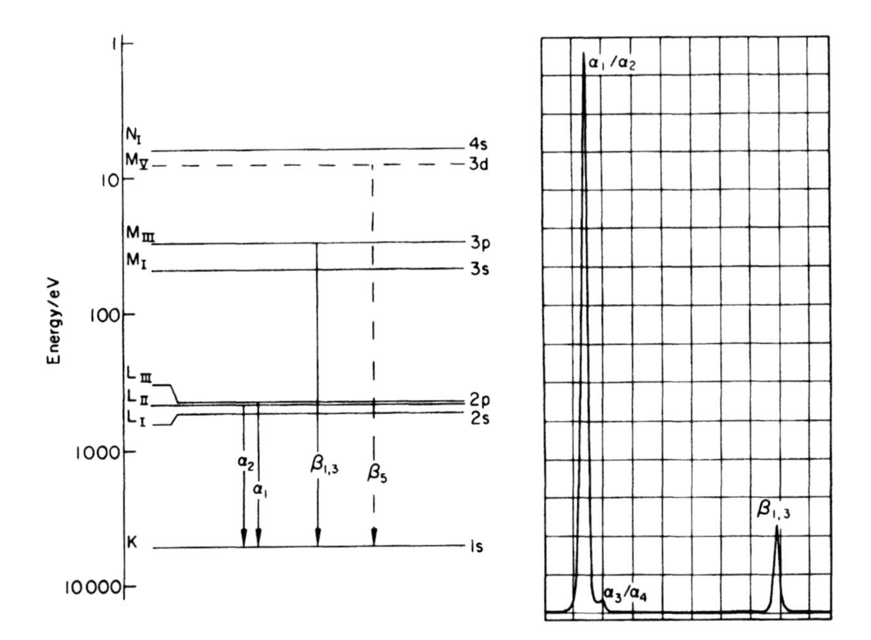

Figure 4.6. *K* spectrum for calcium.

copper and tin spectra is the lower degree of splitting of the $K\alpha_1/K\alpha_2$ doublet in the case of copper. This is because the absolute energy difference between α_1 and α_2 in copper is almost an order of magnitude less than in tin.

4.8.3. The Calcium *K* Spectrum

Calcium is atomic number 20 and, in the ground state, has filled *K*, *L*, 3*s*, 3*p*, and 4*s* levels. The 3*d* level is unoccupied, hence the forbidden $3d \rightarrow 1s$ transition is not observed (see Figure 4.6). Also, in this instance, the energy gap between $^2P^{3/2}$ and $^2P^{1/2}$ is so small that there is no observable splitting of α_1 and α_2. Several other interesting features also begin to appear at this level of orbital occupancy, and these involve the satellite lines to be seen to the high energy (right-hand side of the figure) side of the $K\alpha_1/K\alpha_2$ and, $K\beta_1/K\beta_3$ doublets. These satellite lines become more and more pronounced as the atomic number (that is, the number of electrons) decreases.

4.8.4. The Aluminum *K* Spectrum

Aluminum is atomic number 13 and, in the ground state, has filled *K* and *L* levels with two 3*s* electrons and a single 3*p* electron. Since the $K\beta_1/\beta_3$ line

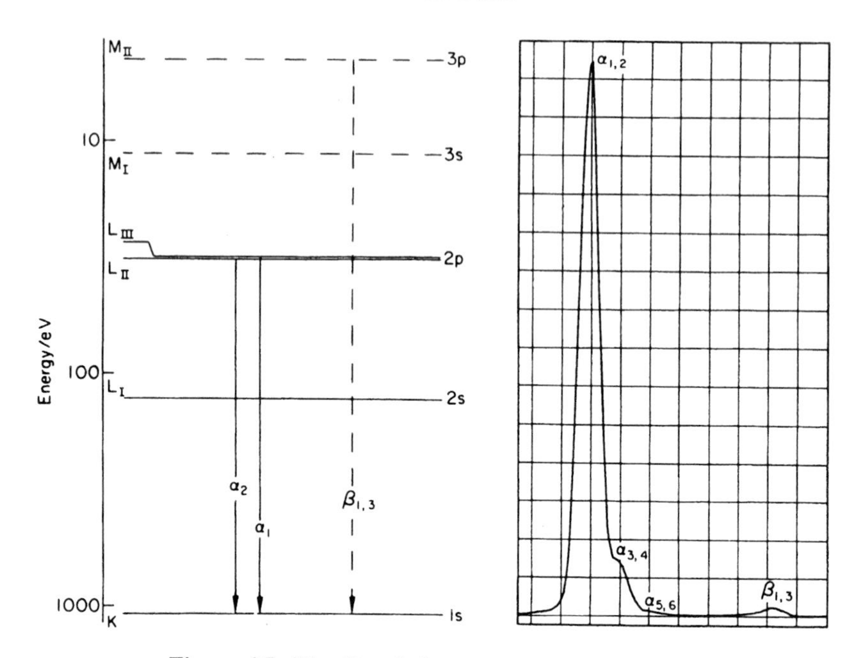

Figure 4.7. The *K* emission spectrum of aluminum.

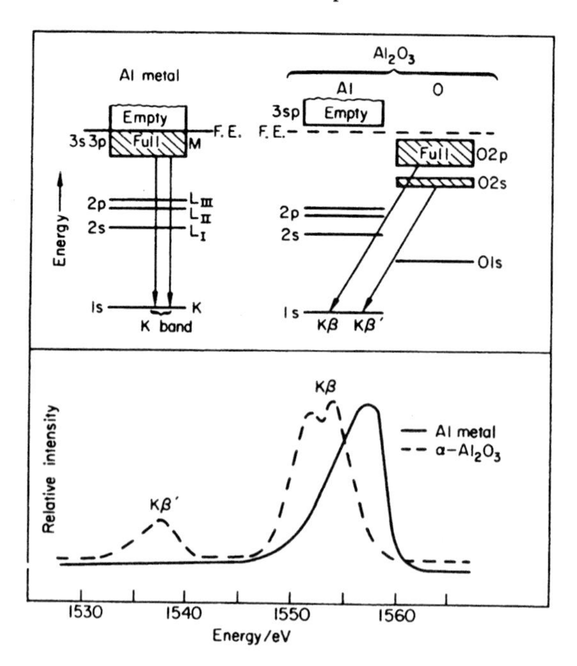

Figure 4.8. The *K* emission bands of aluminum.

comes from a $3p$ transition, only a very weak broad line is observed (see Figure 4.7). The shape of the line is very dependent upon the chemical bonding of the aluminum atoms. Figure 4.8, taken from work by Fischer [7], clearly indicates the effect of bonding and illustrates the reason for the band-like nature of the line which now arises not from the discrete energy level characteristic of an atomic orbital, but rather from the energy band of the molecular orbital of aluminum and oxygen. This change from line to band spectra is very typical of this part of the atomic number region. Also shown on the diagram is the satellite K'_β line.

Returning to Figure 4.7, two satellite doublets, α_3 and α_4, α_5 and α_6, are to be shown on the high energy side of the α_1, α_2 doublet.

Figure 4.9 shows an enlarged view of this portion of the spectrum, and the associated energy level diagram indicates that both satellite doublets occur from double ionizations. It has already been mentioned that removal of two electrons from the same atom is possible, and this is most likely to occur when the electron density is very low, that is, in the case of the lower atomic number elements. Two double ionizations are possible, namely KL_{III} and KL_{II}. In either case, removal of the $2p$ (or $2s$) electron increases the distance and hence the energy gap between $2p$ and $1s$, with the resulting production of two extra satellite pairs. (In fact, a fifth $K\alpha$ satellite is observed between the

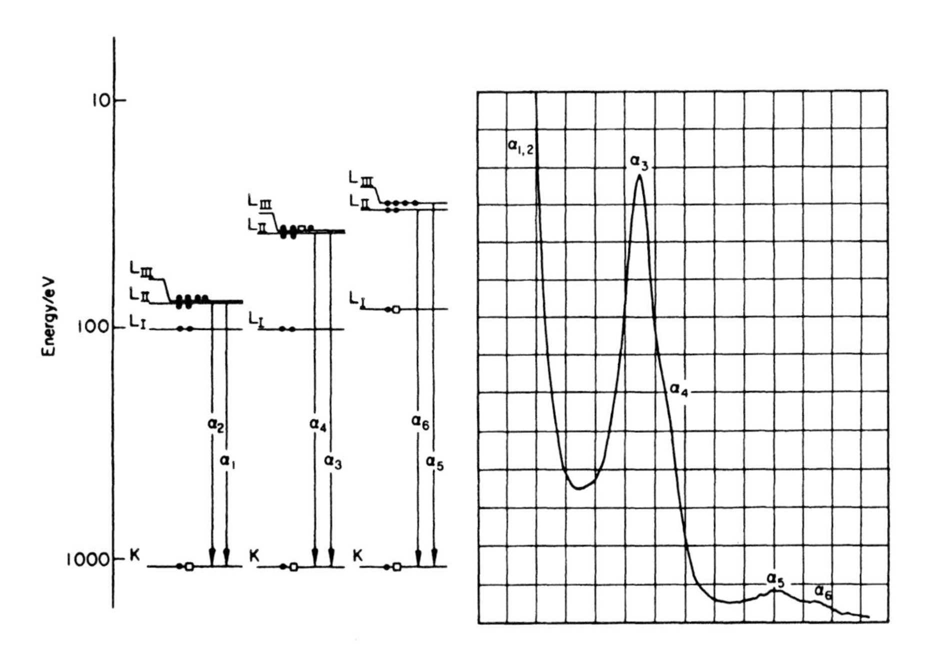

Figure 4.9. *K* satellite lines for aluminum.

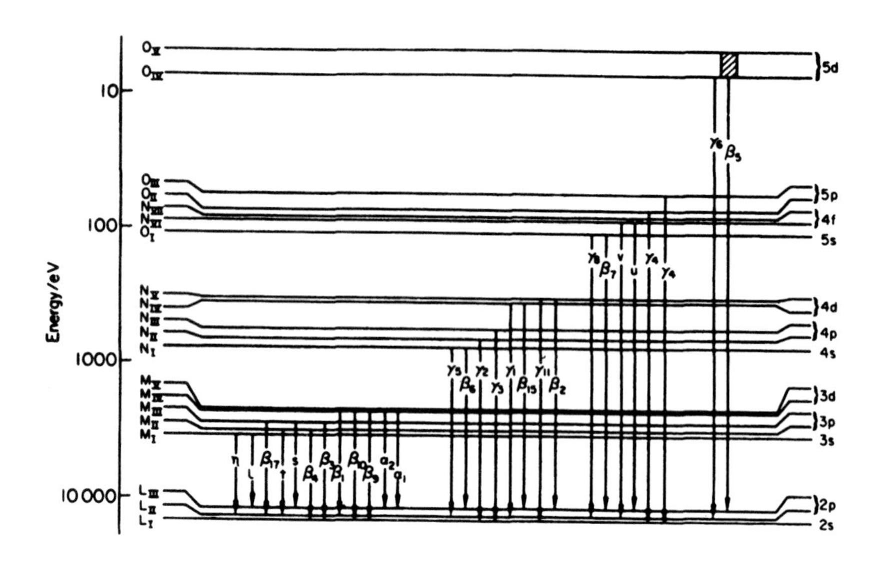

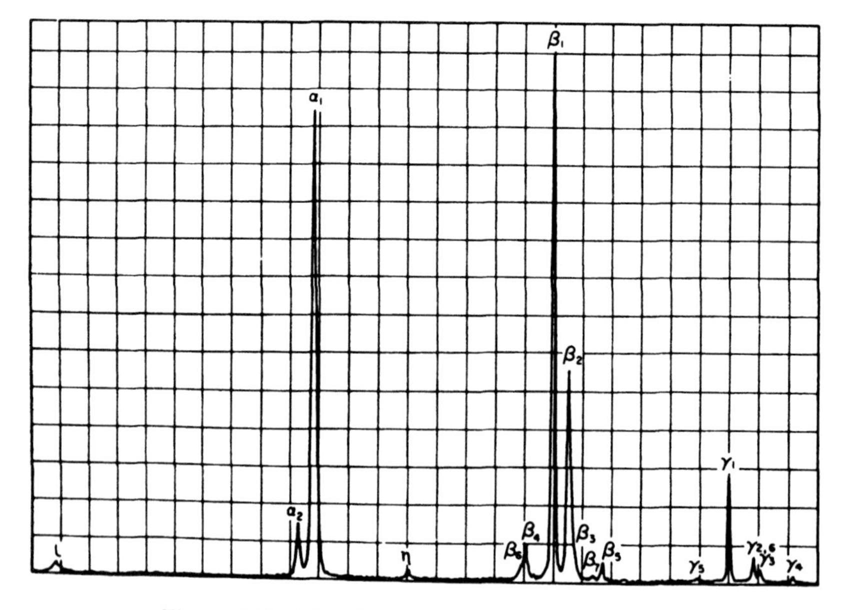

Figure 4.10. The *K* emission spectrum of oxygen.

$K\alpha_2$ and the $K\alpha_3$. This line is the $K\alpha'$ which arises from a more complex dual ionization.)

4.8.5. The Oxygen *K* Spectrum

Oxygen is atomic number 8 and only the 1*s* and 2*s* levels are filled. In the ground state, four 2*p* electrons are present but these will obviously be directly involved in bonding. Figure 4.10 illustrates the oxygen *K* emission spectrum obtained from $LiOH . H_2O$, and clearly shows the band-like nature of the line. Two maxima are seen in the spectrum and these correspond roughly to the upper and lower levels of the 2*p* orbital band. Oxygen has no 3*p* electrons and therefore gives no β lines. The oxygen *K* spectrum is typical of the *K* spectra of the very low atomic number elements. Even lithium, which is atomic number 3, and has the structure $1s^2 2s^1$, gives a *K* spectrum. At first sight, this would seem impossible since there are no 2*p* electrons in the lithium atom. However, it is possible for electrons to be transferred to the vacancy from a molecular orbital arising from a ligand between two dissimilar atoms (e.g., in lithium compounds) or even between two similar atoms. The possible use of the shape and distribution of X-ray emission bands is presently providing considerable insight into the mechanism of chemical bonding [7, 8].

4.9. *L* SPECTRA

L spectra arise following the transference of electrons to fill vacancies in the *L* levels. Since there are three *L* levels, compared with only a single *K* level, there will be a far greater number of possible *L* transitions allowable within the selection rules. *L* spectra are thus more complex than *K* spectra, and between 20 and 30 normal (diagram) lines are observed from the higher atomic number elements. As with the *K* series, a significant number of forbidden transitions and satellite lines are observed. Unlike the *K* series, where satellites arise almost exclusively from dual ionizations directly from primary photon impact, satellites in the *L* series arise mainly from autoionization.

4.9.1. The Gold *L* Spectrum

Gold has atomic number 79 and, in the ground state, has filled *K*, *L*, *M*, *N*, 5*s*, 5*p*, and 5*d* levels plus one 6*s* electron. Figure 4.11 gives the transition diagram for gold and the recorded *L* emission spectrum. Comparison of the transition diagram and the emission spectrum shows the latter to consist of three major groups of lines, the α's, the β's, and the γ's. The α lines plus the

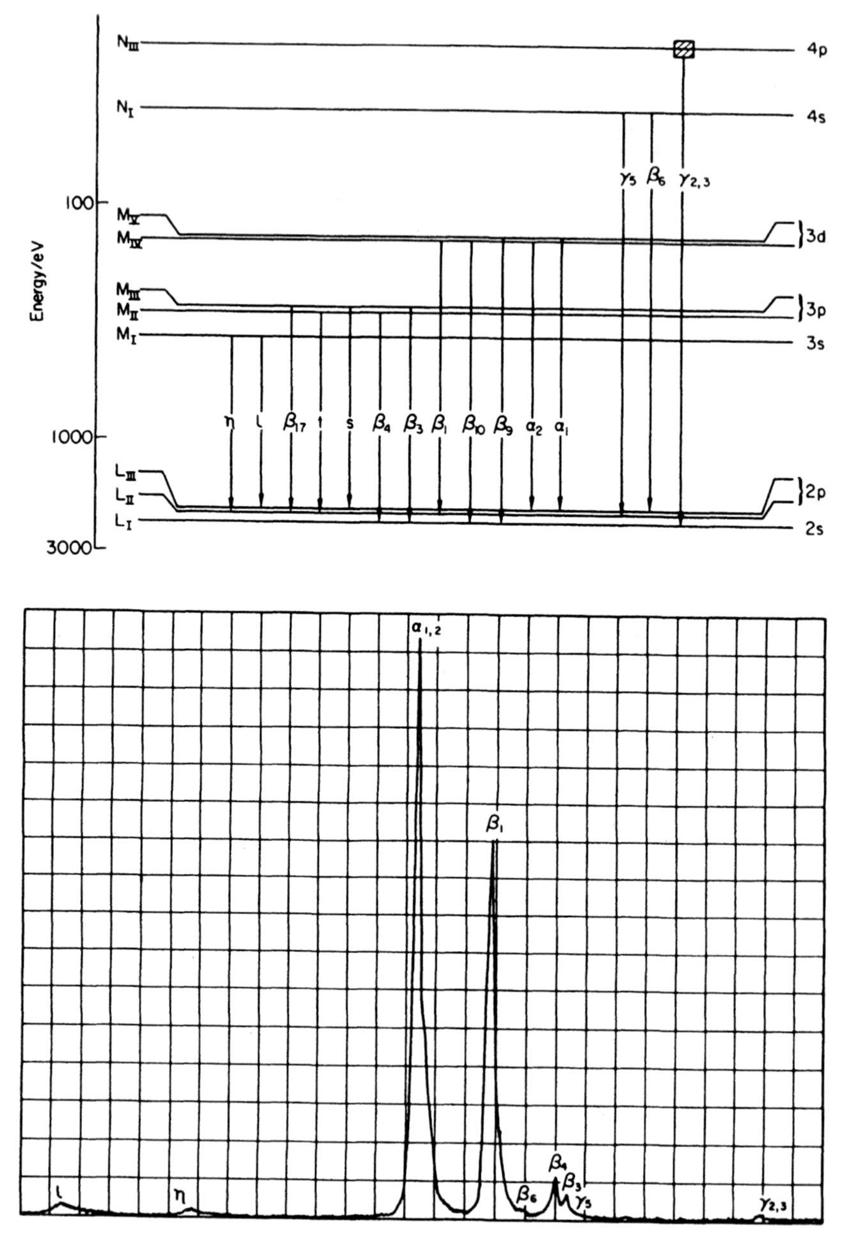

Figure 4.11. The *L* emission spectrum of gold.

majority of the β's come from transitions from the *M* subgroups, whereas the γ lines come mainly from *N* and *O* transitions. The most important exception to this generalization is the β_2 line which is associated with a transition from the $N_v(4d)$ level. One immediately noticeable feature of the *L* spectra from the lower atomic number (about $Z = 40$) elements is the absence of this β_2 line. Other important β lines, which are associated with transitions beyond the *M* levels, are the β_6, β_7, and β_5. Unlike the *K* spectra, the α's are no longer the dominant feature in terms of intensity. The β_1 is frequently as strong, if not stronger (as in this instance) than the α_1. Eight *forbidden* transitions occur in the gold *L* spectrum. These lines are extremely weak and have little analytical significance.

4.9.2. The Strontium *L* Spectrum

The gold spectrum just discussed is typical of most *L* spectra from elements of atomic numbers greater than about 42. Below this, the electron depletion of the upper levels cause significant changes in the emission spectra. Strontium gives an *L* spectrum that is typical of the lower atomic number elements. Strontium is atomic number 38 and, in the ground state, has filled *K*, *L*, *M*, 4*s*, 4*p*, and 5*s* levels. Figure 4.12 shows the transition diagram and the *L* emission spectrum. The marked differences between strontium (Figure 4.12) and gold (Figure 4.11) are very evident. Since strontium has no electrons in the 5*d*, 5*p*, or 4*d* levels, the γ lines are almost nonexistent and

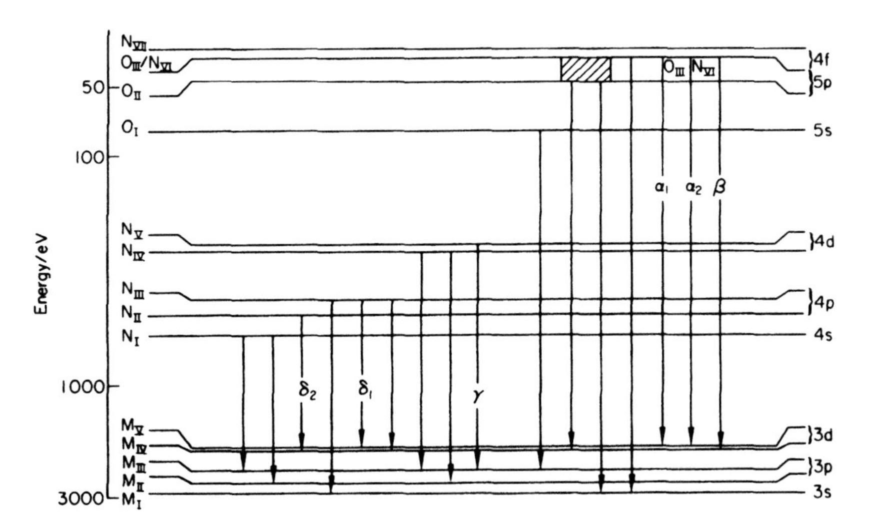

Figure 4.12. The *L* emission spectrum of strontium.

only those associated with the $4p$ level (γ_2 and γ_3) are at all visible. The γ_5 should also be visible but is masked by the β_3 line. Also missing are the β_2 and β_5 lines. The absence of certain lines is by no means the only difference between the two spectra. Due to the absolute differences in the binding energies of the L_{II}, L_{III}, M_{I}, and M_{V} levels, the energy gap $L_{\text{II}} \rightarrow M_{\text{I}}$, ($\eta$) is bigger than that for $L_{\text{III}} \rightarrow M_{\text{V}}$ (α_1) in the case of gold, but is smaller than in the case of molybdenum. The effect is that, whereas the η line occurs on the short wavelength side of the α_1 in the case of gold, it occurs on the long wavelength side in the case of strontium. As in the case of the K spectra (and for the same reason), the α_1 and α_2 lines of the lower atomic number elements are not resolved. The last important feature of the strontium L spectrum that should be noted is the appearance of line broadening on the high energy side of all of the lines.

This is most noticeable in the cases of the $\alpha_{1,2}$ doublet and the β_1 lines, and is very similar to that seen in the K spectra of the lower atomic number elements. Indeed, as in the case of the K spectra, the broadening is due to the occurrence of satellite lines which, in this instance, are resolved neither from their parent lines nor from each other.

4.10. *M* SPECTRA

As might be expected from what has been described previously, M spectra are more complex and more variable than their K and L counterparts. In addition to the large number of levels that may be involved in transitions, it is also found that self-absorption causes significant changes in the emission

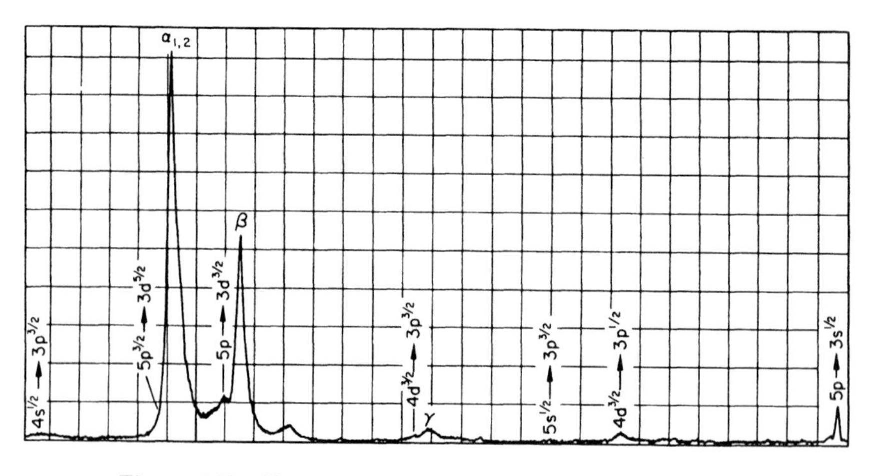

Figure 4.13. Observed transitions in the tungsten M spectrum.

spectrum [9]. Since the measurable wavelength region of most analytical X-ray spectrometers extends only to around 20 Å, the *M* spectra are less often encountered, and then only in the vacuum region (> 3 Å). Nevertheless, the strongest *M* lines are observable by the time the rare earth region is approached (that is, $Z > 57$). Figure 4.13 shows the transition diagram for tungsten. Tungsten is atomic number 76 and, in the ground state, has filled *K*, *L*, *M*, *N*, 5*s*, 5*p*, and 6*s* levels with four electrons in the 5*d* level.

The majority of the *M* lines have not been assigned names, hence most of the lines are identified by their transition states. Some similarity exists between the tungsten *M* spectrum and the strontium *L* spectrum discussed previously. Both spectra are typified by a strong α and relatively strong β emission with weaker lines on the high energy side. Both spectra show line broadening on the high energy side of the α and β lines due, in each case, to the presence of autoionized satellites.

BIBLIOGRAPHY

[1] E. H. S. Burhop, *The Auger Effect and Other Radiationless Transitions*, University Press: Cambridge, 189 pp. (1952).

[2] J. A. Bearden, "X-ray wavelengths," *U.S. Atomic Energy Commission Report NYO-10586*, 533 pp. (1964).

[3] R. Jenkins, R. Manne, J. Robin, and C. Senemaud, "Nomenclature, symbols, units and their usage in spectrochemical analysis, Part VIII, Nomenclature system for X-ray spectroscopy," *Pure and Applied Chemistry*, **63**, 736–746 (1991).

[4] K. D. Servier, *Low Energy Electron Spectroscopy*, pp. 236–237, Wiley-Interscience: New York (1972).

[5] H. Mendel, "A theoretical interpretation of satellite lines in the X-ray emission K spectra of compounds," *Proc. Kon. Ned. Acad. Wetenschappen*, B70, 276 (1967).

[6] K. Siegbahn, et al., *ESCA Atomic, Molecular and Solid State Structure Studied by Means of Electron Spectroscopy*, Almquist and Wiksells: Upsala, 282 pp. (1967).

[7] D. W. Fischer, "Chemical bonding and valence states: non-metals," *Adv. X-ray Anal.*, **13**, 159–181 (1969).

[8] D. J. Fabian, *Soft X-ray Band Spectra*, Academic Press: London, 382 pp. (1968).

[9] D. W. Fischer and W. L. Baun, "The influence of sample self-absorption on wavelength shifts and shape changes in the soft X-ray region: the rare earth M series," *Adv. X-ray Anal.*, **11**, 230–240 (1967).

CHAPTER

5

HISTORY AND DEVELOPMENT OF X-RAY FLUORESCENCE SPECTROMETRY

5.1. HISTORICAL DEVELOPMENT OF X-RAY SPECTROMETRY

X-ray fluorescence spectrometry provides a means of identification of an element, by measurement of its characteristic X-ray emission wavelength or energy. The method allows the quantification of a given element by first measuring the emitted characteristic line intensity and then relating this intensity to elemental concentration. While the roots of the method go back to the early part of this century [1,2,3], it is only during the last 25 years or so that the technique has gained major significance as a routine means of elemental analysis. The first use of the X-ray spectrometric method dates back to the classic work of Henry Moseley in 1912 (see Section 1.3). In Moseley's original X-ray spectrometer, the source of primary radiation was a cold cathode tube in which the source of electrons was residual air in the tube itself, with the specimen for analysis forming the target of the tube. Radiation produced from the specimen then passed through a thin gold window onto an analyzing crystal, where it was diffracted to the detector. At about the same period in time, after the application of the ability to image through opaque objects had demonstrated significant utility, Barkla observed [4] that when elements were irradiated with primary X-rays, the secondary radiation varied in penetration power (as measured by absorbing the rays with foils). Figure 5.1 shows Barkla's classic experiment in which he demonstrated that the total attenuation of the incident X-ray beam was different depending on whether the absorber was placed either between the source and the scatterer, or between the scatterer and detector. From this observation, Barkla concluded that the scatterer was not only scattering, but it was also modifying the beam. He further concluded that hard primary radiation could produce soft secondary radiation. It was Barkla who coined the term *characteristic rays*. He also observed that there were two components to this radiation, a *harder* and a *softer* one (specified by their ability to penetrate the foils). He chose to name them *K* and *L* rays, respectively, starting in the middle of the alphabet, as he assumed that harder and softer radiation would be discovered.

One of the major problems in the use of electrons for the excitation of characteristic X-radiation is that the process of conversion of electron energy

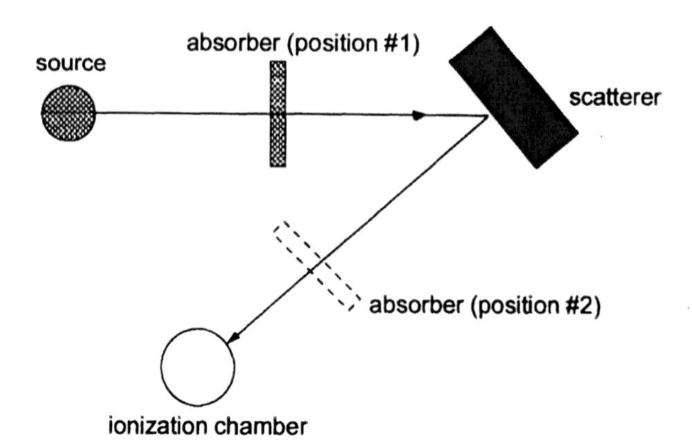

Figure 5.1. Barkla's experiment, which suggested that the scatterer modifies the wavelength distribution of the primary beam.

into X-rays is relatively inefficient; about 99% of the electrons are converted to heat energy. This means, in turn, that it may be difficult to analyze specimens which are volatile or tend to melt. Nevertheless, the technique seemed to hold some promise as an analytical tool and one of the first published papers on the use of X-ray spectroscopy for real chemical analysis appeared as long ago as 1922. In this paper Hadding [5] described the use of the technique for the analysis of minerals. A fact sometimes forgotten is that Moseley not only showed the relationship between wavelength and atomic number but, and perhaps more importantly, he also recognized from the intensity of the impurity elements in his specimens, that this relationship would be a powerful method of chemical analysis, and in fact, proposed that it might lead to the discovery of elements missing from the periodic table (realized in 1923 by the discovery of hafnium by Coster and von Hevesey). The first suggestion to use X-rays rather than electrons for excitation appears to have come from von Heversey [6], who showed that specimens located outside the X-ray tube could be excited to fluoresce their characteristic radiation by the primary X-rays from the tube, making possible the analysis of volatile or low-melting material. Thus, in less than 20 years after Roentgen's work, the basic foundation of X-ray spectrometry, that is, XRF (photon excited) was laid.

The use of X-rays, rather than electrons, to excite characteristic X-radiation, avoids the problem of the heating of the specimen. It is possible to produce the primary X-ray photons inside a sealed X-ray tube under high vacuum and efficient cooling conditions, which means the specimen itself need not be subject to heat dissipation problems or the high vacuum requirements of the electron beam system. Use of X-rays rather than electrons represented the beginnings of the technique of X-ray fluorescence as it is known. The *fluorescence method* was first employed on a practical basis in 1928 by Glocker and Schreiber [7]. Unfortunately, data obtained at that time were rather poor because X-ray excitation is rather inefficient relative to electron excitation, and the detectors and crystals available at that time were rather primitive. Thus the fluorescence technique did not seem to hold too much promise. Widespread use of the technique had to wait until the mid 1940s when X-ray fluorescence was rediscovered by Freidman and Birks [8]. The basis of their spectrometer was a diffractometer that had been originally designed for the orientation of quartz oscillator plates. A Geiger counter was used as a means of measuring the intensities of the diffracted characteristic lines and quite reasonable sensitivity was obtained for a very large part of the atomic number range.

5.2. EARLY IDEAS ABOUT X-RAY FLUORESCENCE

Following Moseley's tragic death in the Dardanelles campaign of World War I [9], the work on the characterization of X-ray spectra was taken up by Siegbahn's group in Sweden. Siegbahn developed vacuum and high resolution spectrometers, and developed the nomenclature system for characteristic X-radiation which is still in use today. By the early 1920s, most of the more common lines had been measured. The period from 1914 to 1932 was, in fact, filled with workers around the world setting up equipment to measure X-ray spectra. There are perhaps only a few milestones of note to list, but they are important for what they accomplished, and for what was overlooked. In 1923, as previously noted, Coster and Von Hevesey discovered $Z = 72$ (hafnium) [10]. It was observed by its characteristic X-ray spectrum emitted by a zircon from Norway. The X-ray fluorescence technique was employed since the material to be analyzed was smeared onto a secondary surface which was bombarded with X-radiation from the anode. Clearly the technique was limited to nonvolatile materials that had some semblance of thermal stability. Since Geiger counters had just been developed, it seemed like a reasonable addition to the instrumentation used to measure the spectra in that device. It was Jönsson in Sweden who first accomplished this [11] and used the technique to measure the spectra of X-ray tubes. Thus the ability to measure X-ray line intensities quantitatively was now a reality. As a tribute to Moseley's insight relative to the use of intensities from impurities in his specimens, it must be noted that, in Australia, Eddy and Laby [12] were measuring parts-per-million trace elements in alloys by XRF Geiger counter spectrometer. In 1932, von Hevesey published the classic text, *Chemical Analysis by X-ray and its*

Applications, and while it is an excellent description of the principles and applications of X-ray analysis, it does neglect to mention the work of de Broglie and of Jönsson, an important omission because of the importance of fluorescent excitation and electronic detection in virtually all instruments to come after. By the end of the 1920s X-ray methods were firmly entrenched.

5.3. REBIRTH OF X-RAY FLUORESCENCE

The period from 1947 to the middle 1960s can be characterized by tremendous advances in the practice of XRF. That is not to say that the more recent 30 years have been devoid of developments, but as the field has become mature, progress becomes slower and less frequent. There was a significant event in 1947 when Friedman and Birks converted an X-ray diffractometer to an X-ray spectrometer for chemical analysis [8], thus combining the advantages of de Broglie's fluorescent excitation and Jönsson's Geiger counter detector. The Friedman/Birks spectrometer was composed of an end-grounded, high-intensity, sealed tungsten-target X-ray tube; a single crystal spectrometer using either (200) NaCl or (220) CaF_2 crystals; a Geiger counter with a special thin window; a collimator laboriously constructed of thin-walled nickel tubing; and a turret device to facilitate changing specimens. To all intents and purposes, the modern X-ray analytical wavelength-dispersive spectrometer uses a version of the same physical geometry, but with more features to add to its versatility and utility.

The Friedman and Birks instrument had certain limitations. Foremost, perhaps, was the low intensity which the Geiger counter could tolerate without becoming saturated. Attempts were made to use multiple detectors to improve this situation, but this was only partially successful, there being room for only a few. With the advent of the gas-proportional and scintillation detectors with a much shorter dead time, counting at much higher rates was made possible. In addition, they were able to discriminate against unwanted energies, enabling the suppression of higher order diffraction from the analyzing crystal and reduce the background. Secondly, the original spectrometer was limited to the wavelengths which could be measured in an air atmosphere, and which could be diffracted with acceptable resolution by the spacings of the crystals available. It was not long before these difficulties were overcome. The crystals used in this prototype instrument, rock salt and fluorite, enabled the measurement of all elements from atomic number 20 (Ca) upward, using either their K or L spectra. While certain natural long-spacing crystals were available (mica and gypsum being notable examples), quality, uniformly diffracting samples were not readily available, and the shorter spacing NaCl was subject to the whim of the atmospheric relative humidity. With the development of synthetic LiF, the (200) cleavage plane became the most efficient analyzing crystal for the 0.5 to 3.5 Å region. More efficient and more stable long-spacing crystals were also being developed [13]. Although a helium atmosphere provided a possible solution to the measurement of long wavelengths, it was difficult to maintain a stable atmosphere, and evacuating the spectrometer to a moderate vacuum was a much more viable solution. Techniques for accomplishing this goal were straightforward, and, when coupled with long spacing crystals being synthesized, instruments became available which were capable of analyzing all elements in the periodic table from atomic number 11 (Na) upwards.

The first commercial X-ray spectrometer became available in the early 1950s and although these earlier spectrometers operated only in air path conditions, they were able to provide qualitative and quantitative information for all elements above atomic number 22 (Ti). Later versions allowed use of helium or vacuum paths that extended the lower atomic number cutoff. Most modern spectrometers allow the determination of elements down to atomic number 9 (F), and with special precautions, even down to atomic number 6 (C). Today, nearly all commercially available X-ray spectrometers use the fluorescence excitation method and employ a sealed X-ray tube as the primary excitation source. Some of the simpler systems may use a radioisotope source, due to cost and/or portability. While electron excitation is generally not used in *stand-alone* X-ray spectrometers, it is the basis of X-ray spectrometry carried out on electron column instruments. The ability to focus the primary electron beam allows analysis of extremely small areas down to a micron or so in diameter. This, in combination with imaging and electron diffraction, offers an extremely powerful method for the examination of small specimens, inclusions, grain boundary phenomena, and so on. The instruments used for this type of work may be in the form of a specially designed electron microprobe analyzer or simply an energy or wavelength dispersive attachment to a scanning electron microscope.

Among major developments in X-ray fluorescence instruments in the early 1960s was the use of lithium fluoride as a diffracting crystal and the use of chromium and rhodium target X-ray tubes that were especially useful for the excitation of longer wavelengths. Multichannel spectrometers, in which many spectrometers were grouped around the specimen, also became available at this time. These devices allowed the simultaneous measurement of many elements, albeit with the loss of some flexibility. The computer-controlled spectrometer became available in the mid-1960s. Probably the most significant development in recent years came in 1970 with the advent of the Si(Li) lithium drifted silicon detector (see Section 5.6). This detector gives a very high energy resolution and provides the means of separating the

X-ray photons coming from an excited specimen without recourse to the use of a relatively inefficient analyzing crystal.

There are many types of X-ray fluorescence spectrometers available on the market today but most of these fall roughly into two categories, wavelength dispersive instruments and energy dispersive instruments. In the wavelength dispersive method, the diffracting property of a single crystal is used to separate the polychromatic beam of radiation coming from the specimen. In the energy dispersive spectrometer, a Si(Li) detector is utilized to give a spectrum of voltage pulses that is directly proportional to the spectrum of X-ray photon energies entering the detector. An electronic voltage level sorter, called a multichannel analyzer, is then used to separate and collect these voltage pulses and store them in terms of their energies. The wavelength dispersive system was introduced commercially in the early 1950s, and probably around 20 000 or so such instruments have been supplied commercially, roughly half of these in the United States. Energy dispersive spectrometers became available in the early 1970s and today there are several thousands of these units in use.

5.4. EVOLUTION OF HARDWARE CONTROL METHODS

Hardware control methods were changing by the 1970s, see Table 5.1. The servomotors of the 1950s and early 1960s were now replaced with the stepper motor, offering a greater level of feedback control. Cumbersome switching and mechanical step-scanning had already been replaced with the use of pegboards. A carefully designed connection board allowed preprogramming of switch sequences, relieving much of the tedium of setting up an analytical sequence. These pegboards were to remain the dominant programming technique for almost a decade. While the pegboard

Table 5.1. Development of Hardware Control Methods.

Year	System	Control
1953	Servomotors	Individually Switched
1955	Servomotors	Mechanical Step-Scanning
1962	Servomotors	Logic Switching (peg-boards)
1968	Steppermotors	Electronic Logic Boxes
1970	Steppermotors	Mainframe Computer
1974	Steppermotors	Electronic Module (NIM, CAMAC)
1975	Steppermotors	Microprocessor
1985	Steppermotors	Personal Computer

solved many problems, it was by no means trivial in its use. The analytical sequences were still complex and the setting up and interpretation of an automatic programmer still required some expertise on the part of the operator. With the introduction of the stepper motor, the use of electronic control boxes soon followed.

While computer control of hardware was used for a couple of years in the early 1970s, it soon became apparent that much was to be gained by the separation of the tasks of hardware control and data processing. In response to this need, several "universal" electronic interface modules were developed. As an example, CAMAC [14] was designed as a universal interface to provide a bridge between high-level instructions from a computer program and various instrumental controllers. This system was used, for example, by Siemens in the SRS X-ray spectrometer system. Early systems used a PDP-11/10 computer, running under RT-11. 28 k words of addressable memory was available and the system supported 2 RK05 disks, each of 2.5 megabytes [15]. The CAMAC module was generally not accepted by United States developers of X-ray instruments and here the NIM module was taken as the standard.

5.5. THE GROWING ROLE OF X-RAY FLUORESCENCE ANALYSIS IN INDUSTRY AND RESEARCH

By 1953 there were a number of automated X-ray spectrometers around, of which the Philips Autrometer was typical. This 25-channel, sequential machine was programmed by a combination of switches, servomotors and mechanical stops, which required many hours of careful mechanical adjustment to set up. By the early 1960s multichannel spectrometers were beginning to appear. Some, like the machine at Avesta Steel, were homemade, but these were soon to be replaced by fast, accurate, and increasingly expensive commercial machines.

The need for fast, reliable, automated analytical instrumentation grew rapidly with accelerated industrial development brought about by World War II. The aircraft industry had developed special high-temperature alloys, and steel making had become a complex art, requiring accurate and fast elemental analysis. Two major contenders came to the forefront in the area of analytical instrumentation, the ultraviolet emission spectrometer and X-ray fluorescence. Because of its ability to measure carbon, the UVE method quickly found a place in the ferrous industry. Similarly, the ability of XRF to measure elements such as sulfur and aluminum meant that the area of nonferrous metallurgy was ideally suited for XRF. The need for high-speed analysis was much greater in the steel industry and this, combined with the nature of UVE, resulted in the development of multichannel UVE spectrometers. In contrast, while high throughput was certainly important in the analysis of brasses, bronzes, etc., the need for high-speed analysis was not so critical. This, combined with the fact that most early manual spectrometers were attachments to diffractometers, meant that almost all early X-ray spectrometers were sequential machines. As we moved into the 1960s, those companies (ARL, Hilger & Watts, etc.) that had been successfully manufacturing simultaneous UVE spectrometers, now started to make X-ray spectrometers which, as might be expected, were also multichannel. Similarly, many companies making XRD systems started to make XRS systems and, hence, sequential machines. By 1970, all of the major manufacturers were supplying computer-controlled spectrometers.

5.6. THE ARRIVAL OF ENERGY DISPERSIVE SPECTROMETRY

The late 1960s and early 1970s marked the arrival of the energy dispersive spectrometer. The Si(Li) solid state detector, originally developed as a sensor for nuclear radiation, was found to be useful for lower-energy X-rays. Thus was born the energy dispersive X-ray spectrometer. It was true that proportional counters and scintillation detectors had been used as energy-dispersive X-ray devices for some purposes, but it was Si(Li) (and somewhat later Ge(Li) and HPGe), with much better energy resolution, that brought energy-dispersive X-ray fluorescence to the forefront of the X-ray analytical community. There are probably more EDXRF instruments in use today around the world than all the wavelength dispersive X-ray fluorescence instruments (using crystal spectrometers) that have ever been employed from the beginning.

Without a doubt, the most significant development of the 1970 time period was the arrival of the Si(Li) detector. The late 1960s and early 1970s saw rapid growth in the use of X-ray spectrometers based on lithium ion drifted Si and Ge solid-state detectors [16]. While the basic concept of drifting Li in single crystals of silicon and germanium had been described long before (in 1960 by Pell [18]), the implementation of techniques capable of providing commercial quantities of Si(Li) detectors had to wait almost another decade. The initial success in the early 1970s was gained mainly in electron column applications. However, the rapid growth in the use of stand-alone EDS systems had a significant impact on the XRF community. The idea of a low-cost, high-speed system, which was apparently unencumbered by problems of sample presentation geometry, seemed almost too good to be true. The count-rate limitations notwithstanding, the sale of stand-alone EDS systems did much to challenge the previously solid bastion of wavelength dispersive spectrometers. The use of the digital computer to replace the relatively expensive hard-wired, multichannel analyzer did much to further enhance the merits of the EDS technique, particularly when software was developed for providing on-line dynamic displays of the data collection process. In later years, the flexibility and sensitivity of the EDS system was further improved by the development of the secondary fluorescer system. The secondary fluorescer system was itself a logical development based on the γ-X radioisotope source.

5.7. EVOLUTION OF MATHEMATICAL CORRECTION PROCEDURES

With the need for greater accuracy, especially in XRF, came the need for matrix correction. Early work at, for example, the British Nonferrous Metals Research Association, employed a table of correction factors which could be applied with a slide rule. From this, grew simple intensity correction models which were simply linear equations requiring only minimal computational power for their solution. As faster computers became available, concentration models evolved, based on powerful matrix inversion and multilinear regression analysis (see Section 12.4 for details of these methods). Very early in this time period, mathematical methods to correct for matrix effects had been developed, which negated the necessity to have a large number of "type standards" for direct conversion of X-ray intensity to concentration. There was a large diversity of these techniques, all loosely defined as "empirical", but they were mostly variations on a theme. They required a well-characterized set of a few standards, from which coefficients could be determined, to account for the effect of each element present in the sample of each other element; some expressions were only applicable over a limited range of concentrations. These coefficients could then be applied in a series of simultaneous equations to solve for the concentration of the unknown specimens. As has been previously discussed, in the early 1950s, Sherman developed a series of intensity/concentration equations which, in principle, required no standard, merely a knowledge of certain "fundamental parameters", absorption coefficients, fluorescence yields, jump ratios, etc. Unfortunately, the complex nature of the relationships required massive computing power, more than was readily available at the time. Sometime later, however, in the mid-1960s, Shiraiwa and Fujino in Japan, and Criss at the Naval Research Laboratory, took advantage of the improvements in computers to program these equations, making it useful to employ them in practical data reduction schemes. The use of these basic equations in a variety of software packages continues today.

5.8. X-RAY ANALYSIS IN THE 1970s

The years immediately following 1970 were critical years in the development of X-ray analytical instrumentation. It was a time of innovation. The recent development of the Si(Li) detector had taken the field of nondispersive analysis to a completely new level, now referred to as energy dispersive analysis. It was a time of instrument automation establishment, with the computer beginning to reveal its potential power. As would be expected, the time sequence in which the automation of data collection and data processing developed closely followed developments in computer hardware and peripherals. While most automated X-ray spectrometers and diffractometers were still using electromechanical control devices, the foundations for the use of the minicomputer, and later the personal computer, were now being put in place. X-ray analysis 30 years ago was rather different from the techniques known today. Layered synthetic microstructure crystals were unknown and lead stearate soap film was used to diffract long wavelengths. The idea of an engineered selectable *d*-spacing analyzing crystal was, of course, well-known, following the work of DuMond and Youtz [19] in 1940 and Dinklage and Frericks in 1963 [20]. But the development and implementation of the necessary technology was still many years away [17]. ESCA and photoelectron spectroscopy were just becoming established as analytical tools, so there was still a great deal of interest in the use of X-ray emission lines from outer orbitals for the study of chemical bonding (see e.g., [21]). Such work is difficult since it pushes both the required long wavelength excitation efficiency and the spectrometer resolution to the very edge of its capability.

In X-ray fluorescence, mathematical correction procedures were, at best, primitive. Although several years had passed since Jacob Sherman showed

Table 5.2. Development of Mathematical Models.

Year	Basis of the Method	Typical Author
1922	Type Standardization	Hadding
1950	Tabular Methods	BNFMRA
1952	Empirical Methods	Gillam and Heal
1955	Fundamental Methods	Sherman
1961	Intensity Correction	Lucas-Tooth/Price
1966	Concentration Correction	Lachance/Trail
1974	Absorption/Enhancement	Rasberry/Heinrich
1980	Separation of 3rd Element Effects	Lachance/Claisse

that it was possible to calculate X-ray intensities from known elemental compositions, few had computers large enough to exploit his ideas.

5.9. MORE RECENT DEVELOPMENT OF X-RAY FLUORESCENCE

The most recent 30 years of this odyssey has seen XRF develop into a very mature analytical tool for both routine quality control in many industries, as well as analytical support for the research laboratory. Both EDXRF and WDXRF have found their place, becoming complementary rather than competitive, in a variety of applications. While the use of XRF as a qualitative tool to provide rough identification of the elemental components of a specimen as a place in the analytical laboratory, it is the quantitative analysis of a wide variety of materials that is the raison d'être for the technique. Although the literature is replete with applications illustrating the usefulness of XRF in industry and the research laboratory, the refinement of the data reduction schemes to improve the quantification of the conversion of X-ray intensity to composition has been an important part of the maturation process.

There are many places in production control where the *empirical* data reduction methods are still employed. Many of the commercial XRF instruments have a variety of software packages available to treat the data, and where the stable of necessary standards can be justified, it is most efficient to use those which take advantage of these standards. Where the required number of standards is not available, the *fundamental parameter* approach is called for. Although, as noted above, standards are not absolutely necessary, the instrumental sensitivity factor *must* be calibrated. It has been suggested that for EDXRF, only one pure element standard is necessary, and for WDXRF, only pure standards are required for each element of interest. However, it is much more satisfying to use at least one multielement standard containing all the analytes. The uncertainties associated with the parameters introduce certain biases into the results, which are minimized by the iterations involving the multielement standard.

It took several years, from the first programming of Sherman's equations, until a usable program, NRLXRF, was made available to the X-ray community [22]. Since that time, there have been many software packages produced using this method of data reduction, many of them proprietary to instrument manufacturers. One of these second generation programs that is not proprietary is NBSGSC [22], a collaboration between the (then) National Bureau of Standards and the Geological Survey of Canada. It is assumed that all of these various incarnations are based on Sherman equations, although there might be other possibilities. A different approach was taken by

Hawthorne and Gardner [23], who applied a Monte Carlo technique to the simulation of the X-ray intensities from homogeneous specimens excited by the continuous and characteristic photons from X-ray tubes.

Techniques used to collect X-ray intensities have become more sophisticated as time has passed. Particularly with the incorporation of computers as an integral component, the details of gathering X-ray data have been reduced to selecting the proper program. Also, the stability of the instruments has reached a point where it is of no concern to the analyst. The primary problem remaining limiting accuracy with which specimens can be analyzed is specimen preparation. Among the most disturbing specimen preparation problems are those dealing with powder specimens, particularly minerals, which can be quite heterogeneous, with components having different physical properties leading to difficulty in producing consistent particle size (see Section 9.3).

BIBLIOGRAPHY

[1] L. S. Birks, "History of X-ray spectrochemical analysis," *American Chemical Society Centenial Volume*, ACS: Washington, DC (1976).

[2] J. V. Gilfrich, "100 years of progress in X-ray fluorescence analysis," *Adv. X-ray Anal.*, **39**, 29–40 (1995).

[3] R. Jenkins, "Evolution of X-ray instrumentation and techniques, 1970–1990," *Adv. X-ray Anal.*, **39**, 13–18 (1995).

[4] C. G. Barkla, "Spectra of the fluorescent Röntgen radiation," *Phil. Mag.*, **22**, 396–412 (1911).

[5] A. Hadding, "Mineral analysis by X-ray spectroscopic methods," *Z. Anorg. Chem.*, **122**, 195–200 (1922).

[6] G. von Hevesey, *Chemical Analysis by X-rays and its Applications*, McGraw-Hill: New York, 333 pp. (1932).

[7] R. Glocker and H. Schreiber, "Quantitative Röntgen spectrum analysis by means of cold cathode excitation," *Ann. Physik.*, **85**, 1089–1102 (1928).

[8] H. Freidman and L. S. Birks, "Geiger counter spectrometer for X-ray fluorescence analysis," *Rev. Sci. Instrum.*, **19**, 323–330 (1948).

[9] J. L. Heilbron, *H. G. J. Moseley, the Life and Letters of an English Physicist*, California Press: London, 312 pp. (1974).

[10] D. Coster and G. von Hevesey, "Missing element of atomic number 72," *Nature*, **111**, 182 (1923).

[11] A. Jönsson, "A study of the intensities of soft X-rays and their dependence on potential," *Z. Phys.*, **43**, 845–863 (1927).

[12] C. E. Eddy and T. H. Laby, "Quantitative analysis by X-ray spectroscopy," *Proc. Roy. Soc. (London)*, **127A**, 20–42 (1930).

[13] B. L. Henke, "Some notes on ultra-soft X-ray fluorescence--10 to 100 Å region," *Adv. X-ray Anal.*, **8**, 269–289 (1965).

[14] "IEEE Standard modular instrumentation and digital interface system (CAMAC) IEEE- Std.583," *Inst. of Electrical and Electronic Engineers Inc.*, New York (1975).

[15] J. R. Rucklidge, "Automatic X-ray fluorescence using CAMAC," *X-ray Spectrom.*, **7**, 57–62 (1978).

[16] R. L. Heath, "The application of high-resolution solid state detectors to X-ray spectrometry--a review," *Adv. X-ray Anal.*, **15**, 1–35 (1972).

[17] T. W. Barbee, "Layered synthetic microstructures (LSM): reflecting media for X-ray optic elements and diffracting structures for study of condensed matter," *Supperlattices Microstruct.*, **1**, 311–326 (1985).

[18] E. M. Pell, "Ion drift in an n-p junction," *J. Appl. Phys.*, **31**, 291–302 (1960).

[19] J. DuMond, and J. P. Youtz, "An X-ray method of determining rate of diffusion in the solid state," *J. Appl. Phys.*, **11**, 357–365 (1940).

[20] J. Dinklage and R. Frericks, "X-ray diffraction and diffusion in metal film-layered structures," *J. Appl. Phys.*, **34**, 2633–2635 (1963).

[21] D. S. Urch, "Chemical bonding effects in X-ray emission spectra--a molecular orbital model," *Adv. X-ray Anal.*, **14**, 250–267 (1970).

[22] G. Y. Tao, P. A. Pella, and R. M. Rousseau, "NBSGSC--a FORTRAN program for quantitative X-ray fluorescence analysis," *NBS Tech. Note 1213*, NIST, Gaithersburg, USA (1985).

[23] A. R. Hawthorne and R. P. Gardner, "Monte Carlo simulation of XRF from homogeneous multi-element samples excited by continuous and discrete energy photons from X-ray tubes," *Anal. Chem.*, **47**, 2220–2225 (1975).

CHAPTER

6

INSTRUMENTATION FOR X-RAY SPECTROMETRY

6.1. INTRODUCTION

An X-ray spectrometer consists of five major portions:

1. An excitation source.
2. A specimen presentation system.
3. A dispersing system.
4. A detection system.
5. A data collection and processing system.

The excitation source is typically a high-powered X-ray tube driven by a high-voltage generator. The specimen presentation system is built in as an integral part of the spectrometer itself. In the case of wavelength dispersive spectrometers, the spectrometer has a means of setting angles (the goniometer) for selectable dispersing crystals. The detection system in wavelength dispersive systems is generally based on a pair of detectors mounted in a tandem configuration. In the case of energy dispersive spectrometers, a high-resolution, solid-state detector is used to fulfill both the dispersing and the detection requirements. In the case of modern spectrometers, based on either wavelength or energy dispersive principles, data collection and analysis systems are typically based on personal computers.

6.2. EXCITATION OF X-RAYS

Several different types of sources have been employed for the excitation of characteristic X-radiation and the most important of these are illustrated in Figure 6.1.

As has been mentioned previously, all of the early work in X-ray spectrometry was done using electron excitation. This technique is still used very successfully today in electron column applications. Electron excitation is not frequently used in classical X-ray fluorescence systems, due mainly to the inconvenience of having to work under high vacuum and problems with

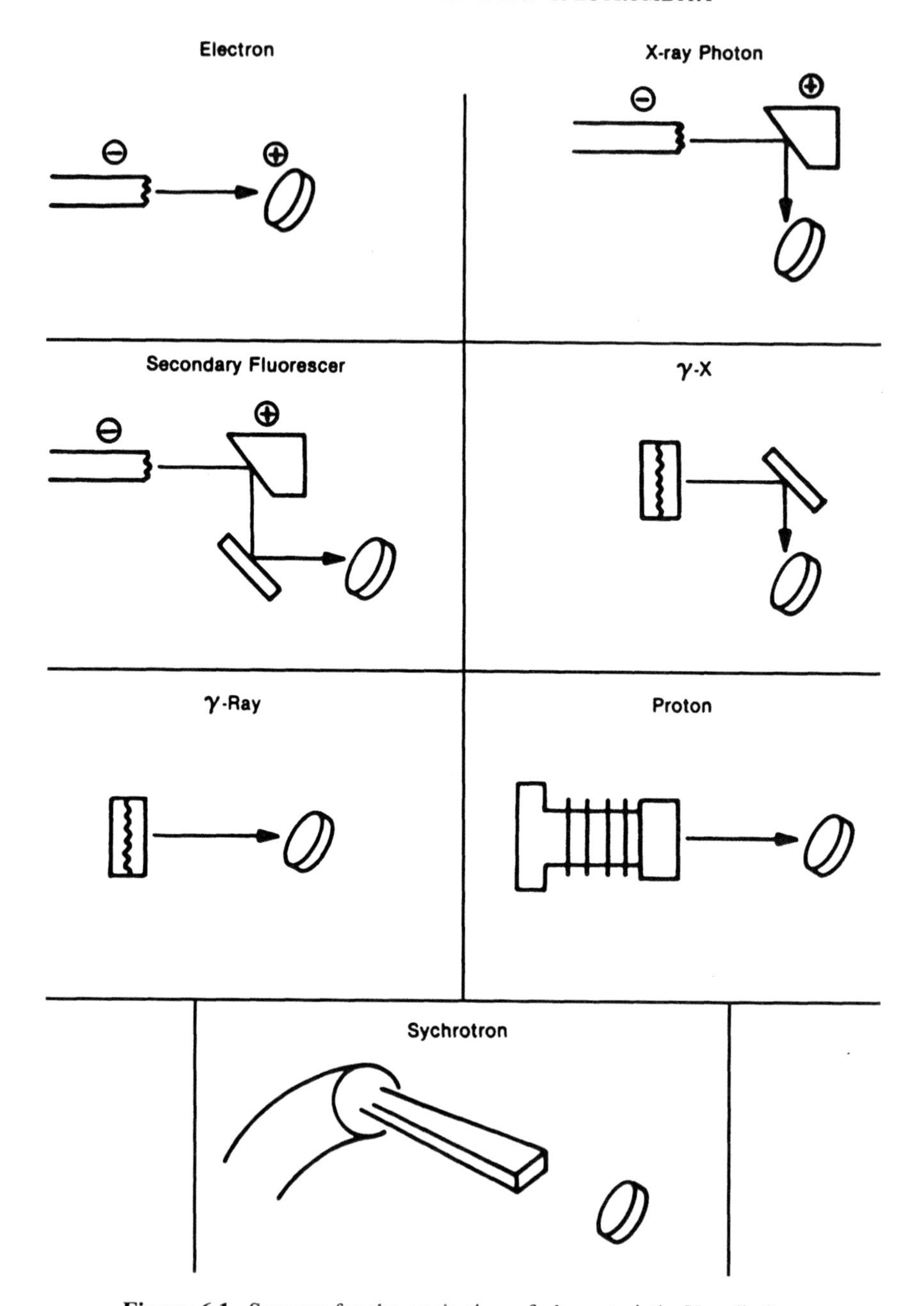

Figure 6.1. Sources for the excitation of characteristic X-radiation.

heat dissipation. By far the most common source today is the X-ray photon source. This source is used in primary mode in the wavelength and primary energy dispersive systems, and in secondary fluorescer mode in secondary target energy dispersive spectrometers. The γ source is typically a radioisotope that is used either directly, or in a mode equivalent to the secondary fluorescer mode, in energy dispersive spectrometry. This configuration is known as the γ-X source and here, γ rays are used to excite X-radiation from a secondary target and this secondary radiation is used to excite the specimen. Typical γ-sources are ^{241}Am, ^{109}Cd, ^{153}Gd, ^{155}Eu, and ^{145}Sm. The γ source and γ-X source are both used in low cost and portable systems, and though the total photon flux is small, they both offer the advantages of high stability, compactness, and low cost [1]. The proton and synchrotron source both offer incredibly high sensitivity, relative to the classical bremsstrahlung source, and each are finding increasing application in areas where high source intensity is an advantage. (see Chapter 5 for further discussion of proton and synchrotron source spectrometers.)

Most conventional wavelength dispersive X-ray spectrometers use a high-power (2–4 kW) X-ray bremsstrahlung source. Energy dispersive spectrometers use either a high-power or low-power (0.5–1.0 kW) primary source, depending on whether the spectrometer is used in the secondary or primary mode. In all cases, the primary source unit consists of a very stable, high-voltage generator, capable of providing a potential of typically 40 to 100 kV, plus a sealed X-ray tube. The sealed X-ray tube has an anode of Cr, Rh, W, Ag, Au, or Mo, and delivers an intense source of continuous and characteristic radiation, which then impinges onto the analyzed specimen, where characteristic radiation is generated. In general, most of the excitation of the longer wavelength characteristic lines comes from the longer wavelength characteristic lines from the tube, and most of the short wavelength excitation, from the continuous radiation from the tube. Since the relative proportions of characteristic to continuous radiation from a target increase with decrease of the atomic number of the anode material, optimum choice of a target for the excitation of a range of wavelengths can present some problems. One way around this problem is to use a dual target tube, and many different varieties of such tubes have been employed over the years. A more recent manifestation of this is a dual anode tube in which the second (low atomic number) material is plated on top of the first (high atomic number) material. At high tube voltages, the electrons penetrate beyond the thin layer of low atomic number material and the output is highly biased toward continuum. At low tube voltages, the electrons dissipate their energy mainly in the low atomic number surface layer with a resulting output biased in favor of longer wavelength radiation. Combinations of Sc/Mo, Cr/Ag, and Sc/W have all proven useful in this regard [2].

In order to excite a given characteristic line, the source must be run at a voltage V_o well in excess of the critical excitation potential V_c of the element in question. The relationship between the measured intensity of the characteristic line I, the tube current i, and the operating and critical excitation potentials is as follows

$$I = K \times i \times (V_o - V_c)^{1.6}, \tag{6.1}$$

where K is a constant. The product of i and V_o represents the maximum output of the source in kilowatts. The optimum value for V_o/V_c is 3 to 5. This optimum value occurs because, at very high operating potentials, the electrons striking the target in the X-ray tube penetrate so deep into the target

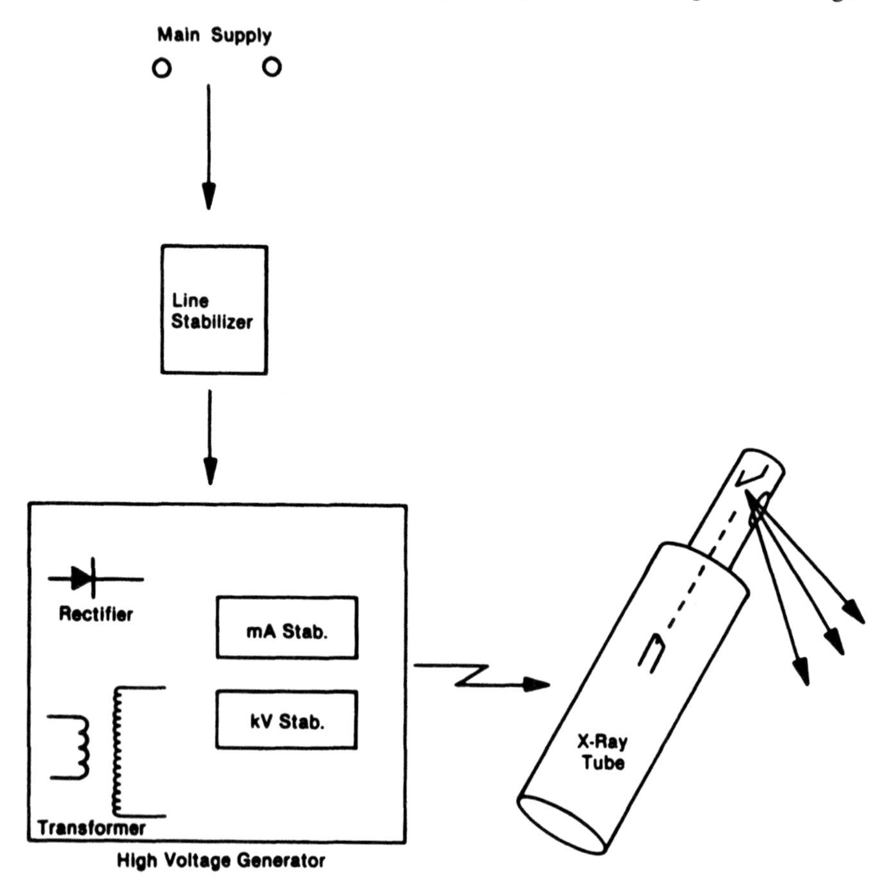

Figure 6.2. Basic components of the primary X-ray photon source using a sealed X-ray tube.

that self-absorption of target radiation becomes significant. Figure 6.2 shows the basic components of a typical high-power source. Power from the main supply is fed via a line stabilizer to the high-voltage generator. Here the voltage is rectified and stabilizing circuits used for both output current and output voltage. The current is fed to the filament of the X-ray tube that is typically a coil of tungsten wire. The applied current causes the filament to glow, emitting electrons in all directions. A portion of this electron cloud is accelerated to the anode of the X-ray tube, that is typically a water-cooled block of copper with the required anode material plated or cemented to its surface. The impinging electrons produce X-radiation, a significant portion of which passes through a thin beryllium window to the specimen. Since it is the intention to eventually equate the value of I for a given wavelength or energy to the concentration of the corresponding analyte element, it is vital that both the tube current and voltage be stabilized to less than a tenth of a percent.

Variations in the photon output from a source are generally referred to as *drift*, and Table 6.1 illustrates the several different categories into which drift components fit. Some drift is inherent to the design of any given system, but most of this is correctable. Such correction measures are necessary, otherwise significant systematic errors may accrue in a given quantitative analysis. Most commercially available high-voltage generators are designed to give very high stability over a relatively short period of time (typically 30–120 minutes). The associated random error due to source instability, is referred to as *short-term drift* and is typically of the order of a tenth of one percent or less. This is the limiting error in any quantitative procedure. There is also a *long-term drift* component that arises from a variety of sources including thermal effects on components, focal spot wander in the tube, etc. The long-term drift is typically 2 to 5 times the magnitude of the short-term drift. *Ultrashort term drift* arises when the stabilization circuits are unable to react quickly enough to short time duration changes in the input power supply. Even a modern solid-state stabilizer may have a reaction time on the order of a few hundred milliseconds. There is also an *ultralong-term* drift component that is due to aging of the X-ray tube, arising mainly to deposition of tungsten

Table 6.1. Different Forms of Drift Encountered in X-Ray Tube Sources.

Form of Drift	Time Duration	Magnitude	Source
Ultralong	Months	1–20%	Aging of X-ray Tube
Long term	Days	0.2–0.5%	Thermal Changes
Short term	30–120 Minutes	$<0.1\%$	Stabilization Circuits
Ultrashort	50-500 Millisecs	0.2–10%	Transients

from the filament onto the inner surfaces of the tube. While ultrashort-term drift can only be eliminated by adding extra stabilization or using an isolation transformer, the effects of both long- and ultralong-term drift can be completely removed using ratio counting. The basis of the ratio counting technique is to measure the counting rate on an instrument reference standard at frequent intervals, during the process of taking counts on specimens being analyzed. The instrument reference is typically a specimen selected from the suite of specimens used in the calibration of the spectrometer for the analytical problem in question. Ideally, the standard should contain all analyte elements, at concentration levels falling somewhere in the middle of the range being covered. By ratioing count rates from the analyte elements in the unknown specimen to those count rates similarly obtained from the instrument reference, most of the effects of instrument and source long-term drift can be eliminated.

6.3. DETECTION OF X-RAYS

In order to adequately define the properties of X-ray detectors it is necessary to define certain terms, which, while in common use, are often misused. *Intensity* [I] is a flux of radiation, measured in number of photons per steradian. It is, in effect, a measure of what comes out of the X-ray tube. The photons are converted to pulses by the detector generating a *count rate* [R], measured in pulses (counts) per second. The detector pulses are integrated for time t giving a *number of counts* [N].

An X-ray detector is a transducer for converting X-ray photon energy into voltage pulses. Detectors work through a process of photoionization in which interaction between the entering X-ray photon and the active detector material produces a number of electrons. The current produced by these electrons is converted to a voltage pulse by a capacitor and resistor, such that one digital voltage pulse is produced for each entering X-ray photon. In addition to being sensitive to the appropriate photon energies, that is, being applicable to a given range of wavelengths or energies, there are other important properties that an ideal detector should possess. These properties are *quantum counting efficiency*, *linearity*, *proportionality*, and *resolution*, and are illustrated in Figure 6.3. In the first part of the figure (a), a photon flux I_o is incident upon the detector, and a fraction I of this incident flux passes through the detector without producing ions within the detector. The quantum counting efficiency is expressed as I/I_o. The property of linearity is illustrated in (b). Here, a number of X-ray photons are entering the detector at a rate of I photons per second, producing a rate R pulses per second. Where R is proportional to I, the detector is said to be linear. Linearity is important

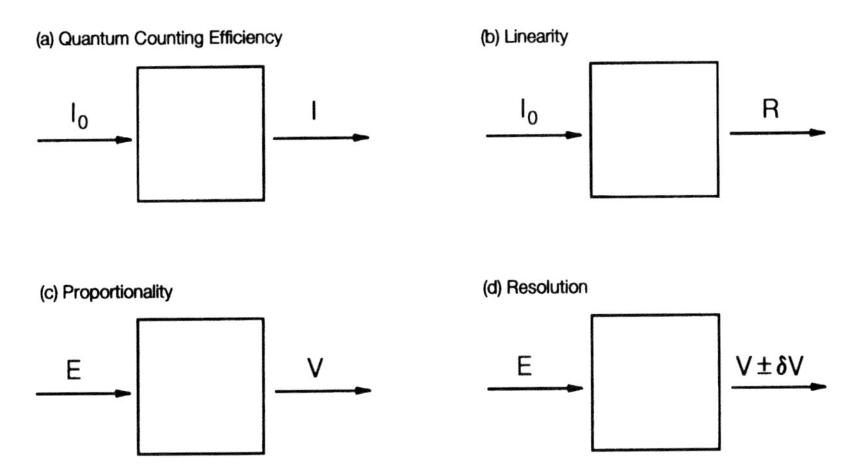

Figure 6.3. Properties of X-ray detectors.

where the various count rates produced by the detector are to be used as measures of the photon intensity for each measured line. Referring to (c), an X-ray photon of energy E enters the detector and a pulse of V volts is produced. Where V is proportional to E, the detector is said to be proportional. Proportionality is needed where the technique of pulse height selection is to be used. The last important detector property is associated with the use of pulse height selection. This property is the detector resolution. As indicated in (d), while each incident photon of energy E gives an average value of V, there is a distribution in the value of V equally to δV. This distribution is a measure of the detector resolution. Pulse height selection is a means of electronically rejecting pulses of voltage levels other than those corresponding to the characteristic line being measured. This technique is a very powerful tool in reducing background levels and the influence of overlapping lines from elements other than the analyte [3]. The properties of proportionality and linearity are, to a certain extent, controllable by the electronics associated with the actual detector. To this extent, while it is common practice to refer to the characteristics of detectors, the properties of the associated pulse processing chain should always be included.

6.3.1. The Gas Flow Proportional Detector

A gas flow proportional counter, shown schematically in Figure 6.4, consists of a cylindrical tube about 2 cm in diameter, carrying a thin (25–50 μm) wire along its radial axis. The tube is filled with a mixture of inert gas and quench gas, typically 90% Argon/10% methane (P-10). The cylindrical tube is

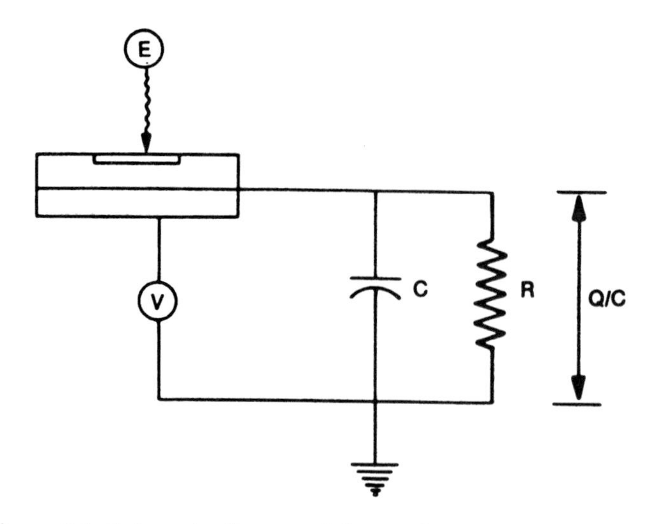

Figure 6.4. Schematic diagram of the gas flow proportional counter.

grounded and about 1400 to 1800 volts is applied to the central wire. The wire is connected to a resistor R shunted by a capacitor C. An X-ray photon entering the detector produces a number n of ion pairs, each comprising one electron and one argon positive ion.

The first ionization potential for argon is about 16 eV, but competing processes, during the conversion of photon energy to ionization, cause the average energy required to produce an ion pair to be greater than this amount. The fraction relating the average energy to produce one ion pair to the first ionization potential is called the Fano factor F [4]. For argon, F is between $1/2$ and $1/3$, and the average energy e_i required to produce one primary ion pair is equal to 26.4 eV. Thus the number of ion pairs produced by a photon of energy E will equal

$$n = \frac{E}{e_i}. \tag{6.2}$$

Following ionization, the charges separate with the electrons moving towards the (anode) wire and the argon ions to the grounded cylinder. As the electrons approach the high field region close to the anode wire, they are accelerated sufficiently to produce further ionization of argon atoms. Thus a much larger number N of electrons will actually reach the anode wire. This effect is called *gas gain*, or *gas multiplication*, and its magnitude is given by M, where M is equal to N/n. For gas flow proportional counters used in X-ray spectrometry, M typically has a value of about 10^5. Thus the final charge Q appearing on the capacitor, for each incident photon, will be equal to $N \times q_e$, where q_e is the elementary charge, equal to 1.60×10^{-19} coulombs. Since capacitance is defined as the ratio of the charge to the potential difference V, the corresponding change in voltage across the capacitor will be equal to

$$V = \frac{nMq_e}{C}. \tag{6.3}$$

Combining Equations 6.3 and 6.4 gives

$$V = \frac{Eq_e}{e_i C} \times M. \tag{6.4}$$

Provided the gas gain is constant, the size of the voltage pulse V produced is directly proportional to the energy E of the incident X-ray photon. In practice, not all photons arising from photon energy E will be exactly equal to V. There is a random process associated with the production of the voltage pulses and the resolution of a counter is related to the variance in the average number of ion pairs produced per incident X-ray photon. The resolution is generally expressed in terms of the full width at half maximum of the pulse amplitude distribution. The theoretical resolution R_t of a flow counter can be derived from

$$R_t\% = \frac{2.36 \times 100}{\sqrt{n}}. \tag{6.5}$$

Combining Equations 6.2 and 6.5 gives

$$R_t\% = \frac{38.3}{\sqrt{E}}. \tag{6.6}$$

6.3.2. The Scintillation Detector

While the gas flow proportional counter is ideal for the measurement of longer wavelengths, it is rather insensitive to wavelengths shorter than about 1.5 Å. For this shorter wavelength region it is common to use the *scintillation counter*. The scintillation counter consists of two parts, the phosphor (scintillator) and the photomultiplier. The phosphor is typically a large single crystal of sodium iodide, that has been doped with thallium. When X-ray photons fall onto the phosphor, blue light photons are produced, where the number of blue light photons is related to the energy of the incident X-ray photon. These blue light photons produce electrons by interaction with a photosurface in the photomultiplier, and the number of electrons is linearly increased by a series of secondary surfaces, called *dynodes*, in the photomultiplier. The current produced by the photomultiplier is then converted to a voltage pulse, as in the case of the gas flow proportional counter. Since the number of electrons is proportional to the energy of the incident X-ray photon, the scintillation counter is a *proportional counter.*

Because of inefficiencies in the X-ray/blue-light/electron conversion processes, the average energy to produce a single event with the scintillation counter is more than a magnitude greater than the equivalent process in the flow counter. For this reason, the resolution of the scintillation counter is much worse than that of the flow counter. The theoretical resolution R_t of the scintillation counter for photons of energy E is given by

$$R_t\% = \frac{128}{\sqrt{E}}. \tag{6.7}$$

6.3.3. The Gas Scintillation Detector

The gas scintillation counter combines the principles of both gas and scintillation counters [5]. In this case, as in the case of the gas proportional counter, the incident X-ray photons first ionize the counter gas. The electrons are accelerated by a voltage across two wire mesh electrodes where electron/atom collisions cause the emission of ultraviolet radiation. This light is measured with a photomultiplier as in the case of the scintillation detector. Because there are fewer losses associated with these processes than with the conventional scintillation counter, the gas scintillation counter gives about a threefold improvement in energy resolution. There are indications that the characteristics of the gas scintillation counter can be improved even further, at which time it may well find a use in conventional wavelength dispersive spectrometers.

6.3.4. The Si(Li) Detector

The Si(Li) detector consists of a small cylinder (about 1 cm diameter and 3 mm thick) of *p*-type silicon that has been compensated by lithium to increase its electrical resistivity [6]. A Schottky barrier contact on the front of the silicon disk produces a *p-i-n* type diode. In order to inhibit the mobility of the lithium ions and to reduce electronic noise, the diode and its preamplifier are cooled to the temperature of liquid nitrogen. By applying a reverse bias of around 1000 V, most of the remaining charge carriers in the silicon are removed. Incident X-ray photons interact to produce a number n of electron hole pairs, as given in Equation 6.2, where, in this instance, e_i is equal to about 3.8 eV for cooled silicon. This charge is swept from the diode by the bias voltage to a charge sensitive preamplifier. A charge loop integrates the charge on a capacitor C to produce an output pulse as in the case of the flow proportional counter, although in this case the M term in Equation 6.3 is equal to unity, since the Si(Li) detector does not have an equivalent property to gas gain.

The resolution R of the Si(Li) detector is given by

$$R = \sqrt{[(\sigma_{noise})^2 + (2.35\sqrt{e_i FE})^2]}. \tag{6.8}$$

The Fano factor for the Si(Li) detector is around 0.12 and the noise contribution about 100 eV. Using a value of 3.8 for e_i, the calculated resolution from Equation 6.8 for Mn $K\alpha$ radiation ($E = 5\,895$ eV) is about 160 eV.

6.3.5. Comparison of X-Ray Detectors

Table 6.2 summarizes some of the characteristics of the three detectors commonly employed in X-ray spectrometry and Table 6.3 compares the resolution of these detectors with that of the crystal spectrometer. As is shown in Equation 6.2, the number of (equivalent) ion pairs produced is directly proportional to the energy of the X-ray photon and inversely proportional to the average energy to produce one ion pair. Going from the Si(Li) detector, to the flow counter, to the scintillation counter, the Average energy to produce an ion pair increases by roughly an order of magnitude each time. Since the resolution of the detector is related to the square root of the number of electrons per photon, the resolution of the three detectors is expected to vary by roughly the square root of 10 or about three times. As shown in Table 6.3(a), the data for Cu $K\alpha$ show such a variation.

Another important practical point shown in Table 6.2 is the actual number of electrons produced, again in this example, for a photon of Cu $K\alpha$. The size

Table 6.2. Characteristics of Common X-Ray Detectors.

Detector Type	Useful Range (Å)	Average Energy for One Ion Pair	Electrons/Photon Cu $K\alpha$		
			Initial	Gain	Final
Gas Flow Proportional	1.5–50	26.4 eV	305	6×10^4	2×10^7
Scintillation	0.2–2	350 eV	23	10^6	2×10^7
Si(Li)	0.5–8	3.8 eV	2116	1	2×10^3

Table 6.3. Resolution of Various Dispersion Devices at the Energy of Cu *Kα* (8.04 keV). Crystal Spectrometer Data based on a 10-cm Long Primary Collimator of 150 μm Spacing and a 5-cm Long Secondary Collimator with 120 μm Spacing.

	Resolution (eV)	Resolution (%)
(a) *Detector Alone*		
Scintillation	3638	45.3
Gas Flow Proportional	1086	13.5
Si(Li)	160	2.0
(b) *Crystal Spectrometer*		
LiF(200)	31	0.39
LiF(220) 1st Order	22	0.27
LiF(220) 1st Order	12	0.15

of the voltage pulse produced by the cathode follower of the detector is proportional to the current produced, that is, to the number of electrons reaching the collector. In a detector that has internal gain, this number of electrons will be the product of the initial number of electrons and the gain of the detector. In the case of the gas flow proportional counter, it is the gas gain, and in the scintillation counter, the photo-multiplier gain. Remember that the Si(Li) detector has no internal gain. The Si(Li) detector has a high number of initial electrons per photon (therefore, a high resolution), but, relative to the other two detectors, has a small number of final electrons. This means that normal external amplification of voltage pulses from the Si(Li) detector cannot be used because this would also amplify noise from the detector. For this reason, a cooled charge sensitive preamplifier is used in Si(Li) spectrometers rather than a simple linear electronic amplifier.

Table 6.3 shows the actual resolution of the detectors used alone, in comparison with a crystal spectrometer. As is shown in the next section, the resolution of a crystal spectrometer is related to a number of factors, of which the *d*-spacing of the crystal and the order of the reflection are the most important. While the resolution of the Si(Li) detector is worse than that of crystal spectrometers in the middle of the usual analytical wavelength region (data here are for Cu $K\alpha$ and as the energy of the analyte photon increases, the resolution difference between detectors alone and crystal spectrometers becomes less), it is sufficient for the separation of most lines. Note that the resolution of both gas flow and scintillation counters is almost never sufficient for use without a crystal spectrometer. (See Section 7.3 for a more detailed discussion of the relative merits of wavelength and energy dispersive spectrometers for the separation of characteristic lines.)

6.3.6. Other Detector Systems

Even though the Si(Li) detector is the most common detector used in energy dispersive X-ray spectrometry, it is certainly not the only one. The higher absorbing power of germanium makes it an alternative for the measurement of high energy spectra, and both cadmium telluride, CdTe, and mercuric iodide, HgI_2, show some promise as detectors capable of operating satisfactorily at room temperature. There are, however, many practical problems in the manufacture of these devices. As an example, in the preparation of CdTe, a Br-Methanol etchant is used and the side product, $CdBr_2$, has been found to poison the surface, causing high leakage current with increase in background noise. This problem has recently been solved and much lower backgrounds are now being reported [7]. The major problem with HgI_2 remains growing crystals of suitable size. The resolution of the detector is, however, promising, making the rather tedious research worthwhile. As an example, a HgI_2 based energy dispersive spectrometer, in which both detector and FET were cooled using a Peltier cooler, has been used in a scanning electron microscope and a resolution of 225 eV (FWHM) obtained for Mn $K\alpha$(5.9 KeV) and 195 eV for Mg $K\alpha$(1.25 KeV) [8].

6.4. WAVELENGTH DISPERSIVE SPECTROMETERS

The function of the spectrometer is to separate the polychromatic beam of radiation coming from the specimen so that the intensities of each individual characteristic line can be measured. A spectrometer should provide sufficient resolution of lines to allow such data to be taken, at the same time providing a sufficiently large response above background to make the measurements statistically significant, especially at low analyte concentration levels. It is also necessary that the spectrometer allow measurements over the wavelength range to be covered. Thus, in the selection of a set of spectrometer operating variables, four factors are important: *resolution, response, background level*, and *range*. Due to many factors, optimum selection of some of these characteristics may be mutually exclusive. As an example, attempts to improve resolution invariably cause lowering of absolute peak intensities.

The wavelength dispersive spectrometer may be a single channel instrument in which a single crystal and a single detector are used for the sequential measurement of a series of wavelengths, or a multichannel spectrometer in which many crystal/detector sets are used to measure many elements simultaneously. In the typical wavelength dispersive spectrometer geometry, a single crystal of known interplanar spacing is used to disperse the collimated polychromatic beam of characteristic wavelengths coming from the sample, such that each wavelength will diffract at a discrete angle. As is shown in Equation 1.8, there is simple relationship between the interplanar spacing of the crystal, the diffracted wavelength, and the diffraction angle. Since the maximum achievable angle on a typical wavelength dispersive spectrometer is around 73°θ, the maximum wavelength that can be diffracted by a crystal of spacing $2d$ is equal to about $1.9d$. In terms of the separating power of a crystal spectrometer, this will be dependant upon two factors, the divergence allowed by the collimators (that to a first approximation determines the width of the diffracted lines), and the angular dispersion of the crystal. The experimental breadth (B_e) of a line in a crystal spectrometer can be expressed as

$$B_e = \sqrt{(B^2 \quad + B^2_{cr})} \tag{6.9}$$

where B_{cc} is the angular aperture of the primary collimator and B_{cr} the rocking curve of the diffracting structure. The wavelength resolution $\Delta\lambda/\lambda$ is a function of B_e and the angle at which the line is diffracted. Thus

$$\Delta\lambda/\lambda = \frac{B_e}{\tan\theta}. \tag{6.10}$$

Typical values for B_e range from 0.1°2θ at low angles to 0.5°at high angles, giving $\Delta\lambda/\lambda$ values in the range 0.002 to 0.02. In terms of energy, this corresponds to approximately 10 to 100 eV.

From the above, it will be clear that there is about one order of magnitude variation in the resolution of a wavelength dispersive spectrometer over the usual wavelength range, depending upon the selection of crystals and collimators. Since mechanical limitations prevent wide selectability of line shape just by selection of collimator divergence, in practice, the resolution of the spectrometer will mainly be a function of the angular dispersion of the analyzing crystal, albeit with some influence of the breadth of the diffracted line profile. The angular dispersion $\Delta d/d$ of a crystal of spacing $2d$ is given by

$$\frac{\Delta d}{d} = \frac{n}{2d \times \cos\theta}. \tag{6.11}$$

From Equation 6.11, the angular dispersion will be high when the d spacing is small. This is unfortunate as far as the range of the spectrometer is concerned because a small value of $2d$ means, in turn, a small range of wavelengths coverable. Thus, as with the resolution and peak intensities, the

Table 6.4. Analyzing Crystals used in Wavelength Dispersive Spectrometer. (For a Comprehensive List of Analyzing Crystals, See [9].)

Crystal	Planes	(2d) Å	Atomic number range	
			K Lines	*L* Lines
LiF	(220)	2.848	>Ti (22)	>La (57)
LiF	(200)	4.028	>K (19)	>Cd (48)
PE	(002)	8.742	Al(13)-K(19)	–
TAP	(001)	26.4	F(9)-Na(11)	–
LSM	–	50–120	Be(4)-F(9)	–

obtaining of high dispersion can only be obtained at the expense of cutting down the wavelength range covered by a particular crystal. In order to circumvent this problem, it is likely that several analyzing crystals will be employed in the coverage of a number of analyte elements. Many different analyzing crystals are available, each having its own special characteristics (see e.g., [9]), but three or four crystals will generally suffice for most applications. Table 6.4 shows a short list of the most commonly used crystals. While the maximum wavelength covered by traditional spectrometer designs is about 20 Å, recent developments now allow the extension of the wavelength range significantly beyond this value (see also Section 8.3).

The actual spectrometer itself, consists of the X-ray tube, a specimen holder support, a primary collimator, an analyzing crystal, and a tandem detector, all mounted on a goniometer. The geometric arrangement of these components is shown in Figure 6.5. A portion of the characteristic *fluorescence* radiation from the specimen is passed via a collimator or slit onto the surface of an analyzing crystal, where individual wavelengths are diffracted to the detector in accordance with the Bragg law. A goniometer is used to maintain the required θ to 2θ relationship between crystal and detector. Typically six or so different analyzing crystals and two different collimators are provided in this type of spectrometer, giving the operator a wide range of choice of dispersion conditions. In general, the smaller the *d*-spacing of the crystal, the better the separation of the lines but the smaller the wavelength range that can be covered. A tandem detector system, comprised of a gas flow counter and a scintillation counter, each with its own collimator, is typically employed. This is used to convert the diffracted characteristic photons into voltage pulses that are integrated and displayed as a measure of the characteristic line intensity. The gas flow counter is ideal for the measurement of longer wavelengths and the scintillation counter is best for short wavelengths.

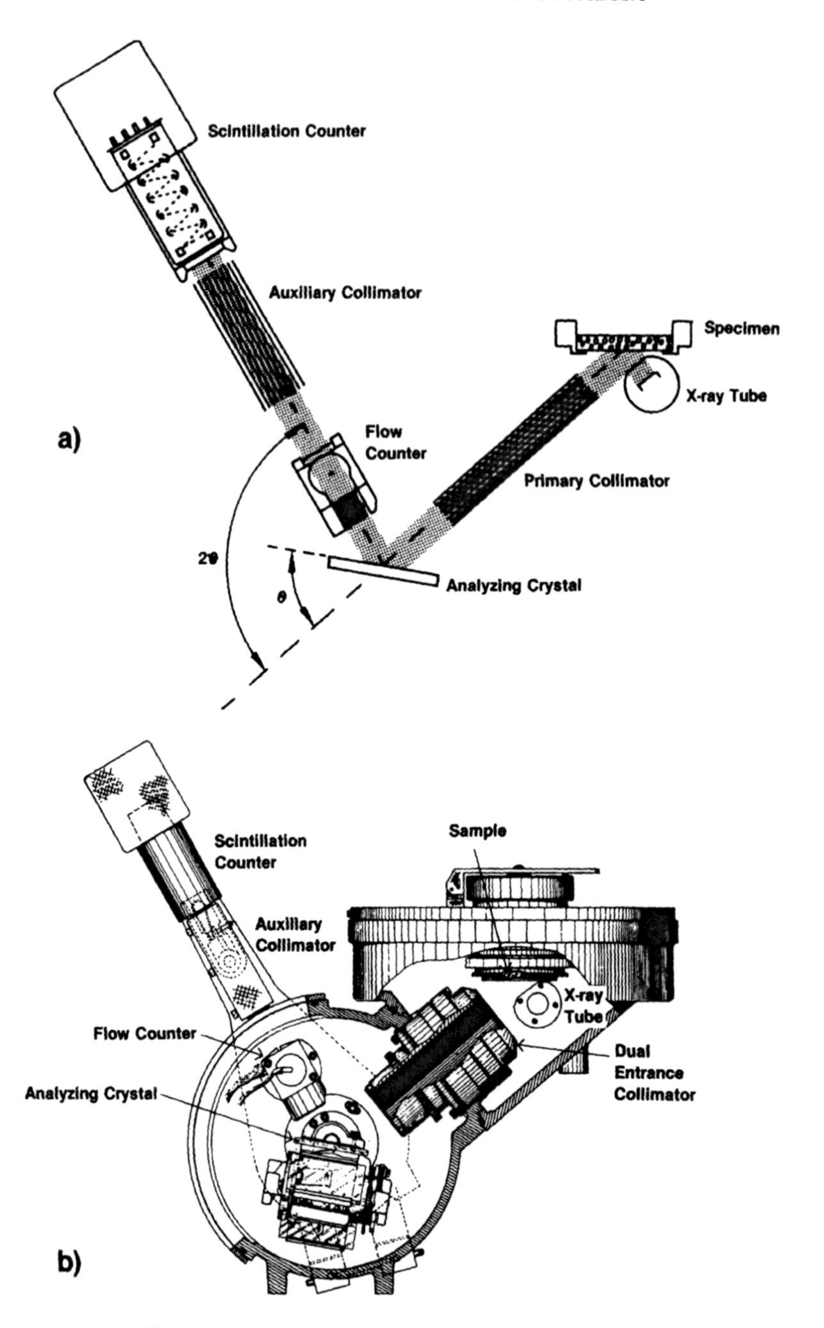

Figure 6.5. The wavelength dispersive spectrometer.

It was recognized some time ago that, since the bremsstrahlung arising from scatter from the primary X-ray source is scattered twice (once by the sample and once by the analyzing crystal), it is plane polarized. Fluorescence radiation from the sample is scattered only once and therefore is not polarized. This can be taken advantage of by arranging the geometry of the source, sample, and analyzing crystal to reduce the amount of scattered radiation from the sample entering the detector [10]. More recently, it has been shown that the use of a curved crystal further increases the

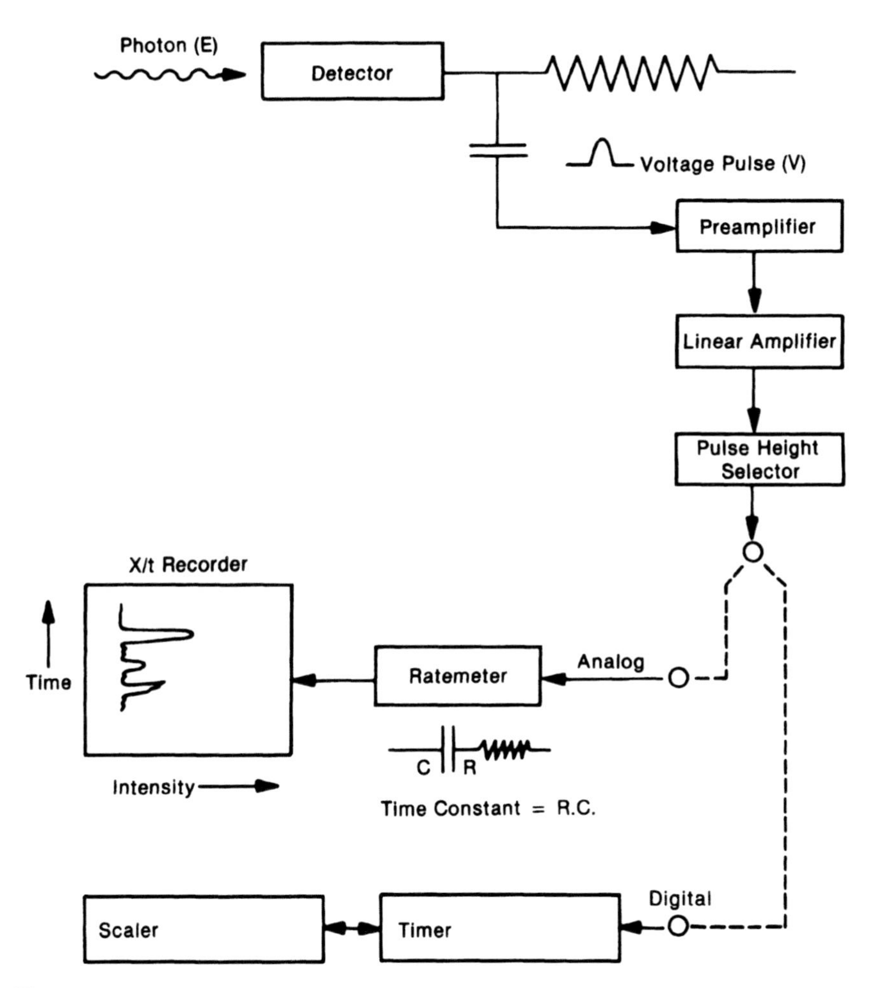

Figure 6.6. Analog and digital instrumentation for the integration and counting of pulses. In fixed time counting, a preset time t generates a stop pulse to the scaler. In fixed counts, a preset count on the scaler generates a stop pulse to the timer.

efficiency of the background reduction, leading to detection limits in the sub-ppm range.

The output from a wavelength dispersive spectrometer may be either analog or digital, and, as illustrated in Figure 6.6, both digital and analog counting equipment is generally available to the user. Pulses from the detector are amplified and passed via the pulse height selector onto one of two circuits. For qualitative work, an analog output is traditionally used and, in this instance, a ratemeter is used to integrate the pulses over short time intervals, typically of the order of a second or so. The output from the ratemeter is fed to an x/t recorder that scans at a speed that is synchronously coupled with the goniometer scan speed. The recorder thus displays an intensity/time diagram that becomes an intensity/2θ diagram. Tables are then used to interpret the wavelengths.

For quantitative work, it is more convenient to employ digital counting, and a timer/scaler combination is provided that allows pulses to be integrated over a period of several tens of seconds and then displayed as count or count rate. In more modern spectrometers a scaler/timer may also take the place of the ratemeter, using a process known as *step-scanning*, illustrated in Figure 6.7. In this instance, the contents of the scaler are displayed on the x axis of the x/t recorder as a voltage level. The scaler/timer is then reset and started to count for a selected time interval. At the end of this time, the timer sends a

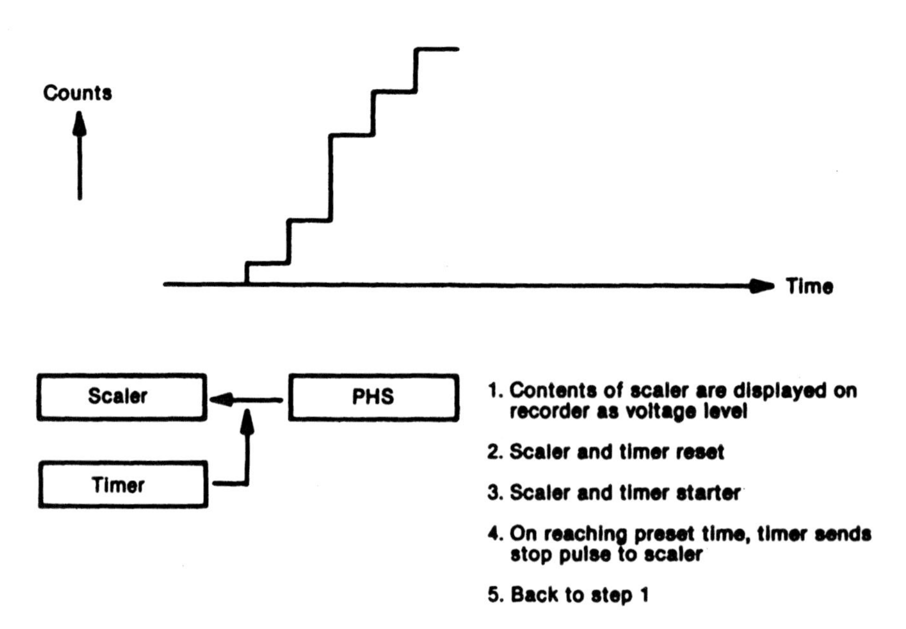

Figure 6.7. Step-scanning to display spectra.

stop pulse to the scaler that now holds a number of counts equal to the product of the counting rate and the count time. The contents of the scaler are then displayed as before, the goniometer stepped to its next position, and the whole cycle repeated. Generally the process is completely controlled by a microprocessor or a personal computer.

6.5. ENERGY DISPERSIVE SPECTROMETERS

Unlike the wavelength dispersive spectrometer, the energy dispersive spectrometer consists of only two basic units, the excitation source and the spectrometer/detection system. In this case, the detector itself acts as the dispersion agent. The spectrometer/detector is typically a Si(Li) detector which, as explained in Section 6.3, is a proportional detector of high intrinsic resolution. A multichannel analyzer is used to collect, integrate, and display the resolved pulses. While properties similar to the wavelength dispersive system are sought from the energy dispersive system, the means of selecting these optimum conditions are very different. Since the resolution of the energy dispersive system is equated directly to the resolution of the detector, this feature is of paramount importance. As indicated in Section 6.3, the value for Si(Li) that should be compared to 10 to 100 eV for the wavelength dispersive system is about 160 eV. (See Chapter 6 for further discussion of the relative merits of wavelength and energy dispersive systems.)

The output from an energy dispersive spectrometer is generally displayed on a CRT and the operator is able to dynamically display the contents of the various channels as an energy spectrum, and provision is generally made to allow zooming in on portions of the spectrum of special interest, to overlay spectra, to subtract background, and so on, in a rather interactive manner. As in the case of modern wavelength dispersive systems, nearly all energy dispersive spectrometers incorporate some form of personal computer that is available for spectral stripping, peak identification, quantitative analysis, and a host of other useful functions

All of the earlier energy dispersive spectrometers were operated in what is called the *primary* mode. These spectrometers consisted simply of the excitation source, typically a closely coupled low-power, end-window X-ray tube, and the detection system. In principle, this primary excitation system offers the possibility of a relatively inexpensive instrument, with two significant advantages over the wavelength dispersive system. The first is the ability to collect and display the total emission spectrum from the sample at the same time, giving great speed in the acquisition and display of data. The second advantage is mechanical simplicity, since there is almost no need at all for moving parts. In practice, however, there is a limit to the maximum count rate that the spectrometer can handle and this led, in the mid-1970s, to the development of the *secondary* mode of operation (see e.g., [11]). In the secondary mode, a carefully selected pure element standard is interposed between primary source and specimen, along with absorption filters where appropriate, such that a selectable energy range of secondary photons is incident upon the sample. This allows selective excitation of certain portions of the energy range, thus increasing the fraction of *useful* to *unwanted* photons entering the detector. While this configuration does not completely eliminate the count rate and resolution limitations of the primary system, it certainly does reduce them.

As has been previously mentioned, within the two major categories of X-ray spectrometers specified, there is a wide diversity of instruments available. The major differences generally lay in the type of source used for excitation, the number of elements that are measurable at one time, the speed at which they collect data, and finally, the price range. All of the instruments are, in principle at least, capable of measuring all elements in the periodic classification from $Z = 9$(F) and upwards. Most can be fitted with multisample handling facilities and all can be automated by use of a personal computer. All are capable of precision on the order of a few tenths of one percent, and all have sensitivities down to the low ppm level. As far as the analyst is concerned, they differ only in their speed, cost, and number of elements measurable at the same time.

BIBLIOGRAPHY

[1] J. R. Rhodes, "Design and application of X-ray emmision analysers using radioisotopes," in "Energy Dispersive X-ray Analysis," *ASTM Special Technical Publication 485*, ASTM: Philadelphia (1971).

[2] J. N. Kikkert and G. Hendry, "Comparison of experimental and theoretical intensities for a new X-ray tube for light element analysis," *Adv. X-ray Anal.*, **27**, 423–426 (1984).

[3] R. Jenkins and J. L. de Vries, *Practical X-ray Spectrometry*, 2nd ed., Chapter 4, MacMillan: London (1970).

[4] U. Fano, "Ionization yield of radiation. II fluctuations of the number of ions," *Phys. Rev.*, **72**, 26–29 (1947).

[5] W. H. M. Ku and R. Novick, "Gas scintillation proportional counters and other low-energy X-ray spectrophotometers," *AIP Conf. Proc.*, **75**, 78–84 (1981).

[6] D. A. Gedcke, "The Si(Li) X-ray energy analysis system: operating principles and performance," *X-Ray Spectrom.*, **1**, 129–141 (1972).

[7] M. Roth and A. Burger, "Improved spectrometer performance of cadmium selenide room temperature gamma-ray detector," *IEEE Trans. Nucl. Sci.*, NS–33, 407–410 (1986).

[8] J. S. Iwanczyk, A. J. Dabrowski, G. C. Huth, J. G. Bradley, J. M. Conley and A. L. Albee, "Low energy X-ray spectra measured with a mercuric iodide energy dispersive spectrometer in a scanning electron microscope," *IEEE Trans. Nucl. Sci.*, NS–33, 355–358 (1986).

[9] E. P. Bertin, *Principles and Practice of X-ray Spectrometric Analysis*, 2nd Ed., Appendix 10, Plenum: New York (1975) .

[10] P. Wobrauschek and H. Aiginger, "The application of linearly polarized X-rays after Bragg reflection for X-ray fluorescence analysis," *Adv. X-ray Anal.*, **28**, 69–74 (1985).

[11] L. S. Birks and H. K. Herglotz, *X-ray Spectrometry* Chapter 2, Dekker: New York (1978).

CHAPTER

7

COMPARISON OF WAVELENGTH AND ENERGY DISPERSIVE SPECTROMETERS

7.1. INTRODUCTION

As has been discussed in earlier chapters, there are many different types of X-ray fluorescence instruments available, based on several different source and dispersion configurations. The most commonly employed instruments are based either on the energy dispersive method or the wavelength dispersive method. Further subdivisions in the energy dispersive instruments include primary or secondary source excitation mode, and in the case of wavelength dispersive instruments, single-channel (sequential) or multichannel (simultaneous). In all cases this may or may not include microprocessor control, and may or may not include a data processing computer. While each of these configurations has clear advantages over its competitors, each also has disadvantages other than the obvious ones of cost and flexibility. Most instrument manufacturers continually strive to develop instruments that offer a good price to performance ratio, and that minimize some of the inherent limitations of a given procedure. This chapter presents some of the advantages and disadvantages of energy and wavelength dispersive based spectrometers.

Wavelength dispersive spectrometers were introduced in the early 1950s and around 30 000 or so such instruments have been supplied commercially, roughly one third of these in the United States. Energy dispersive spectrometers became commercially available in the early 1970s and today there are on the order of 20 000 units in use.

When the first energy dispersive instruments were marketed, the wavelength dispersive method was already an established technique. Also, at this time, the digital minicomputer was being routinely employed in analytical instrumentation both for instrument control as well as data acquisition and processing. It is not surprising, therefore, that nearly all of the early energy dispersive systems incorporated state-of-the-art mini-computers as an integral part of the system. At the same time, the use of graphical display monitors was also becoming commonplace and this too was incorporated into these systems from the very early days. Alongside of its rather conservatively designed wavelength dispersive counterpart, the energy

a) Wavelength Dispersive

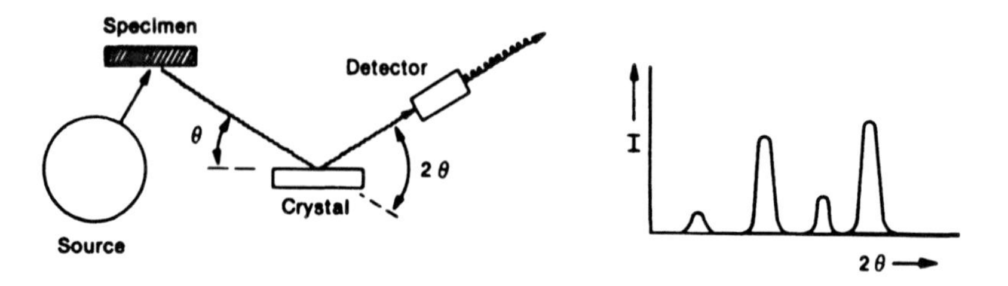

1) Single crystal of fixed 2θ acts as a spectrum analyzer.
2) Scanning 2θ range allows the complete spectrum to be acquired.
3) Selection of single wavelength is achieved by selection of equivalent 2θ value.

b) Energy Dispersive

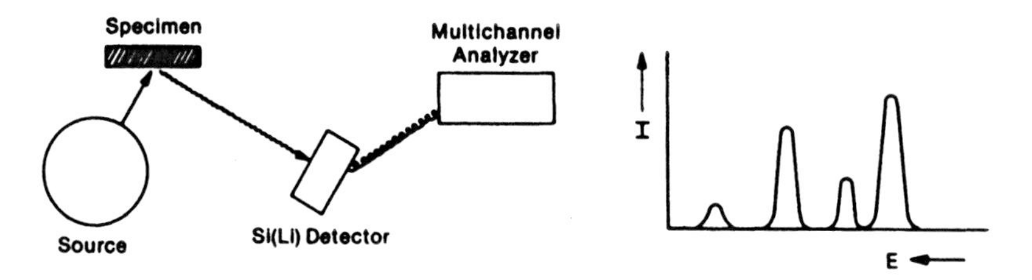

1) Proportional Si(Li) detector gives a distribution of voltage pulses proportional to the spectrum of X-ray photons
2) A multichannel analyzer is used to isolate the voltage pulses into discrete intervals. Consecutive output of the MCA intervals allows complete spectrum to be displayed.
3) Selection of a single energy interval is obtained by selection of appropriate voltage window (i.e., range of channels) on the MCA.

Figure 7.1. The wavelength dispersive spectrometer and the energy dispersive spectrometer.

dispersive system presented a modern state-of-the-art image and many purchasers of those early machines were happy to overlook the count rate and resolution restrictions from which they suffered, in order to take advantage of the rapid data acquisition and display capabilities.

In the design of any spectrometer system, the system designer is confronted with a series of choices and compromises if the end product is to meet the end users' requirements at an acceptable price. This situation is especially important when the spectrometer is to be used for both qualitative and quantitative analysis. High speed of analysis combined with good sensitivity can be obtained for quantitative applications at the expense of flexibility, the multichannel wavelength dispersive spectrometer being a good example of such an end result. Great flexibility can be obtained in qualitative analysis, this time at the expense of sensitivity and some accuracy, as typified by many energy dispersive systems. For most laboratories, there are three basic choices in the selection of an X-ray spectrometer:

1. For *high specimen throughput* quantitative analysis where speed is of the essence, and where high initial cost can be justified, a simultaneous wavelength dispersive system is probably the best.
2. For more *flexibility*, where speed is important but not critical and where moderately high initial cost can be justified, a sequential wavelength dispersive system is probably the best.
3. Where *initial cost* is a major consideration, or where something can be given up in detection limits or accuracy, or where qualitative or semiquantitative analysis is important, an energy dispersive system is probably the best.

For the purpose of comparison, the two major categories of X-ray spectrometers differ mainly in the type of source used for excitation, the number of elements that they are able to measure at one time, the speed at which they collect data, and their price range. All of the instruments are, in principle at least, capable of measuring all elements in the periodic classification from $Z = 9$(F) and upwards, and most modern wavelength dispersive spectrometers can do some useful measurements down to $Z = 6$(C). All can be fitted with multisample handling facilities and all can be automated by use of personal computers. All are capable of precision on the order of a few tenths of one percent and all have sensitivities down to the low ppm level. Single channel wavelength dispersive spectrometers are typically employed for both routine and nonroutine analysis of a wide range of products, including ferrous and nonferrous alloys, oils, slags and sinters, ores and minerals, thin films, and so on. These systems are very flexible but, relative to multichannel spectrometers, are somewhat slow. The multichannel wavelength dispersive instruments are used almost exclusively for routine, high throughput analyses where the great need is for fast accurate analysis, but where flexibility is of no importance. Energy dispersive spectrometers have the great advantage of being able to display information on all elements at the same time. They lack somewhat in resolution over the wavelength dispersive spectrometer, but the ability to reveal elements absent as well as elements present make the energy dispersive spectrometer ideal for general troubleshooting problems. They have been particularly effective in the fields of scrap alloy sorting, forensic science, and the provision of elemental data to supplement X-ray powder diffraction data. In recent years, new wavelength and energy dispersive spectrometers have evolved that offer good price performance characteristics and each of which reflects the latest technology in its own discipline. The secondary target energy dispersive system, as an example, offers good price/performance characteristics, within the mid-price range of instruments. The reaction of the wavelength dispersive spectrometer developers has been to produce instruments in which price has been significantly reduced by sacrificing some of the great speed and sensitivity of the traditional sequential and simultaneous wavelength systems, but that can still outperform energy dispersive type systems in many areas (see e.g., [1]).

7.2. THE MEASURABLE ATOMIC NUMBER RANGE

One of the problems with any X-ray spectrometer system is that the absolute sensitivity (i.e., the measured c/s per %) decreases dramatically as the lower atomic number region is approached (see Section 10.4). The three main reasons for this decrease are the reduction in the fluorescence yield with decrease in atomic number, the decrease in the absolute number of useful long wavelength X-ray photons from a bremsstrahlung source with increase of wavelength, and the increase of absorption effects, generally with increase of the wavelength of the analyte line. The first two problems are inherent to the X-ray excitation process and to constraints in the basic design of conventional X-ray tubes. The third, however, is a factor that depends very much on the instrument design, and, in particular, upon the characteristics of the detector. The detector used in long wavelength spectrometers is typically a gas flow proportional counter, in which an extremely thin, high-transmission window is employed. The detector typically employed in energy dispersive systems is the Si(Li) diode, which has an electrical contact layer on the front surface, typically a 0.02 μm thick layer of gold, followed by a 0.1 μm thick dead-layer of silicon. The absorption problems caused by these two layers becomes most significant for low energy X-ray photons, which have a high probability of being absorbed in the dead layer. Probably the biggest source of absorption loss in the Si(Li) detector is that due to the thin beryllium window that is part of the liquid nitrogen cryostat. A combination of these facts causes a loss in the sensitivity of a typical energy dispersive system of almost an order of magnitude, for the K lines of sulfur ($Z = 16$) to sodium ($Z = 11$). The equivalent number with a gas flow counter would be about a factor of two.

For a number of years, the useful lower atomic number detectable with a typical energy dispersive spectrometer was $Z = 12$. However, newer

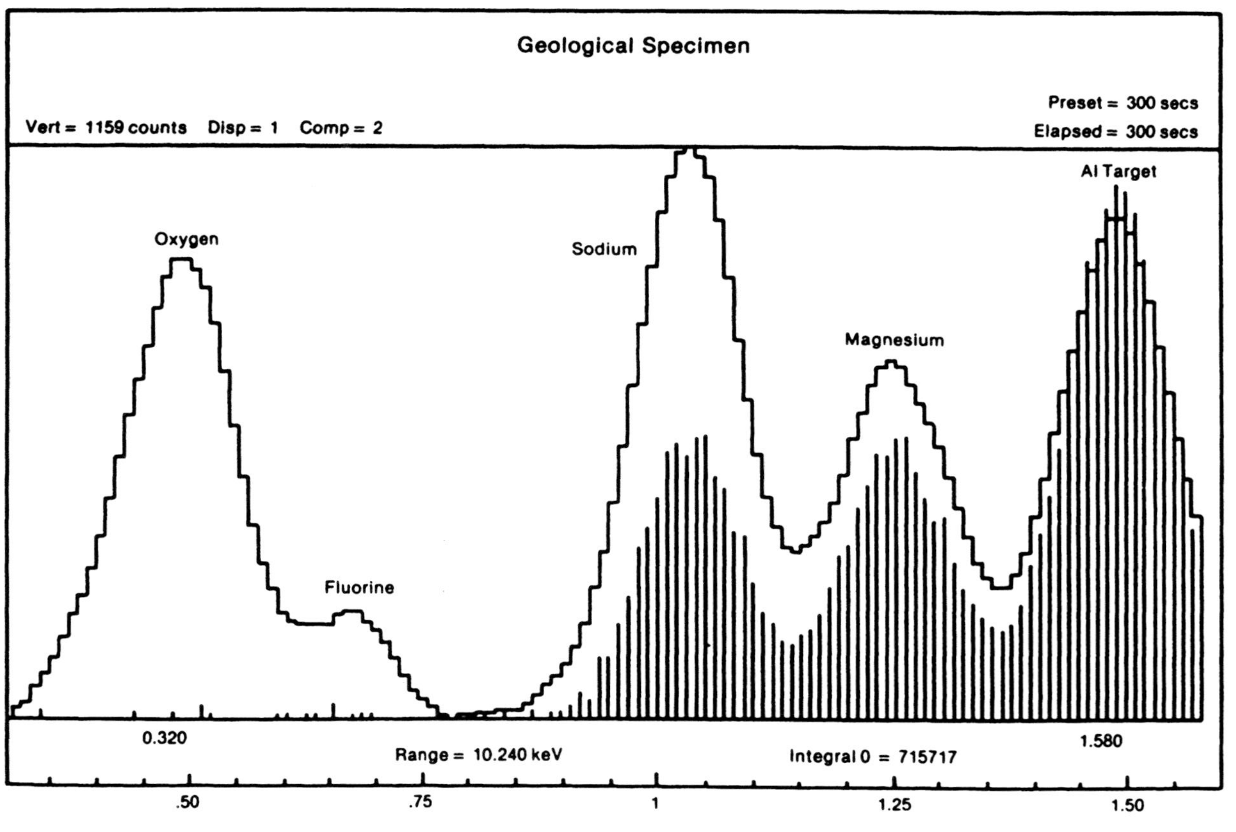

Figure 7.2. Comparison of data obtained with old and new Si(Li) detector technology.

developments in ultrathin windows for the Si(Li) detector [2] now allow measurements down to $Z = 8$ (oxygen). Figure 7.2 shows data published by the Kevex Corporation and compares older (vertical bars) and newer detector technology for the measurement of low atomic number elements in a geological specimen. In comparison, the lower atomic number limit for the conventional wavelength dispersive system is fluorine ($Z = 9$), and, by use of special crystals, this can be extended down to beryllium ($Z = 4$) (see Section 8.3).

7.3. THE RESOLUTION OF LINES

A compromise that must always be made in the design and setup of a spectrometer is that between intensity and resolution, resolution being defined as the ability of the spectrometer to separate lines. In a flat crystal wavelength dispersive system, this resolution is dependant upon the angular dispersion of the analyzing crystal and the divergence allowed by the collimators [3]. In the energy dispersive system, the resolution is dependent only upon the detector and detector amplifier. In absolute terms, the resolution of the wavelength dispersive system typically lies in the range 10 to 100 eV, compared to a value of 150 to 200 eV for the energy dispersive system. Figure 7.3 shows a spectrum obtained on copper, using a state-of-the-

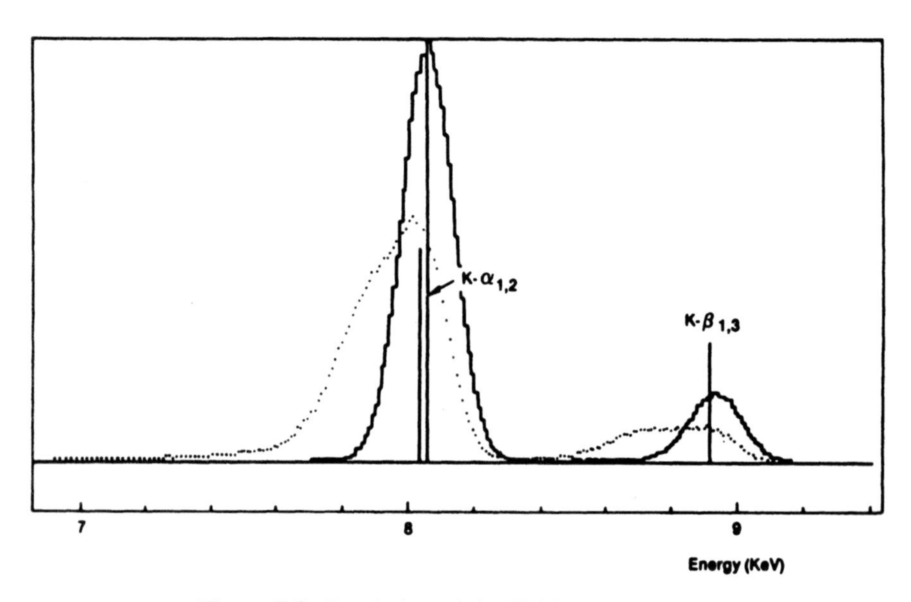

Figure 7.3. Resolution of the Si(Li) spectrometer.

art energy dispersive spectrometer. The figure also indicates the position of the copper $K\alpha$ and $K\beta$ doublets on the same energy scale. The dotted spectrum illustrates the potential shift of the doublets due to Compton scatter. The absolute energy difference between Cu $K\alpha_1$ and Cu $K\alpha_2$ is 21 eV. These two lines could be reasonably well resolved using a wavelength dispersive spectrometer and a LiF(220) analyzing crystal. An added advantage of the wavelength dispersive spectrometer is that the resolution/intensity selection is much more controllable. As an example, when using the wavelength dispersive spectrometer, while Fe $K\alpha$ radiation would normally be measured with a LiF(200) crystal and 150 μm spacing collimator, for those cases where line separation is not a problem, it could also be measured with the LiF(200) crystal and a 450 μm spacing collimator. This latter measurement would give about a factor of three times more intensity, but with worsened resolution. In the case of secondary fluorescer systems, some local modifications to the measured spectrum can be made by use of filters either in the primary beam or in the secondary fluoresced beam and this does offer some flexibility.

7.4. MEASUREMENT OF LOW CONCENTRATIONS

X-ray fluorescence methods are generally reasonably sensitive, with detection limits for most elements in the low ppm range (see also Section 11.5). The lower limit of detection is generally defined as that concentration equivalent to a certain number of standard deviations of the background count rate. This means that three major factors will effect the detection limit for a given element: 1) the sensitivity of the spectrometer for that element in terms of the counting rate per unit concentration of the analyte element; 2) the background (blank) counting rate; and 3) the available time for counting peak and background photons. In comparing the energy and wavelength dispersive type systems, the absolute sensitivity of the wavelength dispersive system is almost always higher than the equivalent value for an energy dispersive system, perhaps by one to three orders of magnitude. This is because modern wavelength dispersive systems are able to handle count rates up to 1 000 000 c/s compared to about 40 000 c/s (for the total output of the selected excitation range) in the case of most energy dispersive systems. The ability of the wavelength dispersive system to work at high counting rates allows use of a high loading at the primary source. A fundamental difference between the wavelength and energy dispersive systems is that in the energy dispersive case, all of the detected radiation falls onto the detector at the same time. Figure 7.4 compares the cases of the energy and wavelength dispersive spectrometers measuring two elements i and j in a specimen, giving intensities I_i and I_j. In this example, the

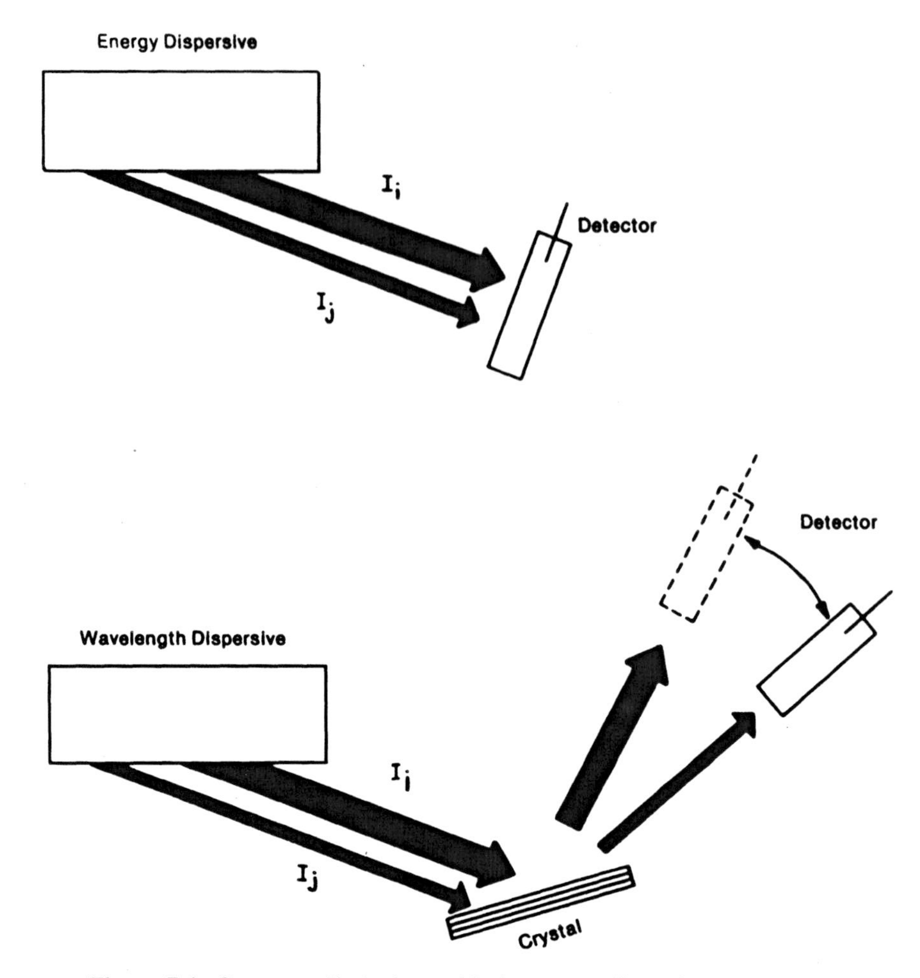

Figure 7.4. Count rate limitations with the energy dispersive spectrometer.

concentration of i is much greater that of j, so that I_i is much greater than I_j. The total radiation flux falling on to the energy dispersive detector is $(I_i + I_j)$, whereas in the case of the wavelength dispersive system, I_i and I_j are measured independently at two different settings of the analyzing crystal. Since the energy dispersive detector is limited on the total radiation flux it can handle, the detection limit of j in i is limited by the total radiation intensity at the detector and, therefore, the concentration of i.

The influence of the background is rather complicated because so many variables come into play. One of the major advantages of the secondary target energy dispersive system over the conventional system is that backgrounds are dramatically lower in the secondary excitation mode, because much of the background in a fluoresced X-ray spectrum comes from scattered primary continuum. The measured backgrounds for the wavelength and energy dispersive systems are very similar in the low atomic number regions of the X-ray spectrum, and for the secondary target energy dispersive system, are lower by up to an order of magnitude in the higher energy (mid-range atomic number) region.

All of these factors taken together lead to detection limits for the secondary target energy dispersive system that are typically significantly worse than the wavelength dispersive spectrometer. The actual detection limit measurable also depends upon the characteristics of the specimen itself, including such factors as absorption, scattering power, and so on.

Table 7.1 shows detection limits reported for trace and major elements in whole rock as reported for primary and secondary target energy dispersive and wavelength dispersive spectrometers [4]. In the table, *PEEDS* stands for *Primary Excitation Energy Dispersive Spectrometry*, *STEDS* for *Secondary Target Energy Dispersive Spectrometry*, and *WDS* for *Wavelength Dispersive Spectrometry*. The trace elements data were obtained directly on the pressed rock powder and the major element data was taken after fusion of the rock with lithium tetraborate. It can be seen that, on average, the secondary target system is better than the primary system by about a factor of between one and

Table 7.1. Detection Limits for Major and Minor Elements in Whole Rock.

Element	PEEDS	STEDS	WDS
Major Elements (% in Fused Samples)			
Na_2O	0.96	0.81	0.16
MgO	0.33	0.13	0.08
Al_2O_3	0.19	0.08	0.032
SiO_2	0.21	0.023	0.050
P_2O_5	0.05	0.023	0.016
K_2O	0.04	0.022	0.008
CaO	0.025	0.034	0.004
TiO_2	0.03	0.008	0.006
MnO	0.015	0.002	0.014
Minor Elements (ppm in pressed powders)			
Rb	5.6	3.0	0.6
Sr	3.5	2.8	0.4
Y	3.5	3.8	0.4
Zr	4.0	2.8	1.1
Nb	4.4	2.8	1.3

four, and the wavelength dispersive data are better again by a further factor of about four.

7.5. QUALITATIVE ANALYSIS

One area in which energy dispersive systems have almost always out-performed wavelength dispersive systems is in qualitative analysis. In wavelength dispersive systems, qualitative analysis has traditionally been performed by synchronously scanning the goniometer at a fixed angular speed, and then using a ratemeter circuit to integrate the digital output from the detector, displaying the output on an x/t recorder. All scanning spectrometers are slow in this sequential angular/intensity data collection mode, not only because the data are taken sequentially, but because, in order to cover the full range of elements, a series of scans must be made with different conditions. In addition to this, scanning at a fixed speed is somewhat inefficient because atomic number varies as a function of one over the square root of the angle in the crystal dispersive system. In effect, this means that in the low atomic number region much scanning time is wasted in scanning angular space that contains no characteristic line data. Although some unique designs have brought about some reduction in data acquisition times [1], this time remains a major limitation of the wavelength dispersive spectrometer.

Angle/wavelength/atomic number tables are readily available to assist in the interpretation of spectra, but this process is still tedious and time consuming. However, count data output from modern wavelength dispersive systems are nearly always digital, and this means that state-of-the-art, spectral processing software can be used to speed up and automate the whole data interpretation process. Good qualitative software packages are available (see Section 10.3) and a complete qualitative analysis can now be performed in about 45 minutes, while retaining all of the inherent resolution and high sensitivity advantages of the wavelength dispersive system.

The output from an energy dispersive system is collected in a simultaneous fashion, either over the complete spectral range, or in selected portions, depending upon the complexity of the specimen and concentration levels involved. The operator is able to dynamically display the contents of the various channels of an energy spectrum as they are acquired. Provision is generally made to interactively allow zooming in on portions of the spectrum of special interest, overlay spectra, subtract background, and so on. Nearly all energy dispersive spectrometers incorporate some form of minicomputer or microprocessor, and software is available for spectral stripping, peak identification, quantitative analysis, and a host of other useful functions.

7.6. GEOMETRIC CONSTRAINTS OF WAVELENGTH AND ENERGY DISPERSIVE SPECTROMETERS

The modern wavelength dispersive system has traditionally tended to be somewhat inflexible in the area of sample handling. The geometric constraints of the conventional system generally stem from the need for close coupling of the sample to X-ray tube distance, and the need to use an air lock of some kind to bring the sample into the working vacuum, usually by means of a multiple position sample carousel. The sample to be analyzed is typically placed inside a cup of fixed external dimensions that is, in turn, placed in the carousel. This presentation system places constraints not only on the maximum dimensions of the sample cup, but also, on the size and shape of samples that can be placed into the cup itself. Primary energy dispersive systems do not require the same high degree of focusing, and to this extent are more easily applicable to any sample shape or size, provided that the specimen will fit into the radiation protected chamber.

In some instances, the spectrometer can even be brought to the object to be analyzed. Because of this flexibility, the analysis of odd shaped specimens has been almost exclusively the purview of the energy dispersive system. In the case of secondary target systems, while the geometric constraints are still less severe than the wavelength system, they are much more critical than in the case of primary systems. This criticality is due not just to the additional mechanical movements in the secondary target system, but also to the limitations imposed by the extremely close coupling of X-ray tube to secondary target. This is probably the reason that most energy dispersive spectrometer manufacturers generally offer a primary system for bulk sample analysis, and retain the secondary target system, generally equipped with a multiple specimen loader, for the analysis of samples that have been constrained to the internal dimensions of a standard sample cup during the specimen preparation procedure.

BIBLIOGRAPHY

[1] R. Jenkins, B. Hammell, and J. A. Nicolosi, "The PN1430, a new low-cost wavelength dispersive spectrometer," *Norelco Reporter*, **32**, 1–8 (1985).

[2] Kevex Corporation, *New Product Bulletin*, Spring (1987), Kevex: Foster City, CA.

[3] R. Jenkins, *An Introduction to X-Ray Spectrometry*, Chapter 4 "Instrumentation," Wiley: London (1976).

[4] R. Jenkins, J. A. Nicolosi, J. F. Croke, and D. Merlo, "Applications of the PW1430 X-ray fluorescence spectrometer," *Norelco Reporter*, **32**, 16–21 (1985).

CHAPTER

8

MORE RECENT TRENDS IN X-RAY FLUORESCENCE INSTRUMENTATION

8.1. THE ROLE OF X-RAY FLUORESCENCE IN INDUSTRY AND RESEARCH

Over the past 30 years or so, the X-ray fluorescence method has become one of the most valuable methods for the qualitative and quantitative analysis of materials. Many methods of instrumental elemental analysis are available today, and among the factors that will generally be taken into consideration in the selection of one of these methods are accuracy, range of application, speed, cost, sensitivity, and reliability. While it is certainly true is that no one technique can ever be expected to offer all of the features that a given analyst might desire, the X-ray method has good overall performance characteristics. In particular, the speed, accuracy, and versatility of X-ray fluorescence are the most important features among the many that have made it the method of choice in laboratories all over the world. Both the simultaneous wavelength dispersive spectrometer and the energy dispersive spectrometers lend themselves admirably to the qualitative and quantitative analysis of solid materials and solutions. Because the characteristic X-ray spectra are so simple, the actual process of allocating atomic numbers to the emission lines is relatively simple, and the chance of making a gross error is rather small. The relationship between characteristic line intensity and elemental composition is also now well understood, and if intensities can be obtained that are free from instrumental artifacts, excellent quantitative data can be obtained. Today, conventional X-ray fluorescence spectrometers allow the rapid quantification of all elements in the periodic table from fluorine ($Z = 9$) and upwards. Recent advances in wavelength dispersive spectrometers have extended this element range down to carbon ($Z = 6$). Over most of the measurable range, accuracies of a few tenths of one percent are possible, with detection limits down to the low ppm level.

8.2. SCOPE OF THE X-RAY FLUORESCENCE METHOD

As was indicated in the previous section, the basis of the X-ray fluorescence technique lies in the relationship between the wavelength (or energy) of the

X-ray photons emitted by the sample element and atomic number Z, as given in Moseley's Law, Equation 1.1. Measurement of wavelength with the crystal dispersive spectrometer, or energy using the energy dispersive spectrometer, allows the identification of the elements present. The intensities of the characteristic lines can then be used to calculate the concentrations of the various elements present. Most commercially available X-ray spectrometers have a range from about 0.4 to 20 Å (40 to 0.6 keV), and this range will allow measurement of the K series from fluorine ($Z = 9$) to lutetium ($Z = 71$), and the L series from manganese ($Z = 25$) to uranium ($Z = 92$). Other line series can occur from the M and N levels but these have little use in analytical X-ray spectrometry. As was discussed in Section 1.3, because of the competing Auger process, the number of vacancies resulting in the production of characteristic X-ray photons is less than the total number of vacancies created in the excitation process. The ratio of the useful to total vacancies is the fluorescent yield, which takes a value of around unity for the higher atomic numbers, to less than 0.01 for the low atomic elements such as sodium, magnesium, and aluminum. This is an important factor in determining the absolute number of counts that an element will give, under a certain set of experimental conditions. It is mainly for this reason that the sensitivity of the X-ray spectrometric technique is rather poor for very low atomic number elements.

There is a wide variety of instrumentation available today for the application of X-ray fluorescence techniques, but for the purpose of this discussion on trends in instrumentation development, it is useful to break X-ray spectrometers down into three main categories:

1. Wavelength dispersive spectrometers
   - Scanning (sequential)
   - Multichannel (simultaneous)
2. Energy dispersive spectrometers
   - Primary excitation
   - Secondary target
   - Isotope excitation
3. Special spectrometers
   - Total reflection (TRXRF)
   - Synchrotron source (SSXRF)
   - Proton induced (PIXE)

Wavelength dispersive spectrometers are by far the most commonly employed, and these systems employ diffraction by a single crystal to separate characteristic wavelengths emitted by the sample. Energy dispersive spectrometers use the proportional characteristics of a photon detector, typically lithium drifted silicon, to separate the characteristic photons in terms of their energies. Since there is a simple relationship between wavelength and energy, these techniques each provide the same basic type of information. The characteristics of the two methods differ mainly in their relative sensitivities and the way in which data are collected and presented. Generally speaking, the wavelength dispersive system is roughly one to two orders of magnitude more sensitive than the energy dispersive system. Against this sensitivity, however, the energy dispersive spectrometer measures all elements within its range at the same time, whereas the wavelength dispersive system identifies only those elements for which it is programmed. To this extent, the energy dispersive system is more useful in recognizing unexpected elements. Although, in principle, almost any high energy particle can be used to excite characteristic radiation from a specimen, a sealed X-ray tube offers a reasonable compromise between efficiency, stability, and cost, and almost all commercially available X-ray spectrometers use such an excitation source. The exception might be in the analysis of very thin specimens where the proton source offers significant advantages [1]. Both wavelength and energy dispersive spectrometers typically employ a primary X-ray photon source operating at 0.5 to 3 kW. As is discussed in Section 10.2, the specimen scatters the bremsstrahlung from the source leading to significant background levels. That tends to become one of the major limitations in the determination of low concentration levels.

The wavelength and energy dispersive spectrometers of the type discussed thus far are typical of what might be found in a modern analytical laboratory. Analysis time required will vary from about 10 seconds to 3 minutes per element. The minimum sample size required is of the order of a few milligrams, although typical sample sizes are probably around several grams. The accuracies obtainable are excellent, and in favorable cases standard deviations of the order of a few tenths of one percent are possible. This accuracy is achievable because the matrix effects in X-ray spectrometry are well understood and relatively easy to overcome. The sensitivity is fair and determinations down to the low parts per million level are possible for most elements. All elements above atomic number 9 (F) are measurable by this technique. The third category considered here is special X-ray spectrometers which, while they may not be generally available to the general user community, do have important roles to play in special areas of application. Included within this category are total reflection spectrometers (TRXRF), synchrotron source spectrometers (SSXRF), and proton induced X-ray emission (PIXE). Two things that each of these three special systems have in common are a very high sensitivity and ability to work with extremely low concentrations and/or small specimens. The TRXRF system makes use of the fact that, at very low glancing angles, primary X-ray photons are almost completely absorbed within thin specimens, and the high background that would generally occur due to scatter from the sample support is absent. The recent development of high intensity synchrotron radiation beams has led to interest in their application for X-ray fluorescence analysis. The high intensity of these beams allows use of very narrow band path monochromators giving, in turn, a high degree of selective excitation. This selectivity overcomes one of the major disadvantages of the classical EDS approach and allows excellent detection limits to be obtained. The proton induced X-ray emission system differs from conventional energy dispersive spectroscopy in that a proton source is used in place of the photon source. The proton source is typically a Van de Graf generator or a cyclotron. Protons in the energy range of about 2 to 3 MeV are typically employed for this type of work. The proton source offers several advantages over the photon source. In addition to being an intense source relative to the conventional photon source, the proton excitation system generates relatively much lower backgrounds. Also, the cross section for characteristic X-ray production is quite large and good excitation efficiency is possible.

8.3. THE DETERMINATION OF LOW ATOMIC NUMBER ELEMENTS

Classically, large single crystals have been used as diffracting structures in the wavelength dispersive spectrometer. The three dimensional lattice of atoms is oriented and fabricated such that Bragg planes form the interatomic $2d$ spacing for the wavelength in question.

Modern spectrometers employ a number of different crystals, each with its specific $2d$ spacing, in order to cover the full wavelength range under conditions of optimum dispersion. The selection of crystals for the longer (>8 Å) wavelength region is difficult, however, mainly because there are not too many crystals available for work in this region. The most commonly employed crystal is probably TAP, thallium acid phthalate ($2d = 26.3$ Å), and this allows measurement of the K lines of elements below atomic number 13 (Al). Unfortunately, while the lower atomic number elements including magnesium, sodium, fluorine, and oxygen can be measured with this crystal, its reflectivity is rather weak, resulting in rather poor sensitivity. In addition, surface deterioration over a period of many months may cause the reflection efficiency to drop by as much as 50%. Several alternatives to single crystals as diffracting structures have been sought over the years, including the use of complex organic materials with large $2d$ spacings, gratings, and specular

a) Natural Single Crystal

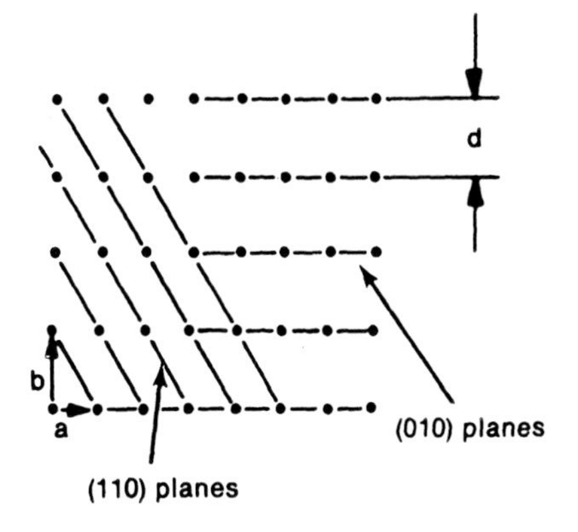

1. Regular three dimensional array of atoms.

   Crystal is oriented and cut in a given crystallographic direction (hkl).
2. Diffraction occurs due to the planes of high atomic density.
3. Diffraction planes correspond to interplanar spacing for selected (hkl), in the example, the (010) set of planes.

b) Multilayer Film

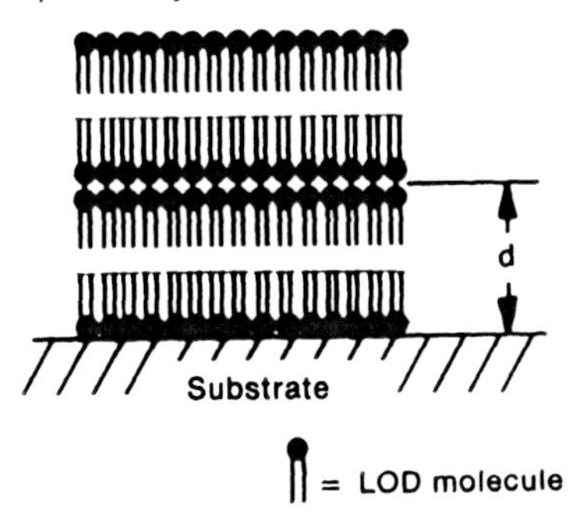

1. Long chain molecules (e.g. lead octadecanate LOD) with heavy atom sites deposited in a specific orientation on to substrate.
2. Diffraction occurs due to high electron density contrast between heavy atoms and rest of molecule.
3. Diffraction planes correspond to twice the length of the chain of the molecule.

c) Layered Synthetic Microstructure

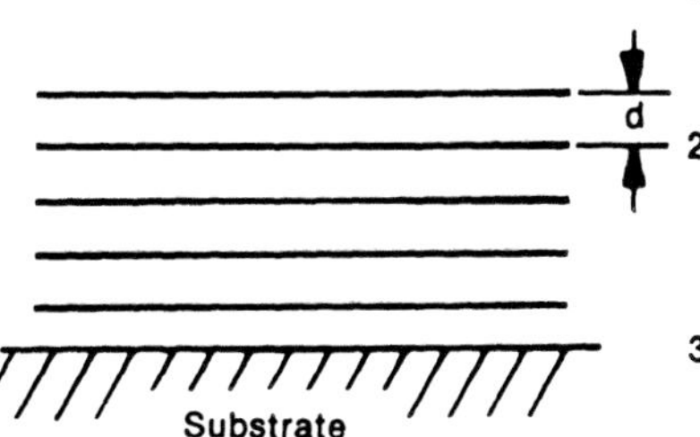

1. Successive layers of low and high atomic number elements deposited onto substrate.
2. Diffraction occurs due to high electron density contrast between low and high atomic number deposits.
3. Diffraction planes correspond to the average distance between successive low and high atomic number layers.

Figure 8.1. Three principal types of diffracting structures.

reflectors, metal disulphides, organic intercalation complexes such as graphite, molybdenum disulphide, mica and clays. Moderate success for long wavelength measurements was achieved by the use of soap films [2] having spacings in the range 80° to 120°. These layered structures are composed of planes of heavy metal cations separated by chains of organic acids. The basis of their usefulness as diffracting structures is the periodic electron density contrast between heavy metal sites and lower density organic material. Lead octadeconate (LOD), with a $2d$ spacing around 100 Å, is one such film that has been used for carbon and oxygen analysis. Unfortunately, this diffracting medium lacks adequate angular dispersion and reflectivity for the elements fluorine, sodium, and higher atomic numbers. In addition to crystals and multilayer films, a third alternative has recently become available and this is the *Layered Synthetic Microstructure*, LSM. LSM's are constructed by applying successive layers of atoms or molecules on a suitably smooth substrate. In this manner, both the $2d$ spacing and composition of each layer are selected for optimum diffraction characteristics. Figure 8.1 illustrates the three principle types of diffracting structures, those fabricated of (a) natural single crystals, (b) multilayer films, and (c) layered synthetic microstructures.

As was discussed in Section 5.9, the first successful attempts to make LSM's were made in 1940. However, while these early workers were partially able to overcome problems of deposition of the layers, the resulting structures were unstable to interdiffusion of the layers. During the development of X-ray normal incidence mirrors, Spiller was able to significantly advance the technology [3] by working with exceptionally smooth substrates and by carefully controlling the layer thickness by in-situ monitoring of X-ray reflectivity. Parallel developments by Barbee and associates made use of sputtering technology in the synthesis of large multilayer mirrors. This work resulted in better control of composition and layer thickness. Both Barbee [4] and Henke [5] presented performance data for these LSM's that by far exceed those obtainable with TAP and LOD. A special feature of LSM's is that, to a certain extent, they can be designed and fabricated to give optimum performance for special applications. Henke [7] described procedures for the detailed characterization of multilayer analyzers that can be effectively coupled to their design, optimization, and application.

Figure 8.2 shows peak diffraction coefficients for LSM's compared with LOD and conventional diffracting structures. The solid line curves exhibit the peak intensity ratios using three commercially available LSM's [8] referred to as PX1, PX2, and PX3. The PX1 LSM has a $2d$ spacing of 50 Å and the PX2 a 2d spacing of 120 Å. These data show factors of about 4.5 times improvement in peak intensities compared to TAP, for the range of elements

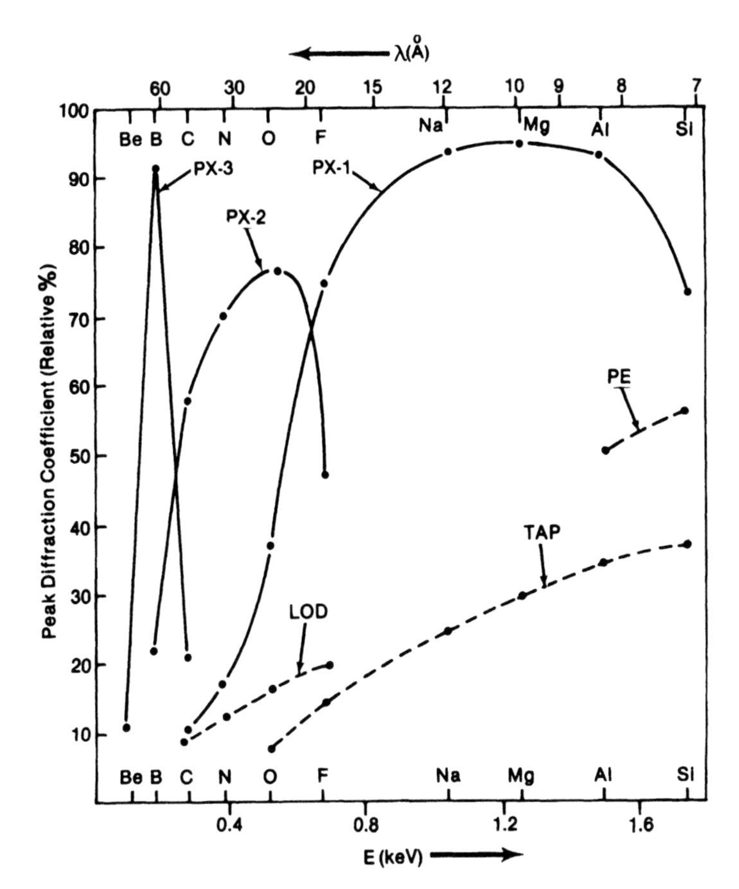

Figure 8.2. Peak Diffraction Coefficients for Penta-Erythritol (PE), Lead Octadecanoate (LOD), and Thallium Acid Phthalate (TAP) Structures (dashed lines), and for Layered Synthetic Microstructures PX1, PX2, and PX3 (solid lines).

Table 8.1. Data Obtained with Low Atomic Number Elements using Layered Synthetic Microstructures. Statistical Lower Limits of Detection Calculated for a Total Counting Time t of 200 Seconds (i.e., $t_b = t/2 = 100$ s. LLD $= (3/m)\sqrt{(R_b/t_b)}$).

Element	Sensitivity M (Counts/Sec/%)	R_b (Counts/Sec)	Matrix	LLD (%)
Be(4)	0.11	16	Be Foil	1.1
B(5)	5	45	Borosilicate Glass	0.4
C(6)	22	200	Coal	0.2

measured. For very low atomic numbers including boron, carbon, nitrogen, and oxygen, the PX2 structure shows peak intensity improvements of about 6 times as compared to LOD.

Table 8.1 shows data obtained for the elements beryllium ($Z = 4$), boron ($Z = 5$), and carbon ($Z = 6$), using LSM's [9], and indicates that detection limits on the order of tenths of a percent are achievable.

8.4. TOTAL REFLECTION X-RAY FLUORESCENCE

One of the major problems that inhibits the obtaining of good detection limits in small samples is the high background due to scatter from the sample substrate support material (see Section 1.6). The suggestion to overcome this problem by using total reflection of the primary beam was made as long ago as 1971 [10], but the absence of suitable instrumentation prevented real progress from being made until the late 1970s [11,12]. Mainly due to the work of Schwenke and his coworkers, good sample preparation and presentation procedures are now available, making *Total Reflection X-ray Fluorescence* TRXRF a valuable technique for trace analysis. The more recent availability of commercial instruments is likely to further enhance the application of this method to a wide variety of problems.

The basis of TRXRF is an interesting property of X-radiation in that, at low glancing angles, incident radiation is no longer scattered, but is absorbed within the specimen or substrate. This effect is called *total reflection*. Figure 8.3 illustrates the principle. The angle at which internal reflection occurs is called the *critical angle* ψ_{crit}. In the top portion of the figure, three scenarios are listed. In each case a beam strikes the scatterer at an angle ψ. In the first case, ψ is less than ψ_{crit} and normal scatter occurs. In the second case, ψ is equal to ψ_{crit} and internal reflection occurs. In the third case, ψ is greater than ψ_{crit} and absorption occurs. The influence on the background is shown in the main body of the figure. Once the critical angle is reached the background drops dramatically.

The TRXRF method is essentially an energy dispersive technique in which the Si(Li) detector is placed close to (about 5 mm), and directly above, the sample. Primary radiation enters the sample at a glancing angle of a few seconds of arc. The sample itself is typically presented as a thin film on the optically flat surface of a quartz plate. In the instrument described by Michaelis et al. [13], a series of reflectors is employed to aid in the reduction of background (see Figure 8.4). Here, a beam of radiation from a sealed X-ray tube passes through a fixed aperture onto a pair of reflectors that are placed very close to each other. Scattered radiation passes through the first aperture to impinge on the sample at a very low glancing angle.

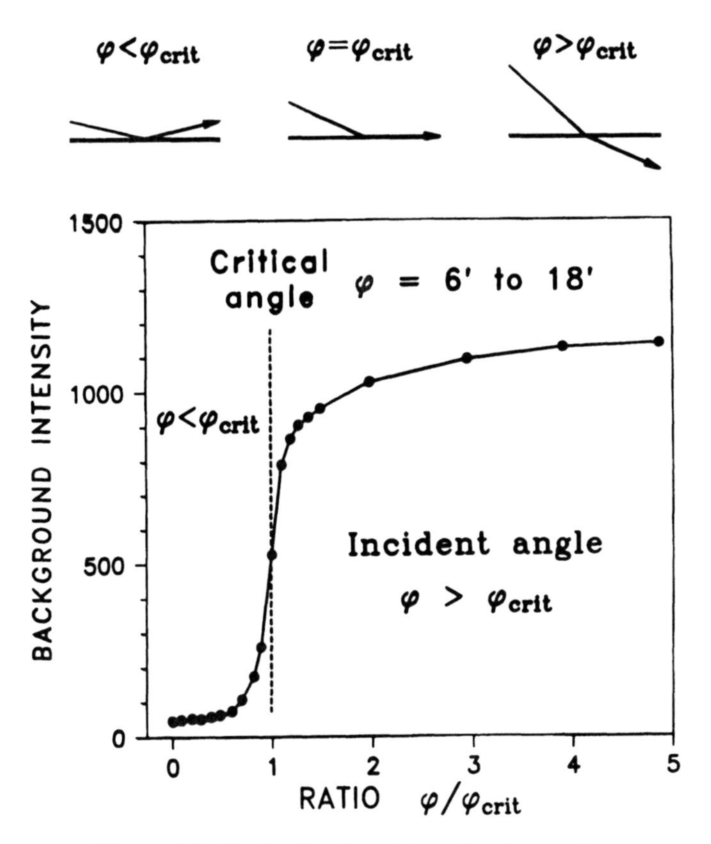

Figure 8.3. Total reflection at low glancing angles.

Because the primary radiation enters the sample at an angle just less than the critical angle for total reflection, this radiation barely penetrates the substrate media. Thus, scatter and fluorescence from the substrate are minimal. Because the background is so low, picogram amounts can be measured or concentrations in the range of a few tenths of a ppb can be obtained in aqueous solutions without recourse to preconcentration [14]. One area in which the TRXRF technique has found great application is in the analysis of natural waters. The concentration levels of, for example, transition metals in rain, river, and sea waters, are normally too low to allow estimation by standard X-ray fluorescence techniques, unless preconcentration is employed (see Section 9.6). Using TRXRF, concentration levels down to less than 10 μg/l are achievable.

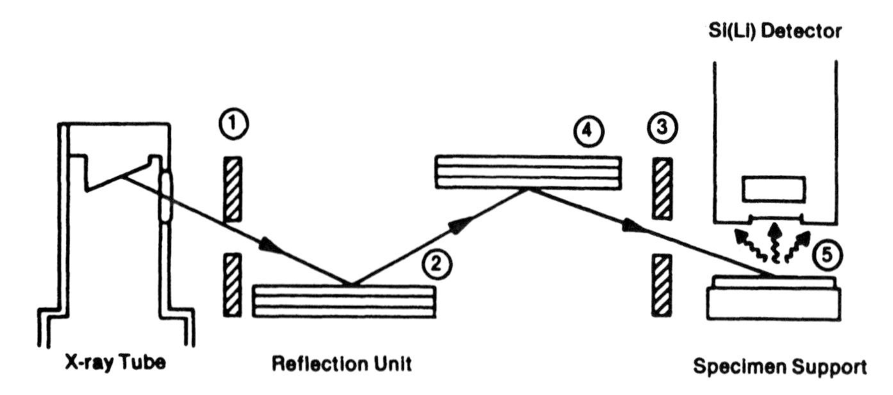

Figure 8.4. Schematic representation of the total reflection X-ray spectrometer: a) First aperture; b) first reflection unit; c) second aperture; d) second reflection unit; and e) specimen.

While the TRXRF method is most applicable to homogeneous liquid samples, success has also been achieved in the application of the method to solids including particulates, sediments, air dusts and minerals. In these instances, the sample is first digested in concentrated nitric acid and then diluted to a calibrated volume with ultrapure water, after the addition of an internal standard. Undissolved material that is still present may be dispersed using an ultrasonic bath before the specimen is taken. In addition to the advantages of ease of specimen preparation and the ability to handle milligram quantities of material, the TRXRF method is also relatively simple to apply quantitatively. Because the specimen is only a few microns thick, one generally does not observe the rather complicated matrix effects usually encountered with thick samples. Thus, the only standard required is to establish the sensitivity of the spectrometer, in terms of c/s per %, for the element(s) in question. This standard is generally added to the sample to be analyzed during the specimen preparation procedure.

8.5. SYNCHROTRON SOURCE X-RAY FLUORESCENCE—SSXRF

In the early 1980s, the availability of intense, linearly polarized synchrotron radiation beams [15] prompted workers in the fields of X-ray fluorescence (see e.g., [16]) and X-ray diffraction (see e.g., [17]), to explore what the source had to offer over more conventional excitation media.

In the synchrotron, electrons with kinetic energies on the order of several billion electron volts (typically 3 GeV), orbit in a high vacuum tube between

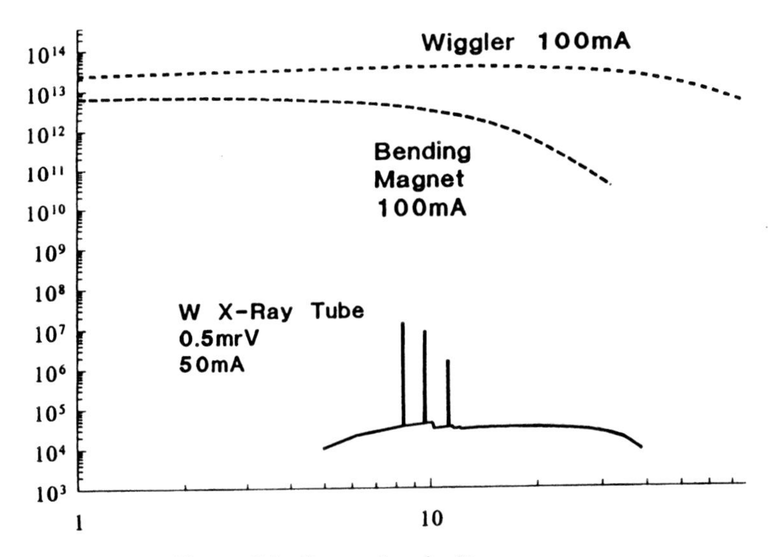

Figure 8.5. Output flux for X-ray sources.

the poles of a strong (about 10^4 Gauss) magnet. A vertical field accelerates the electrons horizontally causing the emission of synchrotron radiation. Thus, synchrotron source radiation can be considered magnetic bremsstrahlung, in contrast to normal electronic bremsstrahlung produced when electrons are decelerated by the electrons of an atom. Figure 8.5 compares the output flux for synchrotron and conventional sealed X-ray tube sources. In the case of both fluorescence and diffraction, it has been found that, because the primary source of radiation is so intense, it is possible to use a high degree of monochromatization between source and specimen, giving a source that is wavelength (and, therefore, energy) tunable, as well as being highly monochromatic. There are several different excitation modes that can be used using SSXRF including, direct excitation with continuum, excitation with absorber modified continuum, excitation with source crystal monochromatized continuum, excitation with source radiation scattered by a mirror, and reflection and transmission modes. Figure 8.6 illustrates the most important of these source configurations. These various schemes have been studied to give optimum signal to noise for various types of specimen [18,19].

During the past decade, much refinement has occurred in the optical arrangements employed in synchrotron based X-ray analysis. Figure 8.7 shows the optical arrangement developed by Michael Hart and his colleagues at the Daresbury Synchrotron Radiation Source (SRS) in the U.K. and, more recently, at the Synchrotron Source Research Laboratory (SSRL) on Long

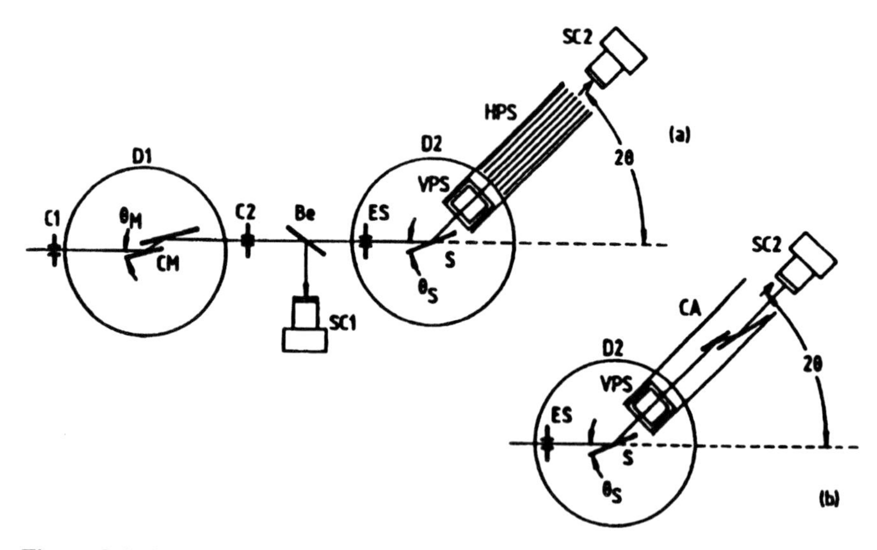

Figure 8.6. General purpose parallel beam X-ray optics developed at SSRL and Daresbury SRS for powder and thin film diffractometry. D1 and D2 represent two different setups. HPS is a long parallel slit; CA is a channel cut analyzer; and CM is a water cooled channel cut monochromator. The specimen diffractometer D2 receives radiation from the monochromator D1 defined by the entrance slit ES. C1 and C2 are Antiscatter slits. SC1 and SC2 are scintillation detectors.

Island, New York [20]. Much of this instrumentation has been designed for X-ray powder diffraction analysis. However, the extreme versatility of the optical arrangements allows application to several different disciplines. In addition to the intensity of the primary beam, several other advantages are available with use of synchrotron source radiation. As examples, synchrotron radiation is highly polarized and the beam has a high degree of parallelism. This in turn leads to extremely narrow line widths.

In the case of EDXRF, the background due to coherent and incoherent scatter can be greatly reduced by placing the detector at 90° to the path of the incident beam and in the plane of polarization. The technique is so sensitive that it has been possible to make studies of concentration profiles at the liquid-gas interface of sulfonated and manganese-neutralized polystyrene dissolved in DMSO [21]. A disadvantage of the SSXRF technique is that the source intensity decreases with time, but this can be overcome by bracketing analytical measurements between source standards and/or by continuously monitoring the primary beam. In addition, problems can arise from the

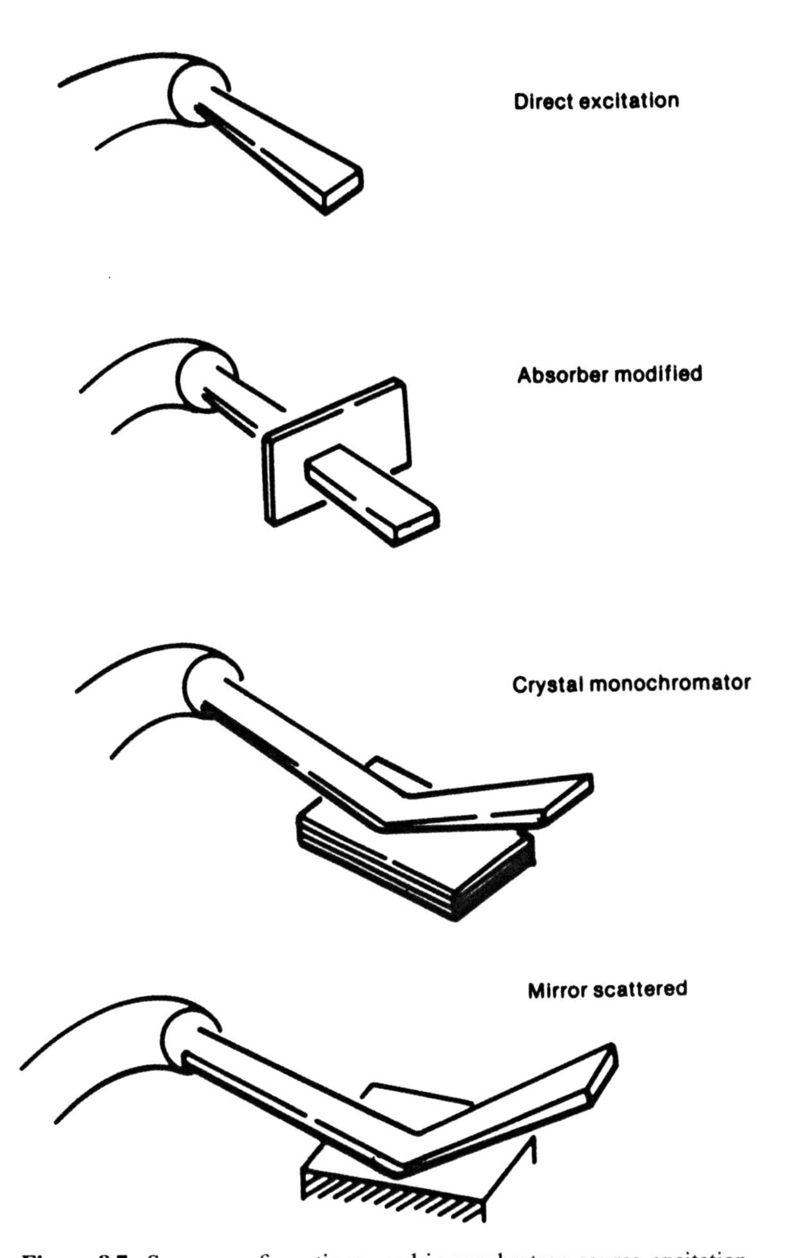

Figure 8.7. Source configurations used in synchrotron source excitation.

occurrence of diffraction peaks from highly ordered specimens. This effect has proven particularly troublesome in the analysis of small quantities (10–100 ppb weight in 20 μm spots) of geological specimens [22]. Giauque et al. have described experiments [23] using the Stanford Synchrotron Radiation Laboratory to establish what minimum detectable limits (MDL's) could be obtained under optimum excitation conditions. Using thin film standards on stretched tetrafluoro-polyethylene mounts, it was found that, for counting times of the order of a few hundred seconds, MDL's on the order of 20 ppb could be obtained on a range of elements including Ca, Ti, V, Mn, Fe, Cu, Zn, Rb, Ge, Sn, and Pb. The optimum excitation energy was found to be about one and a quarter times the absorption edge energy of the element in question. By working with a fixed excitation energy of 18 KeV, the average MDL was found to be about 100 ppb. Synchrotron source radiation EDS is also readily applicable to very small specimen sizes and, as an example, detection limits of less than 1 ppm have been obtained on a 2 μL droplet of air dried whole blood [23]. Other workers have reported detection limits in the range of 0.003 to 0.3 g/cm^2 for a range of elements on thin film filters [15]. In combination with the TRXRF method, using a lapped silicon support, detection limits down to 0.5 ppb have been reported [24].

The intensity of the synchrotron beam is probably four to five orders of magnitude greater than the conventional bremsstrahlung source sealed X-ray tubes. This, in combination with its energy tunability and polarization in the plane of the synchrotron ring, allows very rapid bulk analyses to be obtained on small areas. This factor is especially attractive for the analysis of archeometric samples, for example, ancient ceramics [20]. Because the synchrotron beam has such a high intensity and small divergence, it is possible to use it as a high resolution (about 10 μm) microprobe. Absolute limits of detection, around 10^{-14} g, have been reported using such an arrangement [25]. Synchrotron source X-ray fluorescence has also been used in combination with TRXRF. Very high signal/background ratios have been reported employing this arrangement for the analysis of small quantities of aqueous solutions dried on the reflector, with detection limits of <1 ppb or 1 pg [18].

8.6. PROTON INDUCED X-RAY FLUORESCENCE

While the use of protons as a potential source for the excitation of characteristic X-rays has been recognized since the early 1960s, it is only in more recent years that the technique has come into its own [26]. The *Proton Induced X-ray Fluorescence* PIXE method uses a beam of fast ions (protons) of primary energies in the range 1 to 4 MeV. In addition to the ion accelerator, the system contains an energy defining magnetic deflection field, a magnetic or electrostatic lens along the excitation beam pipe, a high vacuum target chamber for the specimen(s), and an energy dispersive detector/analyzer. The great advantage of the PIXE method over other sources is that it generates only a small amount of background and thus is applicable to very low concentration levels and the analysis of very small samples. As an example, PIXE has been used for monitoring the iron, copper, zinc, bromine, and lead contents of aerosol particulates [27]. This great sensitivity allows integration times as short as 30 to 60 minutes. In another rather unusual application, PIXE has been used for the analysis of a large number of trace elements in wine samples [28].

The use of conventional X-ray fluorescence and PIXE have been compared with special reference to applications in art and archaeology. Together, they seem to offer the museum scientist and archaeologist excellent complementary tools for nondestructive testing [29]. Comparison of the PIXE method has been made with many other spectroscopic techniques for the analysis of 22 elements in ancient pottery [30]. The high sensitivity of PIXE has also been of great use in the field of forensic science. As an example, external beam PIXE was used to verify the presence of lead in the finger bone of a murder victim. The deceased, who had been buried for several years, was known to have suffered a bullet wound in his right hand several years before his death. Analysis of the second right proximal phalanx, using a 1.5 MeV proton beam, showed a unique distribution of lead in an area of metal fragments indicated by X-radiographs. These data confirmed that the lead had come from a gun-shot wound. PIXE has also found a very special area of application in medicine [31]. Applications include in vitro analysis of trace elements in human body fluids and normal pathological tissues, in vivo analysis of iodine in the thyroid, lead in the skeleton, and cadmium in the kidney. Trace elements in blood serum of patients with liver cancer has also been studied by PIXE, and the copper/zinc ratio was found to be significantly higher as compared to normal patients. Serum copper/zinc ratio is potentially useful in the diagnosis and prognosis of liver cancer.

BIBLIOGRAPHY

[1] R. Klockenkamper, et al., "Comparison of different excitation methods for X-ray spectral analysis," *Fresenius Z. Anal. Chem.*, **326**, 105–117 (1987).

[2] B. L. Henke, "Some notes on ultrasoft fluorescence analysis—10 to 100 Å region," *Adv. X-ray Anal.*, **8**, 269–284 (1965).

[3] E. Spiller, "High quality Fabry-Perot mirrors for the ultraviolet," *Optik (Stuttgart)*, **39**, 118–125 (1973).

[4] T. W. Barbee, *Proc. AIP Symposium on Low Energy X-ray Diagnostics*, Monterey, California, pp.131–145 (1981).

[5] B. L. Henke, "Low energy X-ray interactions: photoionization, scattering, specular and Bragg reflection," *AIP Conf. Proc., no. 75, Low energy X-ray diagnostics*, 146–155 (1981).

[6] T. W. Barbee, "Layered synthetic microstructures (LSM) reflecting media for X-ray optic elements and diffraction structures for study of condensed matter," *Superlattices Microstruct.*, **1**, 311–326 (1985).

[7] B. J. Henke, Y. J. Vejio, R. E. Tackaberry, and H. T. Yamada, "The characterization of multilayer analyzers—models and measurements," *Proc. SPIE-Int. Soc. Opt. Eng.*, **563**, 201–215 (1985).

[8] J. A. Nicolosi, R. Jenkins, J. P. Groven, and D. Merlo, "Applications of thin-film multilayered structures to figured X-ray optics," *Proc. SPIE*, **563**, 378–384 (1985).

[9] J. A. Nicolosi, J. P. Groven, and D. Merlo, "The use of layered synthetic microstructures for quantitative analysis of elements boron to magnesium," *Adv. X-ray Anal.*, **30**, 183–192 (1987).

[10] Y. Yoneda and T. Horiuchi, "Optical flats for use in X-ray spectrochemical microanalysis," *Rev. Sci. Instrum.*, **42**, 1069–1070 (1971).

[11] J. Knoth and H. Schwenke, "An X-ray fluorescence spectrometer with totally reflecting sample support for trace analysis at the ppm level," *Fresenius Z. Anal. Chem.*, **291**, 200–204 (1978).

[12] R. Klockenkamper, *Total-Reflection X-ray Fluorescence Analysis*, Wiley: London, 245 pp. (1997).

[13] W. Michaelis, J. Knoth, A. Prange, and H. Schwenke, "Trace analysis capabilities of total-reflection X-ray fluorescence analysis," *Adv. X-ray Anal.*, **28**, 75–83 (1984).

[14] H. Aiginger and P. Wobrauschek, "Total-reflection X-ray spectrometry," *Adv. X-ray Anal.*, **28**, 1–10 (1985).

[15] C. J. Sparks, Jr., *Synchrotron Radiation Research*, H. Winnick & S. Doniach, eds., Plenum Press: New York, p. 459 (1980).

[16] J. V. Gilfrich, et al., "Synchrotron radiation X-ray fluorescence analysis," *Anal. Chem.*, **55**, 187–190 (1983).

[17] W. Parrish, M. Hart, and T. C. Huang, "Synchrotron X-ray polycrystalline diffractometry," *J. Appl. Cryst.*, **19**, 92–100 (1986).

[18] A. Iida, "X-ray fluorescence trace element analysis using synchrotron radiation," *Nippon Kessho Gakkaishi*, **27**, 61–72 (1985).

[19] G. Harbottle, A. M. Gordon, and K. W. Jones, "Use of synchrotron radiation in archaeometry," *Nucl. Instrum. Methods Res. Sect. B*, **B14**, 116–122 (1986).

[20] M. Hart, "Opportunities for materials analysis with next generation synchrotron sources," *Adv. X-ray Anal.*, **35**, 329–332 (1992).

[21] J. M. Bloch, et al, "A glancing angle X-ray fluorescence method to study the concentration profile of a dissolved polymer near an interface," *Brookhaven Natl. Lab. Rep., BNL-51847*, 36–44 (1985).

[22] S. R. Sutton, M. L. Rivers, and J. V. Smith, "Synchrotron X-ray fluorescence: diffraction interference," *Anal. Chem.*, **58**, 2187–2171 (1986).

[23] R. D. Giauque, J. M. Jaklevic, and A. C. Thompson, "Trace element determination using synchrotron radiation," *Anal. Chem.*, **58**, 940–944 (1986).

[24] A. Iida, Y. Goshi, and T. Matsushita, "Energy dispersive X-ray fluorescence analysis using synchrotron radiation," *Adv. X-ray Anal.*, **28**, 61–68 (1984).

[25] W. Petersen, P. Ketelsen, A. Knoechel, and R. Pausch, "New developments of X-ray fluorescence analysis with synchrotron radiation (SYXFA)," *Nucl. Instrum. Methods Phys. Sect. A*, **A246**(1–3), 731–735 (1986).

[26] R. P. H. Garten, "PIXE: possibilities in elemental micro- and trace-analysis," *Trends Anal. Chem.*, **3**, 152–157 (1984).

[27] B. Hietel, F. Schulz, and K. Wittmaack, "Short-term sampling and PIXE analysis of aerosol particulates collected at two different sites in the Munich area," *Nucl. Instrum. Meth. Phys. Res. Sect. B*, **B15**(1-6), 608–611 (1986).

[28] A. Houdayer, P. F. Hinrichsen, J. P. Martin, and A. Belhadfa, "PIXE trace element analysis of a selection of wines," *Can. J. Spectrosc.*, **32**, 7–13 (1987).

[29] K. G. Malmqvist, "Comparison between PIXE and XRF for applications in art and archaeology," *Nucl. Instrum. Meth. Phys. Res. Sect. B*, **B14**, 86–92 (1986).

[30] J. R. Bird, P. Duerden, E. Clayton, D. J. Wilson, and D. Fink, "Variability in pottery analysis," ibid, **B15**, 651–653 (1986).

[31] M. Cesareo, *X-ray Fluorescence in Medicine*, Field Educational Italia: Rome, 239 pp. (1982).

CHAPTER

9

SPECIMEN PREPARATION AND PRESENTATION

9.1. FORM OF THE SAMPLE FOR X-RAY ANALYSIS

There are many forms of sample that can be analyzed by X-ray fluorescence analysis and the form of the "as-received" sample will generally determine the method of pretreatment before analysis. There are three broad categories into which samples will fit [1]:

1. Samples which can be handled *directly* following some simple pretreatment such as pelletizing or surfacing.
2. Samples which require *significant pretreatment*, for example heterogeneous materials.
3. Samples which require *special treatment*, for example radioactive samples.

It is convenient to refer to the material received for analysis as the sample and that which is actually analyzed in the spectrometer as the specimen. While the direct analysis of certain materials is certainly possible, more often than not some pretreatment is required to convert the as-received sample to the analyzed specimen. This step is referred to as specimen preparation. In general, the analyst would prefer to analyze the sample directly, since, if it is taken in the as-received state, any problems arising from sample contamination that might occur during pretreatment are avoided. In practice, however, there are two major constraints that may prevent this ideal circumstance from being achieved, these being sample size and sample heterogeneity. Problems of sample size are frequently severe in the case of bulk materials such as metals, large pieces of rock, etc. Problems of sample heterogeneity will generally occur under these circumstances as well and, in the analysis of powdered materials, sample heterogeneity must almost always be considered.

Table 9.1 lists some typical sample forms and indicates possible methods of specimen preparation. Four types of sample form are listed, bulk solids, powders, liquids, and gases. The as-received sample may be either homogeneous or heterogeneous. In the latter case, it may be necessary to render the sample homogeneous before an analysis can be made. Hetero-

Table 9.1. Forms and Treatments of Samples for Analysis.

Form	Treatment
Bulk Solids	
Homogeneous	Grind to Give Flat Surface
Heterogeneous	Dissolve or React to Give Solution or Homogeneous Melt
Powders	
Homogeneous	Grind and Press into a Pellet
Heterogeneous	Grind and Fuse with Borax
Liquids	
Homogeneous (concentrated)	Analyze Directly or Dilute
Homogeneous (dilute)	Preconcentration
Heterogeneous	Filter to Remove Solids
Gases	
Airborne Dusts	Aspirate Through a Filter to Remove Solids

geneous bulk solids are generally the most difficult samples to handle and it may be necessary to dissolve or chemically react the material in some way to produce a homogeneous solution. Heterogeneous powders are either ground to a fine particle size and then pelletized, or fused with a glass forming material such as borax. Solid material in liquids or gases must be filtered out and the filter analyzed as a solid. Where analyte concentrations in liquids or solutions are too high or too low, dilution or preconcentration techniques may be employed to bring the analyte concentration within an acceptable range. These various specimen preparation techniques will be discussed in detail in succeeding sections of this chapter.

Most laboratory type spectrometers place constraints on the size and shape of the analyzed specimen. In general, the aperture into which the specimen must fit is in the form of a cylinder, typically 25 to 48 mm in diameter, and 10 to 30 mm in height. Although the specimen that is placed in the spectrometer is rather large, because of the limited penetration depth of characteristic X-ray photons, the actual mass of the specimen analyzed is quite small (see Figure 9.1). In the illustration, the specimen is represented as a disk of thickness T, density ρ, and diameter $2r$. As discussed in Section 1.4, an estimation can be made of the approximate path length x, traveled by characteristic X-ray photons, by assuming a certain fraction of absorbed radiation (see Equation 1.5). It was also shown that the penetration depth d is

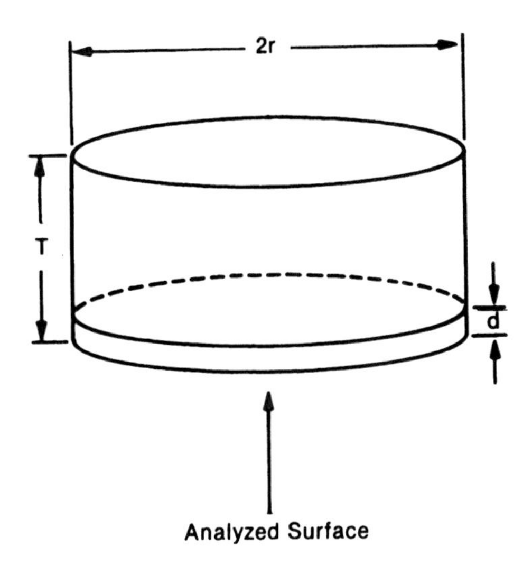

Figure 9.1. Mass of specimen actually contributing to the measured X-ray fluorescence intensity: T = specimen thickness; $2r$ = specimen diameter; d = penetration depth.

related to the path-length and the spectrometer take-off angle ψ_2, as given in Equation 1.6.

Thus the mass of the sample analyzed m_a can be calculated as follows [2]

$$m_a = \pi \times r^2 \times d\rho. \tag{9.1}$$

Combination of Equations 1.5 and 1.6 gives a value for $d\rho$

$$d\rho = \frac{4.6 \times \sin \psi_2}{\mu}. \tag{9.2}$$

Substituting for $d\rho$ in Equation 2 gives

$$m_a = \frac{4.6 \times \sin \psi_2}{\mu} \times \pi r^2. \tag{9.3}$$

Using a value of 35° for ψ_2 and 1.5 cm for r, the approximate value for m_a given in grams is given as

$$m_a = \frac{17.5}{\mu}. \tag{9.4}$$

Table 9.2. Penetration Depths and Amounts of Specimen Contributing to Measured Fluorescence Intensity.

Analyte α-lines	λ (Å)	E (KeV)	Matrix	MAC (cm^2/g)	Depth d(μm)	Mass (mg.)
C *K*	44.0	0.28	Steel	15 000	0.5	0.001
Mg *K*	9.89	1.24	Cement	2 000	5	0.009
U *M*	3.91	3.17	Sandstone	500	20	0.035
Ca *K*	3.36	3.69	Cement	400	25	0.044
Ba *L*	2.78	4.47	Rock	150	65	0.114
Fe *K*	1.94	6.40	Steel	120	35	0.147
Lu *L*	1.62	7.66	Rock	70	140	2.450
Pb *L*	1.18	10.5	Gasoline	3	10 000	6.930
U *L*	0.91	13.6	Sandstone	25	400	0.700
Mo *K*	0.71	17.4	Steel	35	120	0.504
Ba *K*	0.39	32.0	Rock	2	50 000	8.750
Lu *K*	0.23	53.6	Rock	1	100 000	17.5

Table 9.2 lists typical values for various analyte lines in different matrices, that have been calculated using Equation 9.4.

9.2. DIRECT ANALYSIS OF SOLID SAMPLES

While the direct analysis of the sample in the spectrometer would appear to offer significant advantages both in speed and avoidance of contamination in the specimen preparation process, in point of fact, direct analysis is only possible on certain specific occasions. As an example, it is sometimes possible to compress small sample chips into a briquette by high pressure and analyze directly. Although this may give a somewhat uneven surface, it is possible to correct for this by ratioing analytical line intensities to another major sample element line [3]. By far the greatest potential problem in the direct analysis of bulk solids and powders is that of local heterogeneity. As shown in Section 9.1, the actual penetration of X-rays into the specimen is generally rather small and is frequently in the range of a few microns. Where the specimen is heterogeneous over the same range as the penetration depth, what the spectrometer actually analyzes may not be representative of the total specimen. As an example, the data in Table 9.2 represents a series of various

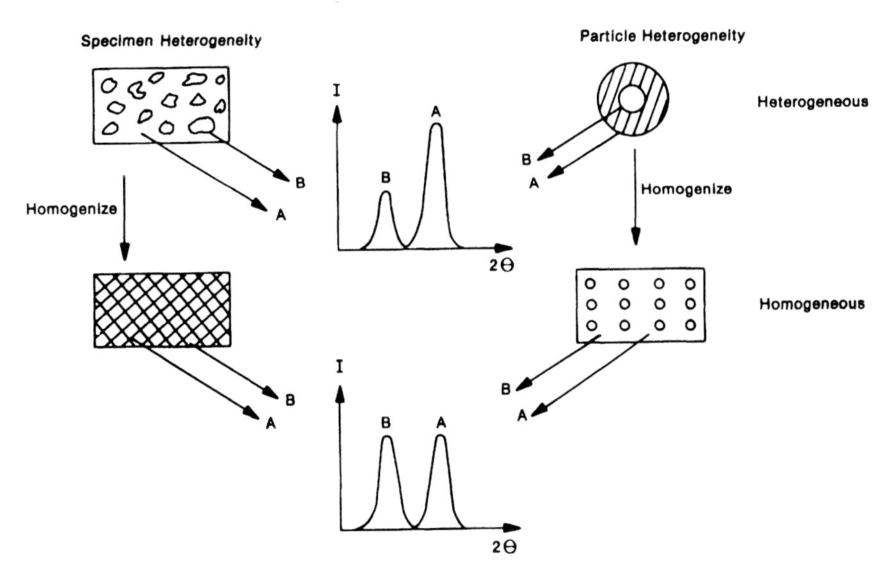

Figure 9.2. Effect of specimen heterogeneity and particle heterogeneity on relative X-ray intensities.

elements in different matrices representing a wide range in mass attenuation coefficients. It can be seen that, for the lower atomic number elements, penetration depths may be on the order of only a few microns, and the mass of specimen actually analyzed is generally much less than 10 g.

The problem in specimen preparation then is to ensure that the relatively thin surface layer actually analyzed is truly representative of the bulk of the sample. This problem may manifest itself either in terms of specimen heterogeneity or particle heterogeneity, as illustrated in Figure 9.2. This represents the measurements of two elements A and B, first in a heterogeneous specimen, then in a heterogeneous particle. In the case of specimen heterogeneity, if the penetration of the X-ray beam is of the same order as the particle size, the measured intensity ratio of lines from the two phases A and B, is quite different than that which would be obtained from a homogeneous sample. A similar effect may be seen if individual particles are heterogeneous. Such an effect can occur, for example, in the case of partially oxidized pyrite (FeS_2) and chalkopyrite ($CuFeS_2$) minerals.

If the penetrating power of the characteristic radiation is sufficiently large, it may be possible to analyze certain materials directly. This factor is particularly important for the analysis of living matter. As an example, measurements have been carried out to determine the amount of in vivo tibia lead using a ^{109}Cd radioisotope source spectrometer [4]. Such a technique allows rapid monitoring of workers with chronic lead exposure and has been determined an acceptable technique for such screening.

9.3. PREPARATION OF POWDER SAMPLES

The most common method of preparing powder samples is first to grind and then to pelletize at high pressure. While grinding is an extremely quick and effective means of reducing the particle size, there is always the potential problem of contaminating the sample during the grinding process. In practice, some materials may be sufficiently soft and homogeneous to allow direct pelletization and analysis. Such a technique is applicable to many pharmaceutical products [5], although it is often necessary to add a small amount of cellulose as a binder. In general, however, the sample is too hard for this technique to be applied, and means of reducing the particles to an acceptable size must be sought. Many studies have been carried out to quantify the magnitude of this problem, not just in X-ray fluorescence but also in other spectroscopic disciplines (see e.g., [6]). Probably the most commonly employed grinding device is the *disk mill* (shatterbox). This device consists of a series of concentric rings, plus an inner solid disk, that are shaken back and forth very vigorously. The actual sample container may be made of hardened steel, agate, or tungsten carbide. These mills are very efficient and are able to reduce powders to less than 325 mesh in a matter of minutes. Unfortunately, they can also be a source of contamination. As an example, Tuff [7] has reported a detailed study of the contamination of a quartzite cobble (99.5% quartz and 0.5% alumina) during crushing and grinding. He reported major pickup of iron, manganese, cobalt, and chromium from the steel jaw crusher.

Probably the most effective means of preparing a homogeneous powder sample is to use the borax fusion method. This method was first proposed by Claisse in 1957 [8], and the principle involves fusion of the sample with an excess of sodium or lithium tetraborate and casting into a solid bead. Chemical reaction in the melt converts the phases present in the sample into glass-like borates, giving a homogeneous bead of controllable dimensions that is ideal for direct placement in the spectrometer. (For a detailed description of the principles of the borax fusion method refer to Reference 34. While manual application of this technique is rather time consuming, a number of automated and semiautomated borax bead making machines are commercially available (see e.g., [9,10]) and these devices are able to produce multiple samples in a matter of minutes. The ratio of sample to fusion mixture is quite critical, since this will not only determine the speed and degree of completion of the chemical reaction, but also the final mass attenuation coefficient of the analyzed bead and the actual dilution factor applied to the analyte elements. Some control is possible over these factors by use of fusion aids (such as iodides and peroxides), and use of high atomic number absorbers as part of the fusion mixture. (Barium and lanthanum salts have been employed for this purpose.) Quantitative comparisons of the variables is not easy because of synergistic effects. Claisse's original method proposed a sample to sodium tetraborate ratio of 100 : 1, and, in 1962, Rose [11] suggested the use of lithium tetraborate in the ratio of 4 : 1. The same authors also suggested use of lanthanum oxide as a heavy absorber. Some years later, Norrish and Hutton [12] concluded that, for whole rock analysis, a ratio of 5.4 : 1, along with added lithium carbonate, was ideal to lower the fusion temperature. More recently, Bower and Valentine [13] published a critical review of these various techniques as applied to whole rock analysis, and have compared the results obtained from different recipes with similar data obtained from pressed pellets.

9.4. DIRECT ANALYSIS OF SOLUTIONS

One of the special problems encountered in the direct analysis of water samples, for example, stems from the need to support the specimen under examination. Most conventional spectrometers irradiate the sample from below, and the support film both attenuates the signal from the longer wavelength characteristic lines and introduces a significant blank. Absorption by air also becomes an important factor for the measurement of wavelengths longer than about 2 Å, and the need to work in a helium atmosphere introduces further attenuation. Thus, in almost all cases, some preconcentration technique is applied to the water sample before analysis. Although the sensitivity obtainable is barely sufficient for direct analysis, a wide range of preconcentration techniques has been developed that brings the concentrations of the required elements well within the range of the system. These preconcentration methods are sufficiently well developed and do not compromise the inherent speed and accuracy of the X-ray fluorescence method. The absence of geometric constraints, in the case of the energy dispersive spectrometer, makes it ideal for the development of special dedicated instruments for trace analysis. A useful outgrowth of this has been the use of particle excited spectrometry PIXE. This technique is discussed in Section 8.6. As mentioned, it offers tremendous potential for the trace analysis of limited amounts of material.

9.5. ANALYSIS OF SMALL AMOUNTS OF SAMPLE

The irradiation area in a typical X-ray spectrometer is on the order of 5 cm^2 and the penetration depth of an average wavelength, about 20 μm. This means that even though 20 g of sample may be placed in the spectrometer, the analyzed volume is still only on the order of 50 mg. The smallest sample that gives a measurable signal above background is at least three orders of magnitude less than this mass, so the lowest analyzable sample is on the order of 0.05 mg, provided that the sample is spread over the full irradiation area of the spectrometer sample cup. Where this is impractical, the smallest analyzable value is increased by roughly the total irradiation area of the cup (about 5 cm^2) over the actual area of the sample that the primary X-ray beam sees. A second important area of materials analysis involving small samples is the investigation of thin films. Although the technique is limited to rather large areas, typically a few mm^2, it does provide useful information about bulk composition of surface films. In the analysis of even large masses of material, it can be easily shown that the actual quantity analyzed is very small. Since penetration depths are on the order of a few tens of microns, and since the irradiation area is typically a few cm^2, the actual weight of sample analyzed is only on the order of a few mg. Table 9.2 lists some examples of typical samples and shows that the actual mass of sample analyzed can range from a few mg to several g, depending on the actual analytical situation. A consequence of this is that, when analyzing small amounts of material, it is far better to spread the sample out over the irradiated area, rather than make a special limited-area sample holder. In fact, excellent sensitivity can be obtained even on mg size samples. A good example of this is found in the analysis of air pollutants collected on filter paper.

It is interesting to compare data obtained with conventional X-ray spectrometers with direct excitation X-ray analysis using electrons, as in electron probe microanalysis (EPMA) or proton induced X-ray emission (PIXE). Using conventional sources, wavelength and energy dispersive spectrometers can achieve MDL's in the 1 to 10 mg/cm^2 range [14]. However, these instruments use an irradiated area of several cm^2 so the absolute mass being detected is on the order of 10^{-7} to 10^{-9} g. The EMPA method can achieve MDL's on the order of 10^{-14} g [15] and the PIXE method about 10^{-10} g [16].

9.6. PRECONCENTRATION TECHNIQUES

While the direct analysis of a solution offers an ideal sample for analysis, all too often the concentration range of the analyte elements is too low to give an adequate signal above background. In these cases, preconcentration techniques must be used to bring the analyte concentration within the sensitivity range of the spectrometer. In principle, one could easily bring an analyte element within the sensitivity range of the spectrometer simply by preconcentration using evaporation. In order to achieve detection limits at the low ppb level, this would mean the evaporation of about 100 mL of water. Although it is not theoretically necessary to take the sample completely to dryness, it is much more convenient to do so because, by this means, a specimen that can be handled is obtained. Unfortunately, taking the sample to dryness does cause experimental problems due to fractional crystallization, splashing of the sample as it reaches dryness, and so on. For these practical reasons, preconcentration by evaporation has not found a great deal of application. On the other hand, the application of evaporation preconcentration techniques does show some promise when combined with special techniques for reducing the relatively high inherent background observed in classical X-ray fluorescence methods. The method of TRXRF discussed in Section 8.4 utilizes the total reflection of X-rays from a highly polished surface. Aiginger and his coworkers [18] have achieved sensitivities down to the ppb level by evaporation of small (about 5 μL) samples of water onto a very flat quartz-glass plate. A thin layer of insulin was used to give a good distribution of the evaporated sample across the surface of the optical flat, and an energy dispersive spectrometer used to measure and analyze the characteristic X-ray emission.

Many different chemical and physical preconcentration techniques have been proposed, including surface adsorption on activated carbon [18] electrodeposition [19], precipitation chromatography [20], liquid/liquid extraction [21], immobilized reagents [22], preconcentration on ion exchange treated polyurethane foam [23], plus a variety of other techniques well known to the analytical chemist. However, by far the most popular, and also the most successful, techniques have been those based on the use of ion exchange resin. The major advantage of most ion exchange methods is that the functional group is immobilized onto a solid substrate providing the potential to batch extract ions from solution. The sample itself can be either the actual exchanged resin or a separate sample containing the appropriate ions reeluted from the resin. The success of the method depends, to a large extent, on the recovery efficiency of the resin that, in turn, is determined by the affinity of the ion exchange material for the ions in question and the stability of complexes present in solution. Preconcentration factors of up to 4×10^4 can be achieved from suitable ion exchange material using around 100 mg of resin [24].

One of the most useful ion exchange resins is Chelex-100. This resin contains iminodiacetic functional groups and chemically acts very similarly to EDTA (ethylenediamine tetra-acetic acid). It shows a good recovery efficiency for a wide range of ions and its chemistry can be predicted from prior experiments with EDTA. However, it is not too successful in the separation of ions from solutions that are high in iron and calcium (e.g., sea water). Many other ion exchange resins have also been used. As examples, Dowex 1X8 has been employed for the determination of cobalt in iron rich materials [25], and Wolfatit RO resin for the separation of trace amounts of gold in cyanide liquors [26]. It is sometimes convenient to use filter paper impregnated with ion exchange resin. This technique has been found to be especially useful in the extraction of low concentrations of rare earth elements [27]. The exchange resin can also be employed as a membrane through which the solution to be analyzed is passed [28]. This is a particularly convenient means of separation for the X-ray fluorescence method because the paper can be mounted directly in the spectrometer for analysis following the separation process. As an example of the use of Chelex-100, 200 mL of water was passed through two Chelex-100 membranes for a period of about 20 minutes. The enrichment factors obtained were in excess of 1000, allowing the separation of potassium, calcium, manganese, cobalt, nickel, copper, zinc, rubidium, strontium, and as chlorides or nitrates at element concentrations in the range 10 ppm to 10 ppb [29].

Another method that has been employed with some success, especially to preconcentrate elements in natural water samples, is the use of coprecipitation. This is a relatively simple method that offers the advantage of giving a fairly uniform deposit that can be easily collected. One of the earlier applications of this technique involved the use of iron hydroxide as a coprecipitant for the determination of iron, zinc, and lead in surface waters [30]. One of the more popular coprecipitants in use today is ammonium pyrrolidine dithiocarbamate (APDC). In the application of this method to the analysis of natural waters, detection limits in the range 0.4 to 1.2 ppb have been claimed for the elements vanadium, zinc, arsenic, mercury, and lead [31]. Other coprecipitants have been described, including the use of iron dibenzyl dithiocarbamate for the determination of uranium at the ppb level in natural waters [32], and polyvinyl pyr-rolidone-thionalide for the determination of iron, copper, zinc, selenium, cadmium, tellurium, mercury, and lead in waste and natural water samples [33].

BIBLIOGRAPHY

[1] V. E. Buhrke, R. Jenkins, and D. K. Smith, *Preparation of Specimens for X-ray Diffraction and X-ray Fluorescence Analysis*, Wiley/VCH, Section 2.1.2 (1997).

[2] ibid., Section 2.7.1.

[3] T. Tokuda, et al., "Improvement of sample preparation for X-ray fluorescence spectrometry in iron and steel analysis," *R&D, Res. Dev. (Kobe Steel Ltd.)*, **35**, 8–11 (1985).

[4] L. J. Somervaille, D. R. Chettle, and M.C. Scott, "In-vivo measurement of lead in bone using X-ray fluorescence," *Phys. Med. Biol.*, **30**, 929–944 (1985).

[5] G. Sanner and H. Usbeck, "Application of X-ray fluorescence to the determination of bromine in drug synthesis with prospects for determining chlorine (sulfur) and fluorine," *Pharmazie*, **40**, 544 (1985).

[6] G. Thompson and D. C. Bankston, "Sample contamination from grinding and sieving determined by emission spectrometry," *Appl. Spectrosc.*, **24**, 210–212 (1970).

[7] M. A. Tuff, "Contamination of silicate rock samples due to crushing and grinding," *Adv. X-ray Anal.*, **29**, 565–571 (1985).

[8] F. Claisse, "Accurate X-ray fluorescence analysis without internal standards," *Norelco Reporter*, **4**, 3–7 (1957).

[9] Claisse Fusion Stirrer Device, Spex Industries, Metuchen, NJ, USA.

[10] G. Willay, "Perl-X and its derivatives in mineral analysis," *Cah. Inf. Tech./Rev. Metall*, **83**, 159 (1986).

[11] H. Rose, I. Adler, and F. J. Flanagan, "Use of La_2O_3 as a heavy absorber in X-ray fluorescence analysis of silicate rocks," *US Geol. Survey Prof. Pap.*, **450–B**, 80–91 (1962).

[12] K. Norrish, and J. Hutton, "X-ray spectrographic method for the analysis of a wide range of geological samples," *Geochim. Cosmochim. Acta*, **33**, 431–453 (1969).

[13] N. Bower and G. Valentine, "Critical comparison of sample preparation methods for major and trace element determination using X-ray fluorescence," X-ray Spectrom., **15**, 73–78 (1986).

[14] L. S. Birks and J. V. Gilfrich, "Evaluation of commercial energy dispersive X-ray analyzers for water pollution," *Appl. Spectrosc.*, **32**, 204-208 (1978).

[15] J. V. Gilfrich et al., "X-ray fluorescence analysis using synchrotron radiation," *Adv. X-ray Anal.*, **26**, 313–323 (1982).

[16] E. M. Johansson and K. R. Akselsson, "A chelating agent—cativated carbon—PIXE procedure for sub-ppb analysis of trace elements in water," *Nucl. Instrum. Methods*, **181**, 221–226 (1981).

[17] P. Wobrauschek and H. Aiginger, "Total-reflection X-ray fluorescence spectrometric determination of elements in nanogram amounts," *Anal. Chem.*, **47**, 852–855 (1975).

[18] P. T. Howe, "Analysis of iodide in ground water by X-ray fluorescence spectrometry after collection as silver iodide on activated charcoal," *At. Energy Can. Ltd., AECL-6444*, 11 pp (1980).

[19] B. H. Vassos, R. F. Hirsch, and H. Letterman, "X-ray microdetermination of Cr, Co, Cu, Hg, Ni and Zn in water using electrochemical preconcentration," *Anal. Chem.*, **45**, 792–794 (1973).

[20] W. P. Zeronsa, G. Dabkowski, and S. Siggia, "Selective separation and concentration of silver via precipitation chromatography," Anal. Chem., **46**, 309–311 (1974).

[21] D. Leyden, et al., "Analytical techniques in environmental chemistry," *Pergamon Ser. Environ. Sci.*, **3**, 469–476 (1980).

[22] D. M. Hercules, et al., "Electron spectroscopy (ESCA). Use for trace analysis," *Anal. Chem.*, **45**, 1973–1975 (1973).

[23] S. Torok, P. Van Dyke, and R. Van Grieken, "Heterogeneity effects in direct XRF analysis of traces of heavy metals preconcentrated on polyurethane foam sorbents," *X-ray Spectrom.*, **15**, 7–11 (1986).

[24] D. E. Leyden, T. A. Patterson, and J. J. Alberts, "Preconcentration and X-ray fluorescence determination of copper, nickel and zinc in sea water," *Anal. Chem.*, **47**, 733–735 (1975).

[25] I. Roelands, "Determination of cobalt in iron-rich materials by X-ray fluorescence spectrometry after solvent and ion-exchange extraction," *Chem. Geol.*, **51**, 3–8 (1985).

[26] H. J. Peter, J. Braun, U. Dietze, and P. Volke, "Trace detection by X-ray fluorescence analysis after enrichment by ion exchange resins," *Z. Chem.*, **25**, 374–375 (1985).

[27] Q. An, "X-ray fluorescence spectrometric determination of trace amounts of RE in ores by ion-exchange paper," *Sci. Appl. Proc. Int. Conf. Rare Earth Dev. Appl.*, **1**, 551–552 (1985).

[28] W. Campbell, E. F. Spano, and T. E. Green, "Micro and trace analysis by a combination of ion exchange resin-loaded papers and X-ray spectrography," *Anal. Chem.*, **38**, 987–996 (1966).

[29] R. E. Van Grieken, C. M. Breseleers, and B. M. Vanderborght, "Ion exchange filter membranes for preconcentration in X-ray fluorescence spectrometric analysis of water," *Anal. Chem.*, **49**, 1326–1331 (1977).

[30] E. Bruninx and E. van Meijl, "Analysis of surface waters for iron, zinc and lead by coprecipitation on iron hydroxide and X-ray fluorescence," *Analyt. Chim. Acta*, **80**, 85–95 (1975).

[31] A. H. Pradzynski, R. E. Henry, and J. S. Stewart, "Determination of selenium in water on the ppb level by coprecipitation and energy dispersive X-ray spectrometry," *Radiochem. Radioanal. Lett.*, **21**, 277–285 (1975).

[32] G. S. Caravajal, K. I. Mahan, and D. E. Leyden, "The determination of uranium in natural waters at the ppb level by thin film X-ray fluorescence spectrometry after coprecipitation with an iron dibenzyldithiocarbamate carrier complex," *Anal. Chim. Acta*, **135**, 205–214 (1982).

[33] R. Panayappan, D. L. Venezky, J. V. Gilfrich, and L. S. Birks, "Determination of soluble elements in water by X-ray fluorescence spectrometry after preconcentration with polyvinylpyrrolidone-thionalide," *Anal. Chem.*, **50**, 1125–1126 (1978).

[34] R. Jenkins, R. W. Gould, and D. A. Gedcke, Quantitative X-ray spectrometry, Chapter 7, Marcel Dekker: London, 484 pp. (1995), 2nd Edition.

CHAPTER

10

USE OF X-RAY SPECTROMETRY FOR QUALITATIVE ANALYSIS

10.1. INTRODUCTION TO QUALITATIVE ANALYSIS

Both simultaneous wavelength dispersive spectrometers and energy dispersive spectrometers lend themselves admirably to qualitative analysis of materials. As shown in Equation 1.1, there is a simple relationship between the wavelength or energy of a characteristic X-ray photon, and the atomic number of the element from which the characteristic emission line occurs. It has also been shown that each element will emit a number of characteristic lines within a given series, *K*, *L*, *M*, and so on. Thus, by measuring the wavelengths, or energies, of a given series of lines from an unknown material, the atomic numbers of the excited elements can be established. Because the characteristic X-ray spectra are so simple, the actual process of allocating atomic numbers to the emission lines is a relatively easy process and the chance of making a gross error is rather small. As a comparison, the procedures for the qualitative analysis of multiphase materials with an X-ray powder diffractometer is much more complex. There are only 100 or so elements, and, within the range of a conventional spectrometer, each element gives, on an average, only half a dozen lines. In X-ray diffraction on the other hand, there are perhaps as many as several million possible compounds, each of which can give 50 or so lines on an average. Similarly, the X-ray emission spectrum is simpler than the ultraviolet emission spectrum. Since the X-ray spectrum arises from a limited number of inner orbital transitions, the number of X-ray lines is few. Ultraviolet spectra, on the other hand, arise from transitions to empty levels, of which there may be many, leading to a significant number of lines in the UV emission spectrum. A further benefit of the X-ray emission spectrum for qualitative analysis is that the effect of chemical combination, that is, valence, is almost negligible because transitions arise from inner orbitals.

10.2. RELATIVE INTENSITIES OF X-RAY LINES

Figure 10.1 shows the strongest lines in the *K*, *L*, and *M* series for the element tungsten ($Z = 74$). It can be seen that the *K* series is dominated by the α_1/α_2

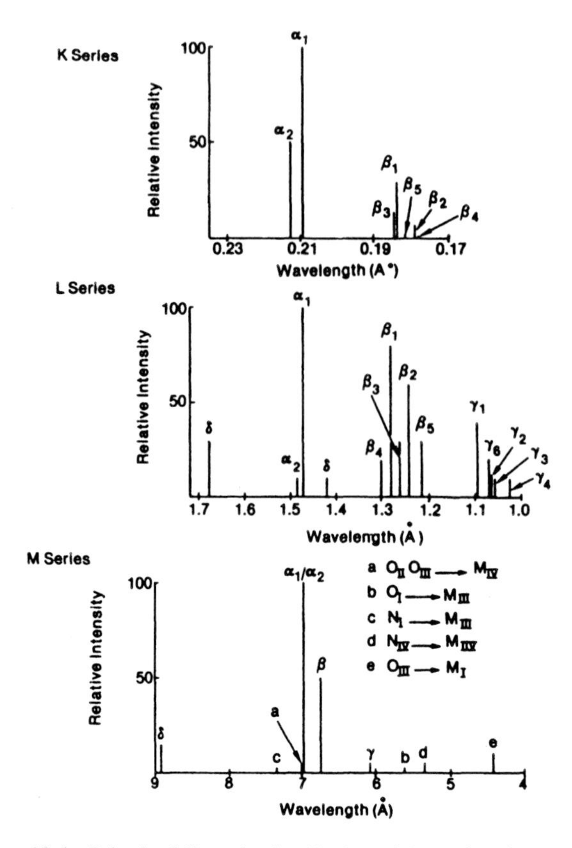

Figure 10.1. Principal lines in the *K*, *L*, and *M* series for tungsten.

doublet, with a group of lines at the short wavelength side of the α's. The *L* series contains three main groups of lines, the α's, the β's, and the γ's. The strongest line is the α_1 and the second strongest the β_1. The *M* series is dominated by the unresolved α_1/α_2 doublet, with the β line the next strongest. While the intensities observed for tungsten are typical of most of the periodic table, there are significant differences in the intensities at the extreme atomic number limits of the various series. As an example, in the *L* series, both the α_2 and the β_1 arise from the 4*d* levels. Since elements of atomic number less than 39 do not have electrons in these levels, these two lines (among others) do not appear in the lower atomic number *L* spectra. By far the greatest differences in spectral intensities appear in the low atomic number *K* spectra.

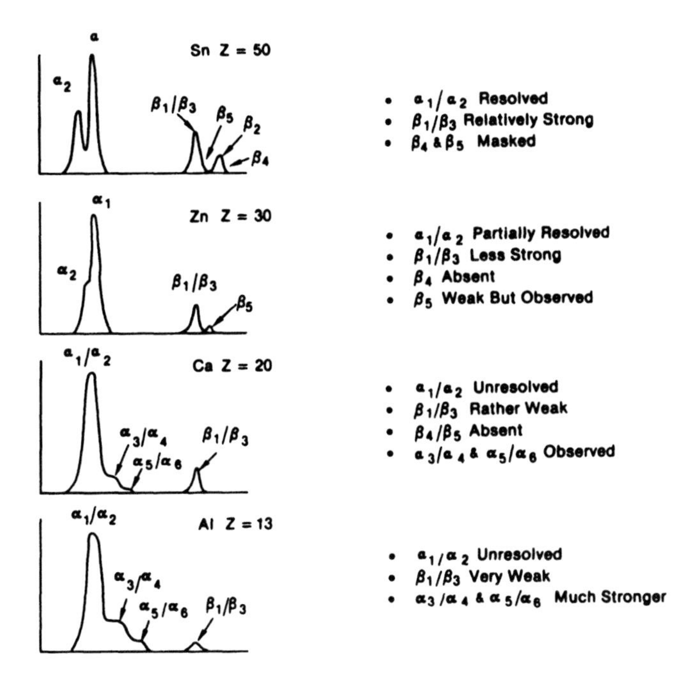

Figure 10.2. Variations in K emission spectra as a function of atomic number.

Figure 10.2 shows the K spectra for four elements, tin ($Z = 50$), zinc ($Z = 30$), calcium ($Z = 20$), and aluminum ($Z = 13$). The tin spectrum is typical of the higher atomic number K series and shows a strong, resolved $K\alpha_1/K\alpha_2$ doublet, a relatively strong $K\beta_1/\beta_3$ doublet that masks the forbidden β_5, and a weaker unresolved $\beta_{2,2'}$ doublet that masks the forbidden β_4. Moving 20 atomic numbers down the periodic table to zinc, leads to the disappearance of the forbidden β_4. The $K\alpha_1/K\alpha_2$ doublet is now only partly resolved and the $K\beta_1/\beta_3$ doublet is somewhat weaker than in the case of tin. Moving another 10 atomic numbers down to calcium leads to a further weakening of the $K\beta_1/\beta_3$ doublet and the complete disappearance of the forbidden $K\beta_5$. However, now the $K\alpha_{3,4}$ and $K\alpha_{5,6}$ satellite doublets appear. Moving a further 7 atomic numbers down to aluminum shows satellites of greater intensity and an even weaker $K\beta_1/\beta_3$ doublet. At very low atomic numbers the transferred electrons almost all come from valence bands and there is an abrupt transition from reasonably sharp line spectra, to band spectra, below atomic number 9. Similar changes in the intensity distribution of spectral lines occur in the longer wavelength L and M series lines.

Figure 10.3 illustrates the origins of peaks and background in a typical experimental X-ray spectrum. The first diagram (a) shows the distribution of

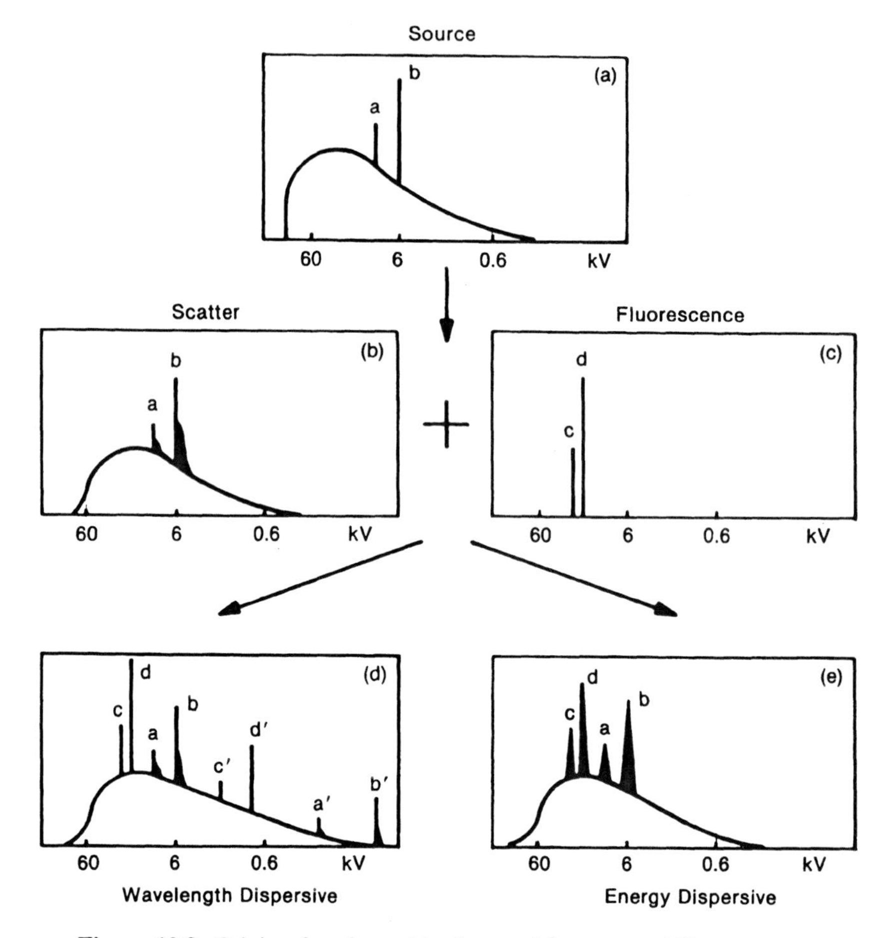

Figure 10.3. Origin of peaks and background in measured X-ray spectra.

the primary exciting radiation, as a plot of intensity versus energy in kV. The primary spectrum contains both continuum and characteristic lines from the source. As shown in diagram (b), scattering effects will cause the primary beam from the source to be scattered by the specimen, both coherently and incoherently. The effect of the incoherent scatter is to shift the continuum toward lower energies and to cause broadening of the characteristic lines, again to the lower energy side of the coherently scattered lines. Diagram (c) shows the positions of the characteristic K lines of an element which is being excited by the primary radiation. In the case of the wavelength dispersive spectrometer, diagram (d), the measured spectrum is a composite of diagrams (b) and (c) with the addition of the appearance of *harmonics* of each of the characteristic lines, regardless of their origin. The harmonics for lines c and d are indicated by c' and d', for the first order, and c'' and d'' for the second order. In the case of EDS (e), the spectrum is the sum of (b) and (c), but with somewhat broadened lines due to the resolution of the Si(Li) detector. Note that all of the characteristic lines are superimposed on top of background. The intensity of the background is mainly dependant upon the scattering power of the sample; the lower the average atomic number, the more efficient the scattering and, therefore, the higher the background. As a consequence, the peak to background ratio of a given line in a measured spectrum is not just dependant on the concentration of the respective element in the specimen, but also on the average atomic number of the specimen. In general, low average atomic number specimens scatter more and absorb less. Thus the background will be high but so too will the peak intensity.

10.3. QUALITATIVE INTERPRETATION OF X-RAY SPECTROGRAMS

Qualitative analysis with an X-ray spectrometer involves identifying each line in the measured spectrogram. As is apparent from the typical intensity distributions shown in Figure 10.4, a preliminary visual study of the spectrogram generally indicates to which line series a given group of lines belongs. The strongest line is generally assumed to be a $K\alpha$, or $L\alpha$, etc. After this strongest line has been assigned to a certain element, all other lines from this element are checked off against a set of tables (see e.g., [1]). The process is then repeated with the next strongest remaining line, and so on, until all significant lines in the spectrogram have been accounted for. As far as the wavelength dispersive spectrometer is concerned, combining Moseley's law Equation 1.1 and Bragg's law Equation 1.8 reveals a rather complex dependance of atomic number on the diffraction angle

$$Z = \sqrt{\left(\frac{1}{K} \times \frac{n}{2d} \times \frac{1}{\sin\theta}\right)}. \tag{10.1}$$

As a consequence, for manual interpretation of spectrograms it is generally necessary to employ tables relating wavelength and atomic number, with diffraction angle for specific analyzing crystals. Some automated wavelength dispersive spectrometers provide the user with software programs for the interpretation and labeling of peaks (see e.g., [2]).

Qualitative analysis with an energy dispersive spectrometer is generally rather simple. A multichannel analyzer stores the distribution of voltage pulses

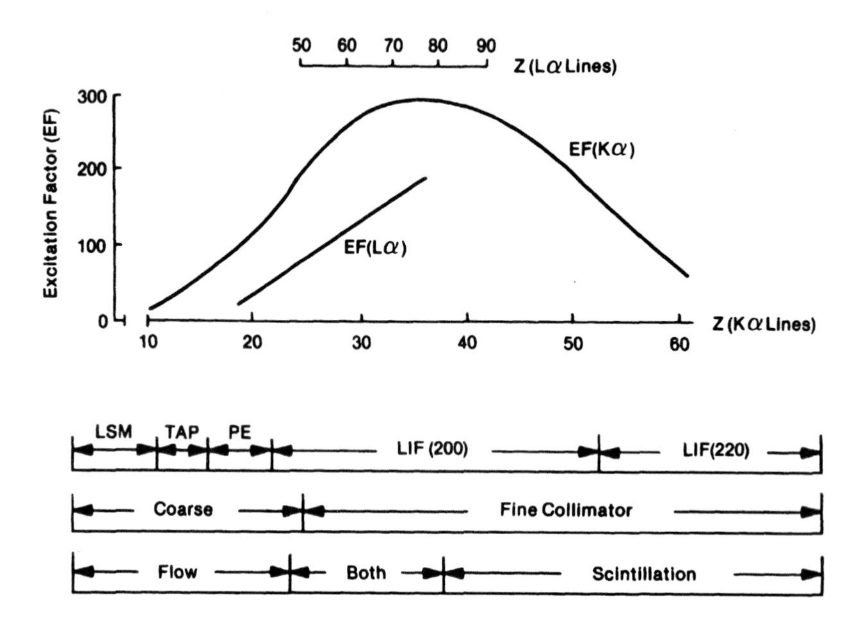

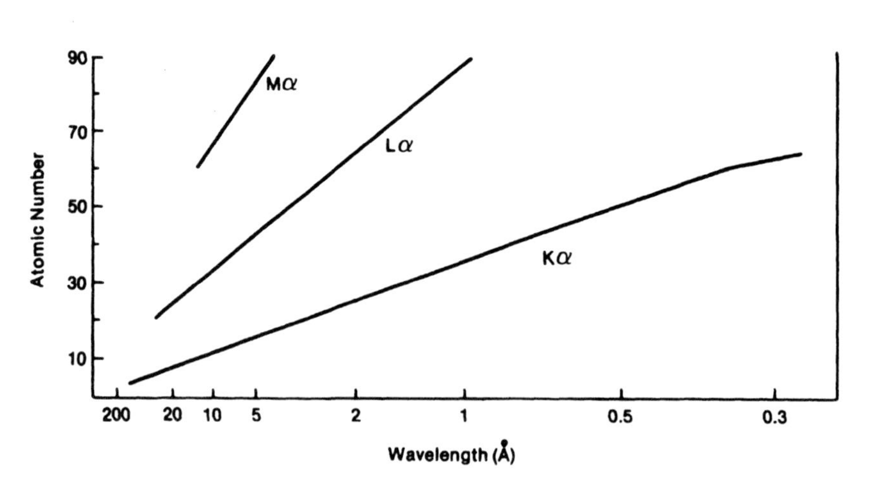

Figure 10.4. Wavelength range and sensitivity of the wavelength dispersive spectrometer.

from a Si(Li) detector, and it is a relatively simple procedure to calibrate the analyzer output directly in terms of photon energy and, from there, to atomic number. One complication that may arise in the interpretation of spectra recorded with an energy dispersive spectrometer is the appearance of artifacts in the acquired spectra. Artifacts may include *sum peaks*, *escape peaks*, and diffraction lines [3]. However, such phenomena are generally recognized by the user, and do not present undue difficulty to the experienced operator.

10.4. SEMIQUANTITATIVE ANALYSIS

The X-ray fluorescence method is a particularly useful tool for application to semiquantitative analysis. A special advantage of this method is that sensitivity does not change dramatically from element to element, but follows a rather smooth curve, with discontinuities where instrument parameters are changed. The lower portion of Figure 10.4 shows the line series covered over the range 0.3 to 200 Å. The center portion of the figure illustrates typical choices of detector, analyzing crystal, and collimator. Within a given set of these parameters, the sensitivity of the spectrometer is determined mainly by two factors, the fluorescent yield for the excited wavelength, and the primary X-ray source overvoltage. As shown in Equation 6.1, provided that the tube current is fixed, the intensity of a given excited X-ray line is proportional to $(V_o - V_c)^{1.6}$. This term is referred to as the overvoltage, and the product of this term and the fluorescent yield is sometimes called the excitation factor. In the upper part of the figure, curves are shown for the excitation factors of the K and L series α lines, as a function of atomic number, for a source at 60 kV. It can be seen that, for the K series, this product varies by about three orders of magnitude over the range of measurable elements. The curve is of the form of a bell, peaking at about atomic number 35. The curve for the L series is almost linear between atomic numbers 50 and 90, with a variation of about two orders of magnitude. Use of such curves allows a rapid semi-quantitative estimation of composition to be made.

BIBLIOGRAPHY

[1] E. W. White and G. G. Johnson, Jr., "X-ray emission and absorption wavelengths and two-theta tables," *ASTM Data Series DS–37A*, ASTM: Philadelphia, USA (1970).

[2] M. F. Garbauskas and R. P. Goehner, "Automated qualitative X-ray fluorescence elemental analysis," *Adv. X-ray Anal.*, **26**, 345–350 (1983).

[3] R. Jenkins, R. W. Gould, and D. Gedcke, *Quantitative X-ray Spectrometry*, Dekker: New York, Chapter 8, Section 4 (1981).

CHAPTER

11

CONSIDERATIONS IN QUANTITATIVE X-RAY FLUORESCENCE ANALYSIS

11.1. CONVERSION OF CHARACTERISTIC LINE INTENSITY TO ANALYTE CONCENTRATION

The great flexibility and range of the various types of X-ray fluorescence spectrometers, coupled with their high sensitivity and good inherent precision, make them ideal for quantitative analysis. In common with all analytical methods, quantitative X-ray fluorescence analysis is subject to a number of random and systematic errors that contribute to the final accuracy of the analytical result. Only by understanding the sources of the errors can they be controlled within reasonable proportions. Like all instrumental methods of analysis, the potentially high precision of X-ray spectrometry can only be translated into high accuracy if the various systematic errors in the analysis process are taken care of. The precision of a wavelength dispersive system, for the measurement of a single, well-separated line, is typically of the order of 0.1%, and about 0.25 to 0.50% for an energy dispersive system. The major source of random error is the X-ray source, that is, the high voltage generator plus the X-ray tube. In addition, there is an error arising from the statistics of the actual counting process. The random error can be significantly worse in the case of the energy dispersive system in those cases where full or partial line overlap occurs. Even though good peak and background stripping programs are available to ameliorate this problem, the statistical limitations of dealing with the difference of two large numbers remains. This problem is probably the biggest hindrance to obtaining precise count data from complex mixtures and this error can reach several percent in worst cases.

A good rule-of-thumb that can be used to estimate the expected standard deviation s at an analyte concentration level C in X-ray fluorescence, is

$$s = K \times \sqrt{(C + 0.1)}, \tag{11.1}$$

where K varies between 0.005 and 0.05. For example, at a concentration level $C = 25\%$, the expected value of s is between about 0.025% and 0.25%. A K value of 0.005 is considered very high quality analysis and a value of 0.05

rather poor quality. The value of K actually obtained under routine laboratory conditions depends upon many factors, but with reasonably careful measurements, a K value of around 0.02 to 0.03 can be obtained. Equation 11.1 is based on the fact that the production of X-rays is a random process; the standard deviation s for the finite number N of pulses measured is equal to $\sqrt{N}$. The number of pulses measured for a particular experiment is the product of the pulse (counting) rate R and the count time t, with a standard deviation given by

$$s = \sqrt{(R \times t)}. \tag{11.2}$$

Provided that all systematic errors are removed, there is a simple straight line relationship between count rate and concentration, that is, $C = K_1 \times R$, where K_1 represents the sensitivity of the spectrometer for the line in question. Substituting for R in Equation 11.2 gives

$$s = K_2 \times \sqrt{(C \times t)}, \tag{11.3}$$

where K_2 is a constant depending on sensitivity and count time. Equation 11.3 has to be extended somewhat because the counting error is not the only random error involved in the measurement. As discussed in Section 6.2, the short term source error is on the order of 0.1% so an additional term is added in Equation 11.3 to allow for this. Addition of this term gives the form of Equation 11.1.

Table 11.1 lists the four main categories of random and systematic error encountered in X-ray fluorescence analysis. The first category includes the selection and preparation of the sample to be analyzed. Two stages are generally involved before the actual prepared specimen is presented to the spectrometer, these being sampling and specimen preparation. The actual sampling is rarely under the control of the spectroscopist, and it generally has to be assumed that the container containing the material for analysis does, in fact, contain a representative sample. The table shows that, in addition to a relatively large random error, inadequate sample preparation and residual sample heterogeneity can lead to very large systematic errors. For accurate analysis, these errors must be reduced by using a suitable specimen preparation method. The second category includes errors arising from the X-ray source, as is discussed in some detail in Section 6.2. As stated in that section, source errors can be reduced to less than 0.1% by use of the ratio counting technique, provided that high frequency transients are absent. The third category involves the actual counting process. As shown in Section 7.3, systematic errors due to detector dead time may be a problem, but these can be corrected either by use of electronic dead time correctors or by some

Table 11.1. Major Sources of Random and Systematic Error.

	Source	Random	Systematic
1.	Sampling	Influenced by Other Sources	–
	Sample Preparation	0–1%	0–5%
	Sample Inhomogeneity	–	0–50%
2.	Excitation Source	0.05–0.2%	0.05–0.5%
	Spectrometer	0.05–0.1%	0.05–0.1%
3.	Counting Statistics	Time Dependant	–
	Deadtime Losses	–	0–25%
4.	Primary Absorption	–	0–50%
	Secondary Absorption	–	0–25%
	Enhancement	–	0–15%

mathematical approach. The fourth category includes all errors arising from interelement effects. Each of the effects listed can give large systematic errors that must be controlled by the calibration and correction scheme. The source of the various interelement effects is discussed in Section 11.4.

11.2. INFLUENCE OF THE BACKGROUND

The background that occurs at a selected characteristic wavelength or energy arises mainly from scattered source radiation. Since scatter increases with decrease in the average atomic number of the scatterer, it is found that backgrounds are much higher from low average atomic number specimens than from specimens of high average atomic number. To a first approximation, the background in X-ray fluorescence varies as $1/Z^2$. Since the spectral intensity from the X-ray source increases quite sharply as one approaches a wavelength equal to one half the minimum wavelength of the continuum, backgrounds from samples excited with bremsstrahlung sources are generally very high at short wavelengths (high energies), again, especially in the case of low average atomic number samples.

As shown in Figure 11.1(a), the measured signal is a distribution of counting rate R as a function of either 2θ angle (wavelength dispersive spectrometers), or as counts per channel as a function of energy (energy dispersive spectrometers). A measurement of a line at peak maximum position P gives a peak maximum counting rate of R_p. In this instance, where

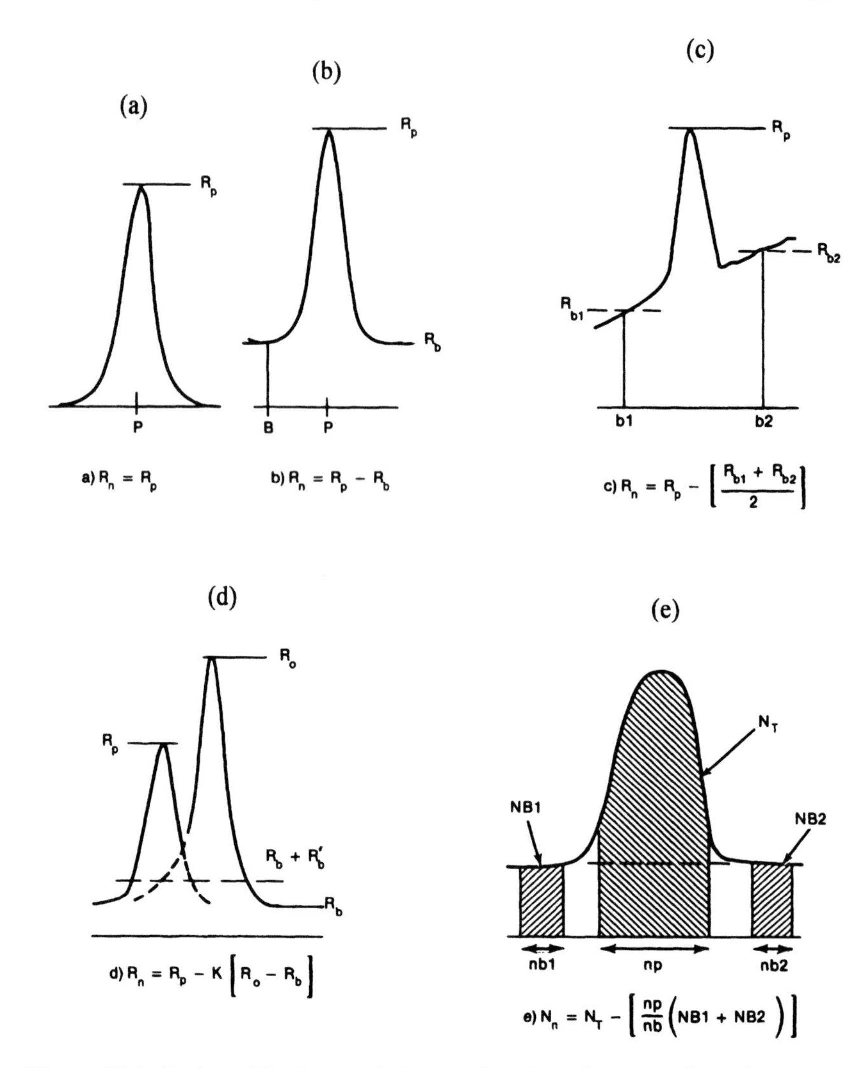

Figure 11.1. Peak and backgrounds in wavelength and energy dispersive spectrometry.

the contribution from the background is insignificant, the analyte concentration is related to R_p. Where the background is significant, Figure 11.1(b), the measured value of R_p now includes a count rate contribution from the background R_b. The analyte concentration in this case is related to the net counting rate $R_n = (R_p - R_b)$. Note from the figure that it is not possible to measure R_b directly. What is generally done is to make a background measurement at a position B, that is close to the peak. The assumption is that the background at B is the same as the background under the peak superimposed on top of a variable background, Figure 11.1(c). In this case it is common practice to measure the background, either side of the peak, at positions B_1 and B_2, giving counting rates of R_{b1} and R_{b2}. The net intensity R_n is now equal to

$$R_n = R_p - \left(\frac{R_{b1} + R_{b2}}{2}\right). \quad (11.4)$$

A further complication occurs when the analyte line is partially overlapped by another line, Figure 11.1(d). In this instance, the measured value of R_p includes a contribution R_b from the background as before, but, in addition, a contribution R_p' from the interfering peak. A line overlap correction must be applied and the assumption is made that the ratio of the peak intensity, of the interfering line to the intensity, at the interfering shoulder of the line, is constant. Thus, the net intensity in this case is

$$R_n = R_p - (R_b + k_1[R_b + k_2 \times R_o])$$

or

$$R_n = R_p - k_1(R_o + [k_2 \times R_b]). \quad (11.5)$$

As illustrated in Figure 11.1(e), in energy dispersive X-ray fluorescence, it is common practice to select a number of channels n_p giving a number of counts N_T, representing a net number of counts N_n on the peak, superimposed on a number of counts B from the background. Two ranges of channels n_{b1} and n_{b2} are then chosen on either side of the peak, giving background counts of N_{b1} and N_{b2} respectively. The background in the selected region is given by

$$B = \frac{n_p}{n_b} \times (N_{B1} + N_{B2}), \quad (11.6)$$

where n_b is the sum of n_{b1} and n_{b2}. The net peak counts are given by

$$N_n = N_T - \left(\frac{n_p}{n_b} \times (N_{b1} + N_{b2})\right). \quad (11.7)$$

11.3. STATISTICAL COUNTING ERRORS

The production of X-rays is a random process that can be described by a Gaussian distribution. Since the number of photons counted is nearly always large, (typically thousands or hundreds of thousands, rather than a few hundred) the properties of Gaussian distribution can be used to predict the probable error for a given count measurement. There will be a random error s_N associated with a measured value of N, this being equal to $\sqrt{N}$. There is a 68.3% probability that a given result will lie within $N \pm \sqrt{N}$, a 95.4% probability that it will lie between $N \pm 2\sqrt{N}$, and a 99.7% probability that it will lie between $N \pm 3\sqrt{N}$. As an example, if 10^6 counts are taken, the 1 σ standard deviation will be $\sqrt{10^6} = 10^3$ or 0.1%. The measured parameter in wavelength dispersive X-ray spectrometry is generally the counting rate and, based on what has been already stated, the magnitude of the random counting error associated with a given datum can be expressed as

$$s(\%) = \frac{100}{\sqrt{N}} = \frac{100}{\sqrt{Rt}}. \tag{11.8}$$

Care must be exercised in relating the counting error (or indeed any intensity related error) with an estimate of the error in terms of concentration. Provided that the sensitivity of the spectrometer in c/s per % is linear, a count error can be directly related to a concentration error. However, where the sensitivity of the spectrometer changes over the range of measured response, a given count error may be much greater when expressed in terms of concentration.

In many analytical situations, the peak lies above a significant background and this adds a further complication to the counting statistical error. An additional factor that must be considered is that, whereas with the scanning wavelength dispersive spectrometer the peaks and background are measured sequentially, in the case of the energy dispersive and the multichannel wavelength dispersive spectrometers, a single counting time is selected for the complete experiment. Thus all peaks and all backgrounds are counted for the same time. To estimate the net counting error in the case of a sequential wavelength dispersive spectrometer, it is necessary to consider the counting error of the net response of peak counting rate R_p, and background counting rate R_b, since the analyte element is only responsible for $(R_p - R_b)$. Equation 11.8 must then be expanded [1] to include the background count rate term

$$s(R_p - R_b) = \frac{100}{\sqrt{t}} \times \frac{1}{\sqrt{R_p} - \sqrt{R_b}}. \tag{11.9}$$

One of the conditions for Equation 11.6 is that the total counting time t must be correctly proportioned between time spent counting on the peak t_p and time spent counting on the background t_b

$$\frac{t_p}{t_b} = \sqrt{\frac{R_p}{R_b}}. \tag{11.10}$$

Several points are worth noting with reference to Equation 11.9. First, where the count time is limited, which is usually the case in most analyses, the net counting error is a minimum when $\sqrt{R_p} - \sqrt{R_b}$ is maximum. Therefore, this expression can be used as a *figure-of-merit* for setting up instrumental variables such as X-ray tube settings, choice of dispersion conditions, and so on. Secondly, it is noted that as R_b becomes small relative to R_p, Equation 11.9 approximates to Equation 11.8. In other words, as the background becomes less significant relative to the peak, its effect on the net counting error becomes smaller. The point at which the peak to background value exceeds 10 : 1 is generally taken as that where background can be ignored completely. A third point to be noted is that, as R_b approaches R_p, the counting error becomes infinite, and this is a major factor in determining the lowest concentration limit that can be detected. This introduces the concept of the lower limit of detection which is discussed in detail in Section 5. In the case of the energy dispersive spectrometer, the peak and background are recorded simultaneously and the question of division of time between peak and background does not arise. However, a choice does have to be made as to what portion of the complete recorded spectrum should be used for the measurement of peak and background [2]. The net counting error s_{net} associated with Equation 11.7 is given by

$$s_{net} = \sqrt{P} + B\left(1 + \frac{n_p}{n_b}\right). \tag{11.11}$$

Where each number of background channels is chosen to be one half of the total number of peak channels, Equation 11.11 reduces to $s_{net} = \sqrt{(P + 2B)}$, or, expressed as a percentage of the peak as

$$s_{net} = \frac{100 \times \sqrt{(P + 2B)}}{P}. \tag{11.12}$$

Equation 11.12 is the formula normally quoted for the percent standard deviation of the net peak intensity.

11.4. MATRIX EFFECTS

In the conversion of net line intensity to analyte concentration, it is necessary to correct for any interelement interactions that may occur. In the case of homogeneous specimens, these interelement effects fall into two broad categories, absorption effects and enhancement effects. Absorption effects include both secondary and primary absorption. Enhancement effects include direct enhancement, involving the analyte element and one enhancing element, plus third element effects, that involve additional element(s) beyond the analyte and enhancer. As shown in Section 1.4, the mass attenuation coefficient is a number that defines the magnitude of the absorption of a certain element for a specific X-ray wavelength. Discontinuities in an absorption curve will occur at wavelengths (energies) corresponding to the binding energy values of the various atomic subshells of the absorbing element. Between these absorption edges, the mass absorption coefficient values vary roughly as $1/Z^3$ where Z is the atomic number of the absorber.

Primary absorption occurs because all atoms of the specimen matrix will absorb photons from the primary source. Since there is a competition for these primary photons by the atoms making up the specimen, the intensity/wavelength distribution of these photons available for the excitation of a given analyte element may be modified by other matrix elements. As an example of this, Figure 11.2 shows three intensity/wavelength distributions of primary source photons being used for the excitation of matrix elements *A* and *B*. The upper part of the figure shows a hatched area under the smooth curve representing the distribution of the primary photons before they strike the specimen. Also shown in the figure are the absorption curves and characteristic lines for elements *A* and *B*. First, considering the excitation of element *A* as shown in the center diagram, it can be seen that the portion of the primary continuum available for the excitation of *A* is indicated by the hatched area between the minimum wavelength of the continuum and the absorption edge of element *A*. Note that the hatched area is less intense than the original primary continuum because of the absorption effect of *B*. Now, considering the excitation of *B* as shown in the lower diagram, as in the case of *A*, the continuum available for the excitation of *B* lies between the minimum wavelength and the absorption edge of *B*, again decreased because of the absorption effect of *A*. However, this time the distribution curve is not completely smooth and has a sharp discontinuity at the wavelength of the absorption edge of *A*. This is because the portion of the continuum to the short wavelength side of the absorption edge of *A* is strongly absorbed and, to the long wavelength side, weakly absorbed, following the shape of the absorption curve of *A* shown in the upper diagram. Thus the effect of primary absorption is to modify that portion of the spectrum most effective in the excitation of a given analyte. The degree of this modification is, in turn, dependent upon all matrix elements present.

Secondary absorption refers to the effect of the absorption of characteristic analyte radiation by the specimen matrix. As characteristic radiation passes out from the specimen in which it was generated, it will be absorbed by all matrix elements, in amounts relative to the mass absorption coefficients of these elements. As an example, referring again to the upper diagram in

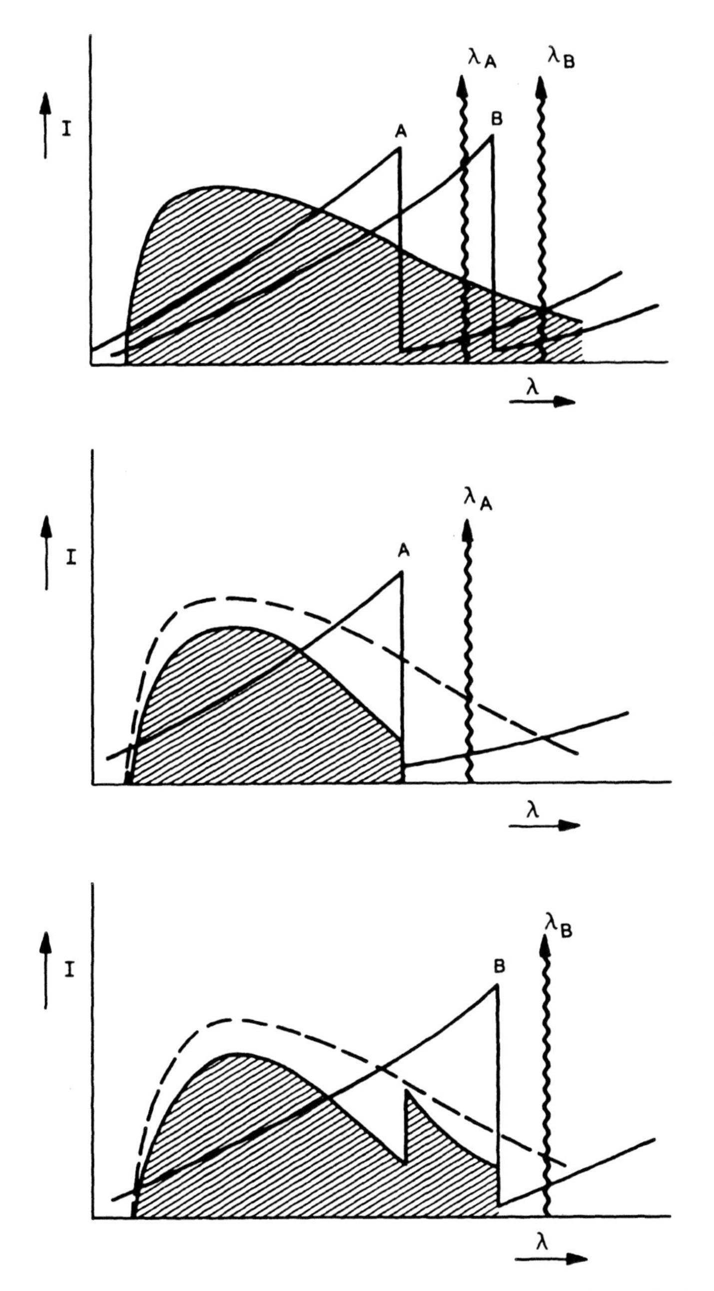

Figure 11.2. Fraction of total primary spectrum available for excitation.

Figure 11.2, note where the absorption curves intersect the characteristic lines of elements A and B.

As far as B is concerned, its characteristic line is intersected at rather low values by both A and B. On the other hand, looking at A, the absorption of B is clearly far greater than that of A itself. Thus, the effect is that element B strongly absorbs element A, and element A weakly absorbs its own radiation. This effect is called secondary absorption. The total absorption of a specimen is dependant on both primary and secondary absorption. The total absorption by element j for an analyte wavelength i is given by the following relationship

$$\alpha_i = \mu_i(\lambda) + A(\mu_i(\lambda_j)). \tag{11.13}$$

The factor A is a geometric constant equal to the ratio of the sines of the incident and takeoff angles of the spectrometer. This factor is needed to correct for the fact that the incident and emergent rays from the sample have different path lengths. The term in the equation refers to the primary radiation. Since most conventional X-ray spectrometers use a bremsstrahlung source, in practice this term is a range of wavelengths although, in simple calculations it may be acceptable to use a single *equivalent* wavelength value [3], where the equivalent wavelength is defined as a single wavelength having the same excitation characteristics as the full continuum.

Looking now at the effects of enhancement, there are a number of routes by which the analyte element can be excited and these are illustrated in Figure 11.3. Figure 11.3(a) shows the direct excitation of an analyte element i by the primary continuum P_1. There may also be excitation of the analyte by characteristic lines from the source, designated in Figure 11.3(b) by P_2. Both continuous and characteristic radiation from the source may be somewhat modified by Compton scatter, and the excitation by this modified source radiation is indicated in Figure 11.3(c) by P_3. Enhancement effects, Figure 11.3(d), occur when a nonanalyte matrix element A emits a characteristic line that has an energy just in excess of the absorption edge of the analyte element. This means that the nonanalyte element in question is able to excite the analyte, giving characteristic photons over and above those produced by the primary continuum. This gives an increased, or *enhanced*, signal from the analyte. An example of this enhancement effect can be seen in the upper diagram of Figure 11.2, where the characteristic line of from A lies to the short wavelength (high energy) side of B and can, therefore, excite B.

The last effect considered here is the so-called *third element effect* and this is shown in Figure 11.3(e). Here, a third element B, is also excited by the source. Not only can B directly enhance i, but it can also enhance A, thus increasing the enhancing effect of A on i. This last effect is called the third

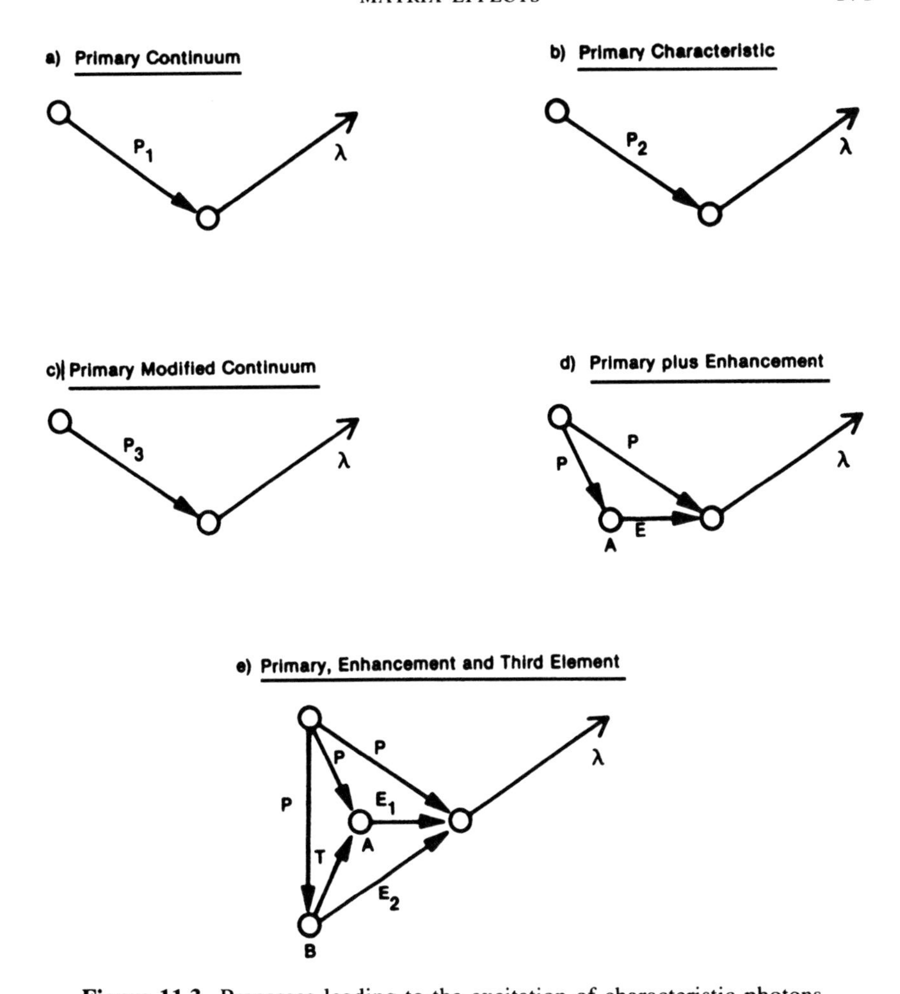

Figure 11.3. Processes leading to the excitation of characteristic photons.

element effect. Table 11.2 shows some of the data published for the chrome/iron/nickel system [4] and illustrates the relative importance of the various excitation routes. Relating these data to Figure 11.3, i is the element chromium, iron is the enhancer A, and nickel is the third element B. The data given are for three different alloy compositions, 25/25/50, 10/40/50, and 40/10/50, nickel/iron/chromium in each case. Note that in each case, roughly 87% of the actual measured chromium $K\alpha$ radiation comes from direct excitation by the source. In the case of direct enhancement by iron, each one

Table 11.2. Relative Importance of Primary Excitation, Enhancement, and Third Element Effect in the Cr/Fe/Ni System.

		Percentage of counts for Chromium $K\alpha$		
Type	Route*	25/25/50	10/40/50	40/10/50
Primary	P	87.5	87.2	87.6
Secondary	E1	6.7	10.6	2.6
Secondary	E2	5.3	2.0	9.4
Third Element	T	0.5	0.2	0.4

percent of iron added increases the chromium radiation by about 0.26%. Direct excitation of chromium by nickel gives about 0.21% increase in chromium radiation per percent nickel added. It can also be seen that the third element effect is much less important, and here the chromium intensity is increased by only about 0.02% per percent of nickel added. From this, it can be predicted that a *fourth* element effect would be negligible.

11.5. ANALYSIS OF LOW CONCENTRATIONS

The X-ray fluorescence method is particularly applicable to the qualitative and quantitative analysis of low concentrations of elements in a wide range of samples, as well as allowing the analysis of elements at higher concentrations in limited quantities of materials. The generally accepted definition for the lower limit of detection is "*that concentration equivalent to two standard deviations of the background counting rate*". From the discussion in Section 11.2, it follows that two standard deviations $2s(N)$ of the total background counts N_b taken will be given (*in counts*) by

$$2s(N) = 2 \times \sqrt{N_b} = 2 \times \sqrt{(R_b \times t_b)},$$

where t_b is the time spent counting on the background. To convert counts to count rate, we divide by time, thus (*in countrate*)

$$2s(R) = \frac{2 \times \sqrt{R_b \times t_b}}{t_b} = 2 \times \sqrt{\frac{R_b}{t_b}}.$$

To convert count rate to concentration, we divide by the sensitivity m (*in concentration*)

$$2s(C) = \frac{2}{m} \times \sqrt{\frac{R_b}{t_b}}.$$

Since two measurements have to be made (peak and background), the error is increased by $\sqrt{2}$ and taking $(2 \times \sqrt{2})$ as ≈ 3, gives the formula for the lower limit of detection *LLD*

$$LLD = \frac{3}{m} \times \sqrt{\frac{R_b}{t_b}}. \tag{11.14}$$

Note that in Equation 11.14, t_b represents one half of the total counting time. The detection limit expression for the energy dispersive spectrometer is similar to that for the wavelength dispersive system, except that t_b now becomes the live time of the energy dispersive spectrometer.

The sensitivity m of the X-ray fluorescence method is expressed in terms of the intensity of the measured wavelength per unit concentration, expressed in c/s per percent. For a fixed analysis time, the detection limit is proportional to $m/\sqrt{R_b}$ and this is taken as a figure of merit for trace analysis. The value of m is determined mainly by the power loading of the source, the efficiency of the spectrometer for the appropriate wavelength, and the fluorescent yield of the excited wavelength. The value of R_b is determined mainly by the scattering characteristics of the sample matrix and the intensity/wavelength distribution of the excitation source.

It is important to note that, not only does the sensitivity of the spectrometer vary significantly over the wavelength range of the spectrometer, but so too does the background counting rate. In general, the background varies by about two orders of magnitude over the range of the spectrometer. By inspection of Equation 11.14, it can be seen that the detection limit is best when the sensitivity is high and the background is low. Both the spectrometer sensitivity and the measured background vary with the average atomic number of the sample. While detection limits over most of atomic number values lie in the low part per million range, the sensitivity of the X-ray spectrometer falls off quite dramatically towards the long wavelength limit of the spectrometer due mainly to low fluorescence yields and the increased influence of absorption. As a result, poorer detection limits are found at the long wavelength extreme of the spectrometer, which corresponds to the lower atomic numbers. Thus the detection limits for elements such as fluorine and sodium are at the levels of hundredths of one percent rather than parts per million. However, detection limits for very low atomic number elements, such as Carbon ($Z = 6$) and Oxygen ($Z = 7$), are very poor and are typically of the order of 3 to 5%.

BIBLIOGRAPHY

[1] R. Jenkins and J. L. de Vries, *Practical X-ray Spectrometry*, 2nd ed., Springer-Verlag: New York, Chapter 5 (1970).

[2] R. Jenkins, R. W. Gould, and D. Gedcke, *Quantitative X-ray Spectrometry*, Dekker: New York, Chapter 4 (1981).

[3] D. A. Stephenson, "A multivariable analysis of quantitative X-ray emission data," *Anal.Chem.*, **43**, 310–318 (1971).

[4] T. Shiraiwa and N. Fujino, "Theoretical calculation of fluorescent X-ray intensities of Nickel, Iron, Chromium ternary alloys," *Bull. Chem. Soc. Japan.*, **40**, 2289–2296 (1987).

CHAPTER

12

QUANTITATIVE PROCEDURES IN X-RAY FLUORESCENCE ANALYSIS

12.1. OVERVIEW OF QUANTITATIVE METHODS

The intensity of a characteristic line from an element in a prepared specimen is related to concentration of the element [1,2,3]. Unfortunately, this relationship is generally not linear. There are two main factors which account for the nonlinearity, matrix effects and specimen heterogeneity effects. An expression can thus be written to equate the concentration C of a given element with these two factors

$$C = K \times R \times M \times S, \tag{12.1}$$

where K is a function of a number of instrumental factors and is generally referred to as the *calibration constant.* R is the measured counting rate from the element in question, M a term describing the matrix effect, and S a term representing specimen effects. While Equation 12.1 is not directly soluble, it does at least serve as a starting point in the development of more practical mathematical relationships.

In an X-ray analytical laboratory, the quantitative method of analysis employed will be typically predicated by a number of circumstances of which probably the four most common are: 1) the complexity of the analytical problem; 2) the time allowable; 3) the computational facilities at the disposal of the analyst; and 4) the number of standards available. It is convenient to break quantitative analytical methods down into two major categories, single element methods and multiple element methods, as shown in Table 12.1. The simplest quantitative analysis situation to handle is the determination of a single element in a known matrix. A slightly more difficult case might be the determination of a single element where the matrix is unknown. As shown in the table, three basic methods are commonly employed in this situation: use of internal standards, use of standard addition, or use of a scattered line from the X-ray source. The most complex case is the analysis of all, or most, of the elements in a sample, about which little or nothing is known. In this case, a full qualitative analysis would be required before any attempt is made to quantitate the matrix elements. Once the

Table 12.1. Quantitative Procedures Employed in X-Ray Fluorescence Analysis.

Single Element Methods:
Internal Standardization
Standard Addition
Use of Scattered Source Radiation
Multiple Element Methods:
Type Standardization
Use of Influence Coefficients
Fundamental Parameter Techniques

qualitative composition of the sample is known, one of three general techniques is typically applied: use of type standardization, use of an influence coefficient method, or use of a fundamental parameter technique. Both the influence coefficient and fundamental parameter technique require a computer for their application.

In X-ray fluorescence analysis of homogeneous specimens, the correlation between the characteristic line intensity of an analyte element and the concentration of that element is typically nonlinear over wide ranges of concentration, due to interelement effects between the analyte element and other elements making up the specimen matrix. As discussed in Section 11.4, matrix effects are made up of various interferences, and these can be rather complex. However, the situation can be greatly simplified in the case of homogeneous specimens, where severe enhancement effects are absent. Here, the slope of a calibration curve is inversely proportional to the total absorption α of the specimen for the analyte wavelength. In this instance, the slope K of the calibration curve is taken as W/I, where I is the line intensity and W the weight fraction of the analyte element.

Thus, the following relationship holds

$$W = I \times \alpha \times K. \tag{12.2}$$

It follows from Equation 12.2 that, where a range of analyte concentrations must be covered, say from a low value W_ℓ to a high value W_h, with corresponding analyte intensities of I_ℓ and I_h, the calibration curve will only be linear if the absorption for the low concentration matrix α_ℓ is equal to the absorption α_h for the high concentration matrix

$$\frac{W_\ell}{W_h} = \frac{I_\ell}{I_h} \times \frac{\alpha_\ell}{\alpha_h} \times \frac{K_\ell}{K_h}. \tag{12.3}$$

The single element techniques, which is discussed in detail later in this chapter, are all methods which reduce the influence of the absorption term in Equation 12.3, generally by referring the intensity of the analyte wavelength to a similar wavelength, arising either from an added standard, or from a scattered line from the X-ray tube. In certain cases, limiting the concentration range of the analyte may allow the assumption to be made that the values of K_ℓ and K_h are nominally the same, that is, the calibration curve is essentially linear. This assumption is applied in the traditional *type standardization* technique. Type standardization was widely employed in the 1960s and 1970s, but now that computers are generally available, it is usually considered more desirable to work with general purpose calibration schemes, which are applicable to a variety of matrix types, over wide concentration ranges.

In 1955, Sherman [4] showed that it was possible to express the intensity/concentration relationship in terms of independently determined *fundamental parameters*. Unfortunately, fundamental type methods require a fair degree of computation and, in the mid-1950s, suitable computational facilities were not available. Table 12.2 lists the typical computer systems which have been used in X-ray fluorescence analysis over the past 30 years or so. In the design of a computer controlled spectrometer, a rough guideline is frequently employed is that the cost of the computer and its associated peripherals should not exceed 20% of the total cost of the spectrometer system. As the cost, power,

Table 12.2. Development of Computers used in Controlled X-Ray Spectrometers Showing Memory and Storage Available. Computer System and Peripherals Available for Roughly 20% the Cost of the Total Spectrometer System.

Approx. Year	Computer Memory	Storage Medium	Program Language
1965	4K	Paper Tape	Assembler
1970	8K	Cassette	Assembler
1973	16K	Floppy Disk (1/3 Mbyte)	High Level/ Assembler
1976	16K	Firm Disk (5 Mbyte)	High Level/ Assembler
1980	32K	Firm Disk (20 Mbyte)	High Level
1985	640K	Firm Disk (40 Mbyte)	High Level
1990	16 M	Firm Disk (100 Mbyte)	High Level GUI Interface

and flexibility of the minicomputers increased, this newer technology was quickly taken advantage of by the spectrometer suppliers. It can be seen from the table that as the computational potential increased, there was a gradual changeover from the use of assembly code to high level languages, typically FORTRAN. Because of the limited computational facilities available in the early 1960s, and because of the clear need for some degree of fast mathematical correction of matrix effects, a number of so-called *empirical correction* techniques were developed which required far less computation and were, therefore, usable by the computers available at the time.

In principle, an empirical correction procedure can be described as the correction of an analyte element intensity for the influence of an interfering element(s), using the product of the intensity from the interfering element line and a constant factor as the correction term [5,6]. Today, this constant factor is generally referred to as an *influence coefficient*, since it is assumed to represent the influence of the interfering element on the analyte. Commonly employed influence coefficient methods may use either the intensity, or the concentration, of the interfering element as the correction term. These methods are referred to as *intensity correction* and *concentration correction* methods, respectively. Intensity correction models give a series of linear equations which do not require much computation, but they are generally not applicable to wide ranges of analyte concentration. Various versions of the intensity correction models [7] found initial application in the analysis of nonferrous metals where correction constants were applied as look-up tables. Later versions [8] were supplied on commercially available computer controlled spectrometers and were used for a wider range of application [9]. The Lachance-Traill model [10] is a concentration model, which, in effect, requires the solving of a series of simultaneous equations by regression analysis or matrix inversion techniques. This approach is more rigorous than the intensity models, and became rather popular in the early 1970s as suitable low-cost minicomputers became available.

12.2. MEASUREMENT OF PURE INTENSITIES

The intensity of an analyte line is subject not only to the influence of other matrix elements, but also to random and systematic errors due to the spectrometer and counting procedure employed. Provided that a sufficient number of counts is taken, and provided that the spectrometer source is adequately calibrated, random errors from these sources are generally insignificant, relative to other errors. Systematic errors from these sources are, however, by no means insignificant, and effects such as counting dead time, background, and line overlap can all contribute to the total experimental error in the measured intensity. A problem may arise in that incorrect conclusions about potential matrix effects may be drawn from data that is subject to instrumental systematic errors. As an example, under a certain set of experimental conditions, a series of binary alloys may give a calibration curve of decreasing slope. The conclusion may be drawn that radiation from the analyte element is enhanced by the other matrix element where, in point of fact, the problem could also be due to dead time loss in the counting circuitry because the count rate is too high. If a correction is applied for an enhancement effect, the procedure will break down if the count rates are changed, for example, by varying the source conditions. Problems of this type can be particularly troublesome in the application of influence correction methods. Unless instrument dependant errors are completely separated from the matrix dependant terms, the instrument effects tend to become associated with the influence correction terms. In practice, this may not be completely disastrous for a specific spectrometer calibrated for a particular application, since the method will probably work, provided that the experimental conditions do not change. However, one major consequence is that it will probably be impossible to *transport* a set of correction constants from one spectrometer to another. Mainly for this reason, it is common practice today to attempt to obtain intensities as free from systematic instrumental effects as possible. Such intensities are referred to as *pure* intensities.

12.3. SINGLE ELEMENT METHODS

12.3.1. Use of Internal Standards

One of the most useful techniques for the determination of a single analyte element in a known or unknown matrix is the use of an internal standard. This technique is one of the oldest methods of quantitative analysis [11] and is based on the addition of a known concentration of an element that gives a wavelength similar to that of the analyte wavelength. Referring to Equation 12.3, if ℓ is replaced by the analyte element x, and h by the internal standard s, provided that the internal standard is selected such that the absorption of the matrix for x and ℓ are about the same, that is, $\alpha_\ell = \alpha_s$, the following relationship is true

$$\frac{W_x}{W_s} = K_i \times \frac{I_x}{I_s}. \tag{12.4}$$

I_x and I_s are the measured intensities of analyte and internal standard wavelengths, and W_s is the known added weight of the internal standard. The value of K_i is equal to K_x/K_s, that is, the ratio of the sensitivities of the spectrometer for the analyte and internal standard elements, respectively. These data can be obtained by separate experiments. Internal standards are best suited to the measurements of analyte concentrations below about 10%. The reason for this limit arises because it is generally advisable to add the internal standard element at about the same concentration level as that of the analyte. When more than 10% of the internal standard is added, it may significantly change the specimen matrix and introduce errors into the determination. Care must also be taken to ensure that the particle sizes of specimen and internal standard are about the same, and that the two components are adequately mixed. If these conditions are not met, the use of the internal standard may well correct for the absorption of the matrix, but it may also introduce problems of heterogeneity, as discussed in Section 9.2.

12.3.2. Standard Addition Methods

Sometimes an appropriate internal standard cannot be found and, in this instance, it may be possible to use the analyte itself as an internal standard. This method is a special case of standard addition, and is generally referred to as *spiking*. Again, referring to Equation 12.3, if ℓ refers to the unknown specimen x, the unknown concentration W_x gives a measured intensity of I_x. By deliberately adding an additional amount of the analyte element, the new intensity of x equals $I_{(x+a)}$. Thus, in the equation, W_h is replaced by $(W_x + W_a)$, and I_ℓ becomes $I_{(x+a)}$. Since the attenuation coefficient is essentially the same for both matrices, before and after addition, and since the slope constants are also identical, it follows that

$$\frac{W_x}{W_x + W_a} = \frac{I_x}{I_{x+a}}. \tag{12.5}$$

Because I_x and I_{x+a} are both measured, and W_a is known, a value for W_x can easily be obtained. The technique of standard addition is especially useful for the determination of analyte concentrations below about 5%. It is also useful in the preparation of secondary standards. However, care must be taken in the application of the method whenever the relationship between analyte concentration and analyte intensity is nonlinear. It is apparent from Equation 12.5 that a linear relationship is assumed between concentration and intensity and, if this is not the case, errors will occur. Since the linearity of the concentration/intensity relationship in a given situation is unsure, it is common practice to repeat the standard addition step at least twice, so that a minimum of three data sets (zero concentration plus the two additions) are available to work with.

12.3.3. Use of Scattered Tube Lines

Chapter 11 shows that the background intensity in a given measurement is always present, due mainly to scatter and, under certain circumstances, it may be possible to pick out a wavelength region in the background and use it as an *internal standard* [12]. It also states that the scattered background B varies approximately as $1/Z^2$, thus

$$BZ^2 = K_1. \tag{12.6}$$

Since the mass absorption coefficient μ varies as the cube of atomic number Z, that is

$$\mu = K_3 \times Z^3, \tag{12.7}$$

and since the slope m of a calibration curve is inversely proportional to the absorption of the specimen matrix, thus

$$m = \frac{I}{C} = \frac{K_2}{\mu}. \tag{12.8}$$

It follows that, by combining Equations 12.7 and 12.8, an expression can be obtained for concentration in terms of the measured peak to background ratio, that is

$$I = \frac{C}{Z^3} \times \frac{K_2}{K_3}. \tag{12.9}$$

Equation 12.9 shows that there is a very high dependance of the slope of the calibration curve on the average atomic number of the specimen in fact, the third power of Z. Combining Equations 12.6 and 12.9 gives

$$\frac{I}{B} = \frac{C}{Z} \times \frac{K_2}{K_1 + K_2}. \tag{12.10}$$

Note from Equation 12.10 that the peak to background ratio is far less dependant (although not independant) on the average atomic number. Thus, while use of a scattered tube line as a standard may not completely overcome a matrix effect, it will generally significantly reduce the effect.

As an example, the data listed in Table 12.3 were obtained from a series of standards that were made to evaluate the use of scattered background as an internal standard for uranium ore analysis [13]. A series of matrices were chosen and 100 ppm of uranium was added to each. The experimentally

Table 12.3. Use of Scattered Background as an Internal Standard in the Analysis of Uranium Ores. Results Given in ppm. The 100 ppm Sample was used as Comparison Reference.

Matrix	c/s/%	Rb	Rp/Rb	Matrix Absorption Coefficient	Z	Uncorrected	Corrected
SiO_2	5.1	390	2.31	7.9	10.5	ppm	ppm
SiO_2/Fe_2O_3	1.9	150	2.15	30.6	17	268	93
Fe_3O_4	0.9	98	1.87	55	21	567	81
$CaCO_3$	2.4	220	2.01	16.9	13	213	91
$KAlSi_3O_6$	3.8	304	2.23	12.2	12	134	97

observed c/s per % for uranium is listed in the second column of the table and it can be seen that, because of the wide range in matrix absorption coefficients, the sensitivity varies by about a factor of six. There is a similarly wide variation in the background count rate, this time due to the large difference in the average atomic number Z. If one simply attempted to use the first sample, that is the SiO_2, as the standard, and ratiod the observed count rate on each of the other four matrices to it in order to obtain a measure of the concentration, values would widely vary from the expected value of 100 ppm. These results are listed in Table 12.3 as *uncorrected*. However, it follows from Equation 12.10 that if the values of K_1, K_2, and (if used) K_3 really are constant, and the atomic number term Z is ignored, the concentration should be proportional to the peak to background ratio, thus

$$C = K_4 \times \frac{I}{B}. \qquad (12.11)$$

Again, using SiO_2 as a reference standard, a new set of concentrations are obtained, listed as *corrected* in the table. While these corrected data are not exactly equal to 100 ppm, they are, in most cases, probably good enough for routine work, especially when one considers that the standards used represent the extremes of likely ore composition.

12.4. MULTIPLE ELEMENT METHODS

12.4.1. Type Standardization

As has been previously stated, provided that the total specimen absorption does not vary significantly over a range of analyte concentrations, enhancement effects are absent, and the specimen is homogeneous, a linear relationship will be obtained between analyte concentration and measured characteristic line intensity. Where these provisos are met, type standardization techniques can be employed. It is also clear from previous discussion, that by limiting the range of analyte concentration to be covered in a given calibration procedure, the range in absorption can also be reduced. Type standardization is probably the oldest of the quantitative analytical methods employed, and the method is usually evaluated by taking data from a well-characterized set of standards and, by inspection, establishing whether a linear relationship is indeed observed. Where this is not the case, the analyte concentration range may be further restricted. The analyst of today is fortunate in that many hundreds of good reference standards are commercially available [14]. While the type standardization method is not without its pitfalls [1], it is nevertheless extremely useful and is especially good as a quality control application when a finished product is being compared with a desired product.

Special reference standards may be made up for particular purposes, and these may serve the dual purpose of instrument calibration as well as establishing working curves for analysis. As an example, two thin glass film standard reference materials specially designed for calibration of X-ray spectrometers are available from NIST (formerly The National Bureau of Standards) in Washington, as Standard Reference Materials (SRM) 1832 and 1833 [15]. They consists of a silica-base film deposited by focused ion-beam coating onto a polycarbonate substrate. SRM 1832 contains aluminum, silicon, calcium, vanadium, manganese, cobalt, and copper. SRM 1833 contains silicon, potassium, titanium, iron, zinc, and rhodium. The standards are especially useful for the analysis of particulate matter [16].

12.4.2. Influence Correction Methods

Lachance [17] has suggested dividing influence coefficient correction procedures into three basic types, *fundamental, derived*, and *regression*. Fundamental models [18,19] are those which require starting with concentrations then calculating the intensities. Derived models are those which are based on some simplification of a fundamental method [10,20], but which still allow concentrations to be calculated from intensities. Regression models are those which are semiempirical in nature [8,21], and which allow the determination of influence coefficients by regression analysis of data sets obtained from standards (see e.g., [22]). [All regression analysis includes terms for W (or concentration C); an intensity (or intensity ratio) term I; an instrument dependent term which essentially defines the sensitivity of the spectrometer for the analyte in question, and a correction term, which

Table 12.4. Commonly Employed Influence Correction Models.

Linear Model

$\frac{W_i}{R_i} = K_i$

Lachance/Traill (1966)

$\frac{W_i}{R_i} = K_i + \sum_j^n \alpha_{ij} W_j$

Claisse/Quintin (1967)

$\frac{W_i}{R_i} = K_i + \sum_j^n \alpha_{ij} W_j + \sum_j^n \gamma_{ij} W_j^2$

Rasberry/Heinrich (1974)

$\frac{W_i}{R_i} = K_i + \sum_j^n \alpha_{ij} W_j + \sum_{k=j}^n \beta_{ik} \frac{W_k}{1+W_i}$

Lachance/Claisse (1980)

$\frac{W_i}{R_i} = K_i + \sum_j^n \alpha_{ij} W_j + \sum_j^n \sum_{k/=j}^n \alpha_{ijk} W_j W_k$

corrects the instrument sensitivity for the effect of the matrix.] The general form is as follows

$$W = I\langle instrument\rangle[1 + \Sigma(model)]. \tag{12.12}$$

The different models vary only in the form of the correction term. Table 12.4 shows several of the more important of the commonly employed influence coefficient methods. All of these models are concentration correction models in which the product of the influence coefficient and the concentration of the interfering element are used to correct the slope K_i of the analyte calibration curve. The Lachance-Traill model [10] was the first of the concentration correction models to be published. Some years after the Lachance-Traill paper appeared, Heinrich and his coworkers at the National Bureau of Standards, suggested an extension to the Lachance-Traill approach [21] in which absorbing and enhancing elements are separated as α and β terms. These authors suggested that the enhancing effect cannot be adequately described by the same hyperbolic function as the absorbing effect. A thorough study [18] of Lachance-Traill coefficients, based on theoretically calculated fluorescence intensities, shows that all binary coefficients vary systematically with composition. Both the Claisse-Quintin [20] and Lachance-Claisse [23] models use higher order terms to correct for so-called *crossed effects*, which include enhancement and third element effects. These models are generally more suited for very wide concentration range analysis.

In all of these methods, one of three basic approaches is used to determine the values of the influence coefficients, following the initial measurement of intensities using a series of well-characterized standards. The first approach is to use multiple regression analysis techniques to give the best fit for slope, background, and influence coefficient terms. Alternatively, the same data set can be used to graphically determine individual influence coefficients. As an example, in the case of the Lachance-Traill equation, for a binary mixture a/b the expression for the determination of a would be

$$\frac{W_a}{I_a} = 1 + \alpha_{ab} \times W_b. \tag{12.13}$$

By plotting data from a range of analyzed standards in terms of W_a/R_a as a function of W_b, a straight line should be observed with a slope of α_{ab} and an intercept of unity. This approach is especially useful for visualizing the form of the influence coefficient correction [24]. Thirdly, the influence coefficient can be calculated using a fundamental type equation based on physical constants (see Section 12.4).

Several useful extensions to the basic models have been proposed. As example, a FORTRAN program (NBSGSC) for quantitative X-ray spectrometry has been described by Tao et al. [25]. The NBSGSC program is able to handle pressed or fused samples. This ability to process data from fused samples is due to the fact that, in mathematical correction processes using influence coefficients, since ($\Sigma w = 1$) giving $(n + 1)$ equations, one term can be dropped (this is usually α_{ii}). This fact can be used to avoid analyzing all of the matrix, for example α_x for ignition loss, and so on. However, such an approach may change the other α's.

The major advantage of using the influence coefficient methods is that a wide range of concentration ranges can be covered using a relatively inexpensive computer for the calculations. A major disadvantage is that a large number of well-analyzed standards may be required for the initial determination of the coefficients. However, where adequate precautions have been taken to ensure correct separation of instrument and matrix dependant terms, the correction constants are transportable from one spectrometer to another and, in principle, need to be determined once.

While influence coefficient methods have gained a good deal of popularity, workers still sometimes encounter difficulties in their application, and there are a number of typical reasons why they may fail [26]. The most common of these are:

1. Failure to adequately separate instrumental and matrix dependant effects.

2. Poor judgment on the part of the analyst as to whether or not a correction term should really be included.
3. Poor technique on the part of the analyst in the determination of the influence coefficients.
4. Poor quality and/or range of calibration standards.
5. Inadequacy of the regression analysis program used in the determination of the coefficients.
6. Application of the technique in cases where the specimens are insufficiently homogeneous.

12.4.3. Fundamental Methods

Since the early work of Sherman [4], there has been a growing interest in the provision of an intensity/concentration algorithm which would allow the calculation of the concentration values without recourse to the use of standards. Sherman's work was improved upon first by the Japanese team of Shiraiwa and Fujino [27], and later by the Americans, Criss and Birks [28,29], with their program NRLXRF [30]. The same group also solved the problem of describing the intensity distribution from the X-ray tube [31]. The problem for the average analyst in the late 1960s and early 1970s, however, remained finding sufficient computational power to apply these methods. In the early 1970s, de Jong suggested an elegant solution [32] in which he proposed the use of a large mainframe computer for the calculation of the influence coefficients, then use of a small minicomputer for their actual application, using a concentration correction influence model. While software packages were then available for minicomputers, a drawback remained in their application systems which used a modified primary excitation spectrum. Most fundamental quantitative approaches in use today employ measured or calculated continuous radiation functions in the calculation of the primary absorption effect. Where sharp discontinuities or *breaks* in this primary spectrum occur, as in the case of the energy dispersive system, the calculation becomes very complicated. This is probably the reason why many energy dispersive based quantitative procedures not covered by type standardization tend to favor simpler, but less accurate, methods based on scattered tube lines.

BIBLIOGRAPHY

[1] R. Jenkins, R. W. Gould, and D. A. Gedcke, *Quantitative X-ray Spectrometry*, 2nd ed., Dekker: New York, 484 pp. (1995).

[2] R. Tertian and F. Claisse, *Principles of Quantitative X-ray Fluorescence*, Heyden: London, 385 pp. (1982).

[3] G. R. Lachance and F. Claisse, *Quantitative X-ray Fluorescence Analysis*, Wiley: New York, 402 pp. (1994).

[4] J. Sherman, "Theoretical derivation of fluorescent X-ray intensities from mixtures," *Spectrochim. Acta*, **7**, 283–306 (1955).

[5] M. J. Beattie and R. M. Brissey, "Calibration method for X-ray fluorescence analysis," *Anal. Chem.*, **26**, 980–983 (1954).

[6] E. Gillam and H. T. Heal, "Some problems in the analysis of steels by X-ray fluorescence," *J. Appl. Phys.*, **2**, 353–358 (1952).

[7] H. J. Lucas-Tooth and B. J. Price (1961), "A mathematical method for investigation of interelement effects in X-ray fluorescent analysis," *Metallurgia*, **54**, 149–152, (1961)

[8] H. J. Lucas-Tooth and C. Pyne, "The accurate determination of major constituents by X-ray fluorescent analysis in the presence of large interelement effects," *Adv. X-ray Anal.*, **7**, 523–541 (1964).

[9] R. Jenkins, J. de Klerck, and S. van Gelder, "Use of the computer controlled spectrometer for the analysis of elements with matrix corrections based on intensities," *Philips S.& A.E. Bulletin* FS-28, Philips: Almelo (1970).

[10] G. R. Lachance and R. J. Traill, "A practical solution to the matrix problem in X-ray analysis," *Can. Spectrosc.*, **11**, 43–48 (1966).

[11] L. S. Birks, "X-ray Spectrochemical Analysis, 2nd ed.," *Interscience Chemical Analysis Series*, Volume **11**, 143 pp. (1959).

[12] G. Anderman and J. W. Kemp, "Scattered X-rays as internal standards in X-ray emission spectroscopy," *Anal. Chem.*, **30**, 1306–1309 (1958).

[13] G. W. James, "Parts-per-million determinations of uranium and thorium in geologic samples by X-ray spectrometry," *Anal. Chem.*, **49**, 967–969 (1977).

[14] Lists of commercially available calibration standards for use in X-ray Spectrometry were published in *The International Journal of X-ray Spectrometry*, Volumes **6**, **7** and **8** (1977–1979), Wiley/Heyden: London.

[15] P. A. Pella, D. E. Newbury, E. B. Steel, and D. H. Blackburn, "Development of National Bureau of Standards thin glass films for X-ray fluorescence spectrometry," *Anal. Chem.*, **56**, 1133–1137 (1986).

[16] T. Hayaska, Y. Shibata, Y. Inoue, H. Hayashi, and Y. Kurosawa, "Determination of multielements in suspended particulate matter by X-ray fluorescence spectrometry," *Kawasaki-shi Kogai Kenkyusho Nenpo*, **12**, 13–23 (1985).

[17] G. R. Lachance, private communication.

[18] R. Tertian, "Mathematical matrix correction procedures for X-ray fluorescence analysis. A critical survey," *X-ray Spectrom.*, **15**, 177–190 (1986).

[19] R. Rousseau, "Fundamental algorithm between concentration and intensity in XRF analysis," *X-ray Spectrom.*, **13**, 115–125 (1984).

[20] F. Claisse and M. Quintin, "Generalization of the Lachance-Traill method for the correction of the matrix effect in X-ray fluorescence analysis," *Can. Spectrosc.*, **12**, 129–134 (1967).

[21] S. D. Raspberry and K. F. J. Heinrich, "Calibration for interelement effects in X-ray fluorescence analysis," *Anal. Chem.*, **46**, 81–89 (1974).

[22] W. N. Schreiner and R. Jenkins, "A non-linear least squares fitting routine for optimizing empirical XRF matrix correction models," *X-ray Spectrom.*, **8**, 33–41 (1979).

[23] G. R. Lachance and F. Claisse, "A comprehensive alpha coefficient algorithm," *Adv. X-ray Anal.*, **23**, 87–92 (1980).

[24] G. Lachance, *Introduction to Alpha Coefficients*, Corporation Scientifique Claisse Inc: Sainte-Foy, Quebec, 189 pp. (1984).

[25] G. Y. Tao, P. A. Pella, and R. M. Rousseau, "NBSGSC— a FORTRAN program for quantitative X-ray fluorescence analysis," *NBS Tech. Note 1213*, National Institute for Standards and Technology, Gaithersburg, USA (1985).

[26] R. Jenkins, "A review of empirical influence cofficient methods in X-ray spectrometry," *Adv. X-ray Anal.*, **22**, 281–292 (1979).

[27] T. Shiraiwa and N. Fujino, "Theoretical calculation of fluorescent X-ray intensities of nickel, iron, chromium ternary alloys," *Bull. Chem. Soc. Japan.*, **40**, 2289–2296 (1987).

[28] J. W. Criss and L. S. Birks, "Calculation methods for fluorescent X-ray spectrometry — Empirical coefficients vs fundamental parameters," *Anal. Chem.*, **40**, 1080–1086 (1968).

[29] J. W. Criss, "Fundamental parameter calculation on a laboratory microcomputer," *Adv. X-ray Anal.*, **23**, 93–97 (1980).

[30] J. W. Criss, L. S. Birks, and J. V. Gilfrich, "A versatile X-ray analysis program combining fundamental parameters and empirical coefficients," *Anal. Chem.*, **50**, 33–37 (1978).

[31] J. V. Gilfich and L. S. Birks, "Spectral distribution of X-ray tubes for quantitative X-ray fluorescence analysis," *Anal. Chem.*, **40**, 1077–1080 (1986).

[32] W. K. de Jongh, "X-ray fluorescence analysis applying theoretical matrix correction. Stainless steel," *X-ray Spectrom.*, **2**, 151–158 (1973).

CHAPTER

13

APPLICATIONS OF X-RAY METHODS

13.1. COMPARISON OF X-RAY FLUORESCENCE WITH OTHER INSTRUMENTAL METHODS

Since the early growth of instrumental methods of analysis in the 1950s and 1960s, there has always been a conceived need to develop a general purpose method of analysis. This means a single technique that can be applied to the analysis of any combination of elements in any matrix, yielding all of the information the analyst would like to have, and giving the ultimate in precision and accuracy. During the past 40 years or so, a number of methods have been developed which, though initially seemed to satisfy all of the needs of the analyst, still fell short in certain areas. In point of fact, an analytical technique is almost never *completely* made obsolete by a newer one. Like most other branches of science, development in analytical instrumentation is generally evolutionary rather than revolutionary. While there may be significant improvements in, for example, detectors or sources, that bring about dramatic improvement, in the long run, techniques more often than not, finish up complementing one another rather than replacing one another. All of this can create a problem for the analyst who has to choose from the vast array of techniques available, and who has to choose what is best for the particular application.

The techniques currently available for the quantitative determination of elements include [1,2]: 1) atomic absorption; 2) atomic fluorescence; 3) flame emission spectrometry; 4) mass spectrometry; 5) X-ray fluorescence; 6) electrochemistry; 7) emission spectrometry; and 8) nuclear and radiochemical analysis. Nearly all of these techniques are based either on the study of radiation emission or radiation absorption by a sample. Techniques that involve study of radiation emission include the X-ray fluorescence method, both in the wavelength and energy dispersive modes, and ultraviolet emission. The fundamental differences between X-ray and UV techniques arise mainly because of the wavelength of the excited radiation. In X-ray methods, the radiation characteristic of the atom arises from transitions from atomic orbitals from the inner shells of the atom. This means that the characteristic wavelengths are largely free from bonding effects, and spectra are rather simple. On the other hand, longer wavelength X-rays are difficult to disperse and even more difficult to quantitatively detect, and there can be significant problems in the analysis of elements of atomic number less than ten. Ultraviolet emission arises from electron transitions to outer atomic orbitals, and while the spectra are more complex, it is much easier to disperse, focus, and detect UV and visible light than X-rays. The techniques that involve absorption of discrete wavelengths include flame and furnace atomic absorption (AA) spectroscopy and inductively coupled plasma (ICP) spectroscopy. The key differences between AA and ICP methods stem from the temperature at which the sample can be held during the absorption process. In general, a technique that offers better dissociation of the element gives absorption data that are much freer from interferences than one where the element is still tightly bound to its neighbors. Better dissociation generally means higher temperatures, thus plasma sources clearly hold advantages over flames. Sometimes, advantages may be gained by combining two techniques. Examples might include the use of X-ray excited UV emission (UV/XRF), gas chromatography/mass spectroscopy (GC/MS), and, more recently, inductively coupled plasma/mass spectrometry (ICP/MS).

The UV emission spectrometer based on arc and spark sources has been the workhorse of the instrumental elemental analysis laboratory for more than four decades. The wide range of application of the UV emission method, in terms of range of elements analyzable, sensitivity, speed, and accuracy, have made this the method of choice in thousands of laboratories all over the world. A major drawback, however, has always been the problem of the introduction of the sample into the excitation system. This problem has spawned an incredible range of sample pretreatments and introductions including ultrasonic nebulization, microliter flow injectors, wire loop micro-furnaces, furnace atomizers, electrothermal vaporization, and a whole host of others.

Flame AA was developed in the early 1960s and today is one of the most useful of the instrumental methods for elemental analysis. There are few interferences, and what few there are can be relatively easily controlled and standardization is usually simple. Elements that are not easily determined by the flame AA method include the more refractory elements that are only partly dissociated in the flame, such as boron, vanadium, tantalum, and tungsten, and elements which have their resonance lines in the far UV, including phosphorus, sulfur, and the halogens. Better dissociation can be achieved by use of plasmas rather than flames, and the inductively coupled plasma (ICP) method is the most commonly used by far. An additional advantage is that detection limits are better by about a factor of three comparing ICP with flame AA. The third commonly used AA method is the graphite furnace technique. On a relative basis, furnace AA has better detection limits than Flame AA or ICP by one or two orders of magnitude.

On an absolute mass basis, it is generally better than either of the other two methods by as much as three orders of magnitude. While a major drawback in graphite furnace AA has been the complex matrix effects, these problems are now better understood, and methodologies are being developed to control or overcome them. Unfortunately, furnace determinations are slow and are typically single element. Analysis times typically run to several minutes.

While not too many detailed studies have been reported comparing these various techniques with X-ray fluorescence, in general, sensitivities and detection limits are of the same order. As an example, in a comparative study between energy dispersive X-ray spectrometry and AA spectrometry for the determination of metals in suspended particulates, detection limits for manganese, iron, nickel, copper, zinc, and lead were found to be 0.01 to 0.05 $\mu g/cm^2$. The AA spectrometry concentration data were about 35% lower than the equivalent XR spectrometry data, and this was attributed to the incomplete solution of solid matter before analysis by AA spectrometry [3]. Similarly, in a comparative study between X-ray spectrometry and ICP-AE spectrometry, 5 to 10% precision has been reported for samples which included soils and grasses [4].

The traditional use of the mass spectrometer has involved analysis in the gaseous phase, and has been mostly restricted to volatile organic liquids. However, a dramatic change has taken place in the last several years which now allows study of the chemistry and analysis of biological substances in condensed polar phases. New fields of plasma and laser desorption, secondary ion bombardment of the liquid surface layer have been developed with, for example, energetic primary ions in the technique of LSIM spectrometry. While the sensitivity of these so-called *soft ionization techniques* cannot yet compete with techniques such as gas chromatography/mass spectrometry (GC/M), indications are that new developments in ion optical systems may eventually bring the sensitivity of the technique down from the existing nanomole range to the picomole sample size.

While features such as accuracy, sensitivity, and cost are probably vital in the selection of a technique, one area of growing importance is that of automation. A technique that is easily automated will find far greater acceptance than one that cannot be easily automated. Perhaps the best example of this is found in atomic absorption. Flame atomic absorption units are not easily automated for the analysis of simultaneous elements in a cost-effective manner, mainly because of the need to change light sources. As a result, almost all AA instruments are designed for the analysis of single elements. This is, of course, not to say that flame AA absolutely cannot be automated. For example, there are units available that can analyze six elements in 50 samples in 30 to 40 minutes. However, this might not be the method of choice here, considering that, with the modern ICP system, which does not need to change sources between different elements, perhaps 40 to 50 elements per minute can be analyzed. In addition to the question of the ease and cost of automation, another constraint to be considered is the form and size of the sample needed for the analysis. Some techniques are ideally, but not exclusively, suited to the analysis of liquids and solutions, flame AA and ICP being good examples. Others, such as XRF, are ideally, but not exclusively, suited to the analysis of solid samples.

13.2. APPLICATION OF X-RAY SPECTROMETRY

The great flexibility and range of the various types of X-ray fluorescence spectrometers, coupled with their high sensitivity and good inherent precision, make them ideal for quantitative analysis. Single channel wavelength dispersive spectrometers are typically employed for both routine and nonroutine analysis of a wide range of products, including ferrous and nonferrous alloys, oils, slags and sinters, ores and minerals, thin films, and so on. These systems are very flexible but, relative to multichannel spectrometers, are somewhat slow. Multichannel wavelength dispersive instruments are used almost exclusively for routine, high throughput, analyses, where the great need is for fast accurate analysis, but where flexibility is of little importance. Energy dispersive spectrometers have the great advantage of being able to display information on all elements at the same time. They lack somewhat in resolution over the wavelength dispersive systems, but also find great application in quality control, troubleshooting problems, and so on. They have been particularly effective in the fields of scrap alloy sorting, forensic science, and the provision of elemental data to supplement X-ray powder diffraction data.

An area where X-ray techniques are finding increasing application is in the analysis of pollutants. For many years, the sensitivity of the fluorescence method was barely sufficient for the direct measurement of contaminant levels at the ppm level, and preconcentration methods had to be employed. More recent developments have allowed sensitivities that are quite sufficient for the sample masses typically encountered in the analysis of trace metals in air and water samples, and X-ray fluorescence methods are finding increasing application in this area [5]. A good example of the wide applicability of the X-ray fluorescence method is in the paper industry, where it has been used for a broad range of applications including the analysis of lime and lime related compounds, calcite, dolomites, silicates, silica glass sand, gypsum, MnO_2 and iron ores, wood treated with pentachlorophenol and ammonical copper arsenate, sulfur, chlorine, bromine, and titanium in paper, pulp, pitch and asphalt, and, evaluation of paper printability and ink holdout [6]. Because of the nondestructive feature of X-ray analysis, the technique has found special usefulness in the field of archeometry. Examples include the examination of the chemical composition of medieval durable blue soda glass from York Minster in England. This study showed three distinct compositional groups, indicating in turn, three probable sources of the glass [7]. Energy dispersive spectrometer studies of early American pottery has indicated details of manufacturing location and time period [8]. A study of old porcelain revealed that the chemical composition of the ceramic body, the glaze and the various colors, could be used as a means of authenticating 18th and 19th century pieces [9]. Similar energy dispersive methods have been used for authenticating fine Chinese porcelain from the Ming (1348–1644) and Qing (1644–1911) dynasties [10]. It has been reported that all Chinese porcelains from Kangxi (1662–1722), up to the time of World War II, have barium contents in the range 100 to 130 ppm. After this time, the barium content varied from 60 to 700 ppm with only a few pieces in the 100 to 130 ppm range. Thus, it is possible to rapidly identify most modern reproductions [11].

There are certain areas of elemental analysis that are still unique to X-ray fluorescence. As an example, nondispersive XRF has been used for the analysis of technetium in solutions by use of the Tc $K\alpha$ line. At this time, there are no good chemical methods for the analysis of technetium [12]. Direct X-ray fluorescence methods are also being used for the study of the short-lived, super-heavy elements. Another unique application of the fluorescence technique is in the direct analysis of biological samples. As an example, radioisotope XRF has been used for the analysis of biological samples such as those provided by biopsy of human aorta or tissue section. A collimated ^{109}Cd source of 10 to 20 mCi strength gave a sensitivity of about 10^{15} atoms, and is being used for tumor growth studies [13] and studies involving atherosclerosis [19].

In addition to the many hundreds of papers that are published annually describing specific applications of the technique [20,21], and a journal specifically devoted to the fluorescence method [22], several texts have been devoted to specific areas of application. These include geology [23], medicine [24], metallurgy, and art [25].

13.3. COMBINED DIFFRACTION AND FLUORESCENCE TECHNIQUES

Like all analytical techniques, both X-ray powder diffraction and X-ray fluorescence have their advantages and disadvantages. Table 13.1 indicates the more important features of the two techniques. In terms of the range of

Table 13.1. Features of X-Ray Spectrometry and Diffraction

Feature	Spectrometry	Diffractometry
Range	Z > 5	All Ordered Materials
Speed	Secs to Mins	Mins to Hours
Precision	0.1%	0.25%
Accuracy	0.1 to 1.0%	0.5 to 5%
Sensitivity	low ppm	0.1 to 2%
Cost	$50,000 to 250,000	$25,000 to 125,000

application, X-ray fluorescence analysis allows the quantification of all elements in the periodic table from fluorine ($Z = 9$) and upwards. Accuracies of a few tenths of one percent are possible and elements are detectable, in most cases, to the low ppm level. X-ray powder diffractometry is applicable to any ordered (crystalline) material and, although much less accurate or sensitive than the fluorescence method, is almost unique in its ability to differentiate phases. The techniques differ widely in terms of speed of analysis. The modern multichannel wavelength dispersive spectrometer is able to produce data from 20 to 30 elements in less than one minute. An energy dispersive spectrometer, or even a relatively simple wavelength dispersive spectrometer, is able to perform a full qualitative analysis on an unknown sample in less than 30 minutes. The diffraction technique, on the other hand, can take as much as an order of magnitude longer. Even with the most sophisticated computer-controlled powder diffractometers available today, diffraction experiments are invariably very time consuming. A similar difference also exists in the sensitivities of the two methods. Whereas the fluorescence method is able to measure signals from as little as one part per million of a given element, the diffraction technique is often hard put to measure one percent.

In terms of precision and accuracy, the spectrometry method out performs the diffraction technique again, perhaps by as much as an order of magnitude. The diffraction technique is, at best, a rather insensitive, slow technique, giving somewhat poor quantitative accuracy. On the other hand, it will be realized that the information given by the X-ray diffraction method is unique, and no other technique is able to provide such data. This is not true of the X-ray fluorescence method, since, as discussed in Section 13.1, there are many other techniques now available to the analytical chemist for the quantification of elements. The fluorescence and diffraction techniques are, to a large extent, complementary, since one allows accurate quantification of elements to be made, and the other allows qualitative and semiquantitative estimations of the way in which the matrix elements are combined to make up the phases in the specimen to be made.

As discussed in Chapter 10, both the simultaneous wavelength dispersive spectrometer and the energy dispersive spectrometers lend themselves admirably to the qualitative and semiquantitative analysis of materials. Because the characteristic X-ray spectra are so simple, the actual process of allocating atomic numbers to the emission lines is a relatively simple process and the chance of making a gross error is rather small. There are only 100 or so elements and, within the range of a conventional spectrometer, each element gives, on an average, only half a dozen lines. As a comparison, the procedures for the qualitative analysis of multiphase materials with an X-ray powder diffractometer are much more complex. In diffraction, on the other hand, there are perhaps as many as several million possible compounds, each of which can give fifty or so lines on average. However, a combination of spectrometric and diffraction data can often be used to greatly simplify what would otherwise be a complex and time consuming analysis. By judicious use of both fluorescence and diffraction data, a *material balance* can often be established. The fluorescence data is first used to establish the higher concentration elements and then the diffraction data used to suggest how these elements might be present in the analyzed specimen.

13.4. PORTABLE XRF EQUIPMENT

Two of the disadvantages of classical X-ray spectrometers are their large size, and the need to present a specimen that has defined size limitations. Because of the frequent need to carry out fast, in situ measurements, on odd-shaped specimens, much development has been devoted to the provision of portable systems which can be transported and used in the field. One of the earliest of these was the Norelco Portospec [14]. This instrument worked with a 40 kV 60 μA transmission gold target tube and was able to measure all elements above vanadium ($Z = 23$) with detection limits of the order of 1%. The Portospec was a wavelength dispersive system, in that it included a minature goniometer, complete with an analyzing crystal. What was then known as *nondispersive analysis* (today, this would be akin to *energy dispersive analysis*) was also employed using a radioisotope source and a scintillation detector [15]. Such systems were extremely compact and later developments lead to the development of small spectrometers which would actually fit inside a borehole [16]. The problem in designing a compact, light-weight spectrometer is invariably related to the need for a powerful enough source to give reasonable X-ray intensities. In more recent years, there have been significant improvements in the design of portable X-ray spectrometers, leading to versatile instruments which are finding increasing use in, for example, the monitoring of hazardous wastes [17,18].

13.5. ON-STREAM ANALYSIS

The area of process analytical chemistry deals with the problems of the provision of qualitative and quantitative information about a given chemical process [26]. Such data may be collected off-line, at-line, or on-line, and X-ray fluorescence is applicable in all three of these areas. This section deals mainly with the truly *on-line* applications of X-ray fluorescence. The use of conventional X-ray spectrometry for on-line process control has long been recognized [27] and, at the present time, there are something like 200 units in regular use in mineral benefication plants in various parts of the world. Conventional X-ray spectrometry, similar to that already described in previous chapters, has been used for on-stream application in which the normal fixed sample is replaced by a slurry cell. Special purpose isotope excited systems have also been used with a great deal of success. Probably the major use of on-stream X-ray fluorescence analysis is in mineral extraction, and applications fall roughly into three categories.

- Analysis of flowing slurries, found in most ore dressing and mineral benefication plants.
- Analysis of dry flowing solids, for example, crushed ores on conveyor belts or dry raw cement mix en route to the kiln.
- Analysis of drill cores and insides of boreholes.

In each of the three areas, but especially in the first, the major disadvantage of the on-stream analyzer system based on a conventional X-ray source/crystal dispersion arrangement, is that its high inherent cost limits the size of the installation to one or, at most, two units. This in turn requires some extremely sophisticated plumbing arrangements for the sample streams, leading to introduction of delays in getting the analysis to the control center. A cycling system is generally employed to allow sampling of multiple lines, but even allowing for certain priority channels, the cycle time can be on the order of 3 to 15 minutes. Although this may be satisfactory in some installations, the growing trend towards use of on-line computer control requires almost immediate use of analytical data, and delays on the order of minutes are unacceptable. In addition, there are special problems in the determination of the lower atomic number elements ($Z < 20$) partially due to the attenuation of the fluorescent signal by the plastic windows of the flow cell, but even more due to particle effects which arise because individual particles of solid in the slurry may be heterogeneous over the same order of magnitude as the penetration depth of the X-ray beam.

The potential use of radioisotope sources for on-stream analysis is also well established and a survey by Watt [28] includes an impressive list of applications and feasibility studies involving the use of radioisotopes for the on-stream analysis of slurries. The problems successfully handled thus far include the determination of elements including calcium, copper, zinc, niobium, molybdenum, barium, and lead in flotation feeds and slurries. A great advantage of the radioisotope system is its low cost and the fact that sources can be obtained or specially fabricated for almost any special application. Unfortunately the photon yield is generally much lower than with X-ray tube excited sources which in turn precludes the use of the relatively inefficient crystal dispersive spectrometers. Energy dispersive spectrometers have played an important role in this area and additional techniques, including use of filters and secondary sources, have done much to enhance their applicability. Typical of these special sources is the γ-X source [29]. The γ-X source comprises three essential parts: the γ source itself (the isotope ^{153}Gd is a frequently used material), the target material in the shape of a cone, and a lead shield. The γ source excites K X-radiation from the target cone, and this secondary X-radiation is then directed onto the sample under analysis.

Looking now at the second area, the conventional X-ray spectrometer, modified for the analysis of a flowing sample, has found some success in the analysis of dry flowing solids. As an example, use of continuous sampling on sticky tape, followed by monitoring with an energy dispersive spectrometer, has yielded some success, and detection limits of a few mg/cm^2 have been reported in the determination of heavy metals in fly ash from waste incineration [30]. For many materials, however, the particle size effect, referred to previously, remains a major problem. However, a good deal of success has been achieved by use of successive sampling rather than continuous irradiation. In continuous irradiation, a sample depth equivalent to tens of microns is continuously irradiated, for an integration time related to the number of count required, this time being typically on the order of 60 seconds. Following a short delay for the data readout, the process is then repeated. In successive sampling, a sample is taken at time intervals, this time being the time required to prepare and analyze the sample, again typically of the order of 200 seconds. In some instances, the sample may be taken continuously and an aliquot taken for analysis, but in all cases, a homogeneous specimen of about 2 to 5 grams is analyzed. In the first case, one relies on homogeneity over the cross section of the stream, but this is not a critical feature in the second method, where a larger analyzed volume is taken and homogenized before analysis. The successive sampling technique, combined with high pressure pelletizing or borax fusion of the sample, followed by conventional X-ray fluorescence spectrometry, has been applied with great success, for example, to the analysis of cements.

The third area concerns the analysis of drill cores and the insides of drill holes. Data has been reported on the direct analysis of drill core samples using PIXIE [31], as well as more conventional energy dispersive X-ray fluorescence [32]. Special instruments based on the use of semiconductor counters have been used for certain applications in the analysis of the insides of drill core holes, specifically for the determination of high atomic number elements in low grade ores. For example, Burkhalter [33] describes a system for the determination of gold and silver in a silicon matrix at concentration levels down to 20 ppm. In this example, an annular source of ^{125}I was used in combination with a silicon semiconductor counter for the determination of silver, and a germanium detector for the analysis of the higher energy gold K spectrum. In another example, a precision of 0.1% has been obtained in the determination of iron in dolomite borehole samples, using radioisotope X-ray fluorescence, based on an energy dispersive technique using ^{109}Cd as a source [34]. This precision is sufficient as specified for oil and gas exploration.

BIBLIOGRAPHY

[1] W. Slavin, "Flames, furnaces plasmas — How do we choose," *Anal. Chem.*, **58**, 589A (1986).

[2] *Analytical Chemistry Fundamental Reviews*, April (1986).

[3] T. Hayasaka, Y. Shibata, Y. Inoue, H. Hayashi, and Y. Kurosawa, "Comparison of atomic absorption and energy dispersive X-ray fluorescence for the determination of metals in suspended particulates," *Kawasaki-shi Kogai Kenkyusho Nenpo*, **12**, 5–12 (1985).

[4] P. P. Coetzee, P. Hoffmann, R. Speer, and K. H. Liesser, "Comparison of trace element determination in powdered soil and grass samples by energy dispersive XRF and by ICP-AES," *Fresnius Z. Anal. Chem.*, **323**, 254–256 (1986).

[5] *X-ray Fluorescence Analysis of Environmental Samples*, T. G. Dzubay, Ed., Ann Arbor Science: Michigan, 310 pp. (1977).

[6] V. Kocman, L. Foley, and S. C. Woodger, "The use of rapid quantitative X-ray fluorescence analysis in paper manufacturing and constructional materials industry," *Adv. X-ray Anal.*, **28**, 195–202 (1985).

[7] G. A. Cox and K. J. S. Gillies, "The X-ray fluorescence analysis of medieval durable blue soda glass from York Minster," *Archeometry*, **28**, 57–61 (1986).

[8] P. L. Crown, L. A. Schwalbe, and J. R. London, "Evaluating the variability of south western ceramics with X-ray fluorescence spectrometry," *Adv. X-ray Anal.*, **28**, 169–176 (1985).

[9] A. Stiegelschmitt and G. Tormandl, "Study of old porcelain with reference to its art historical classification," *Sprechsaal*, **118**, 974–978 (1985).

[10] C. T. Yap, "Chinese porcelain: genuine or fake?," *Phys. Bull.*, **37**, 214–215 (1986).

[11] C. T. Yap and S. M. Tang, "Quantitative XRF analysis of trace barium in porcelains by source excitation," *Appl, Spectrosc.*, **39**, 1040–1042 (1985).

[12] G. A. Akopov, V. V. Berdikov, E. A. Zaitsev, B. S. Iokhin, and A. P. Krinitsyn, "Possible determination of technicium concentration in solutions by means of X-ray fluorescence," *Soviet Atomic Energy*, **60**, 87–89 (1986).

[13] A. S. Frank, M. K. Schauble and I. L. Preiss, "Trace element profiles in Murine Lewis Lung Carcinoma by radioisotope induced X-ray fluorescence," *J. Amer. Pathol.*, **122**, 421–432 (1986).

[14] D. C. Miller, "The Norelco portable spectrometer (Portospec)," *Adv. X-ray Anal.*, **3**, 57–67 (1959).

[15] J. R. Rhodes and T. Furata, "Applications of a portable radioisotope X-ray fluorescence spectrometer to analysis of minerals and alloys," *Adv. X-ray Anal.*, **23**, 19–25 (1979).

[16] M. W. Springett, "Portable X-ray fluorescence spectrometers and their use in an underground exploration program for tin," *Adv. X-ray Anal.*, **11**, 249–274 (1967).

[17] G. A. Raab, C. A. Kuharic, W. H. Cole III, R. E. Enwall, and J. S. Duggan, "The use of field-portable X-ray fluorescence technology in the hazardous waste industry," *Adv. X-ray Anal.*, **33**, 629–637 (1989).

[18] M. Bernick. P. F. Berry, G. R. Voots, G. Prince, J. B. Ashe, J. Patel, and P. Gupta "A high resolution portable XRF HgI_2 spectrometer for field sceening of hazardous metal wastes," *Adv. X-ray Anal.*, **35B**, 1047–1053 (1991).

[19] I. L. Preiss, T. Ptak, and A. S. Frank, "Trace element profiles of biological samples using radioisotope X-ray fluorescence," *Nucl. Instrum. Methods Phys. Res. Sect. A*, **A242**(3), 539–543 (1986).

[20] *Analytical Chemistry, Fundamental Reviews*, published on even years, American Chemical Society.

[21] *CA SELECTS — X-ray Analysis & Spectroscopy*, bi-monthly, Chemical Abstracts Service, Columbus, Ohio, USA.

[22] *International Journal of X-ray Spectrometry*, published bimonthly, Wiley/Heyden: London, (1971 to date).

[23] I. Adler, *X-ray Emision Spectrography in Geology*, Elsevier: Amsterdam, 258 pp. (1966).

[24] R. Cesareo, *X-ray Fluorescence (XRF and PIXIE) in Medicine*, Field Educational Italia, 238 pp. (1982).

[25] H. K. Herglotz L. S. Birks, *X-ray Spectrometry*, Dekker: New York, 518, pp. (1978).

[26] J. B. Callis, D. L. Illman, and B. R. Kowalski, "Process analytical chemistry," *Anal. Chem.*, **59**, 624A–637A (1987).

[27] R. Jenkins and J. L. de Vries, "Use of radioisotopes for on-stream analysis," *Canadian Spectr.*, **16**, 3–8 (1971).

[28] J. S. Watt, "Current and potential applications of radioisotope X-ray and neutron techniques of analysis in the mineral industry," *Australasian I. M.&M. Proceedings*, 69–77 (1970).

[29] J. A. Hope and J. S. Watt, "Alloy analysis and coating weight determination using γ-ray excited X-ray sources," *Int. J. Appl. Rad. Isotopes*, **16**, 9–14, (1965).

[30] K. E. Lorber, "Monitoring of heavy metals by energy dispersive X-ray fluorescence spectrometry," *Waste Management Res.*, **4**, 3–13 (1986).

[31] L. E. Carlsson and K. R. Akelsson, "Application of PIXE and XRF to fast drill core analysis in air," *Adv. X-ray Anal.*, **24**, 313–321 (1981).

[32] R. J. Arthur, J. C. Laul, and N. Hubbard, "Energy dispersive X-ray fluorescence analysis of intact salt drill cores," *Adv. X-ray Anal.*, **28**, 189–194 (1984).

[33] P. G. Burkhalter, "Detection limit for silver by energy-dispersive X-ray analysis using radioisotopes," *Instruments in Industry and Geophysics*, **20**, 353–362 (1969).

[34] F. S. Tham and I. L. Preiss, "Precision measurement of iron concentration in dolomites using radioisotope X-ray fluorescence," *J. Radioanal. Nucl. Chem.*, **99**, 133–144 (1986).

INDEX

Italic page numbers indicate figures; the letter "t" indicates tables; *See also* indicates related topics or more detailed topic breakdowns.

Lightning Source UK Ltd.
Milton Keynes UK
UKOW041522180313

207827UK00001B/15/A

begin to create the sense of new forces to left of Labour. There is every possibility of providing a political input into a deepening discontent that finds an expression in the workplaces and on the streets. That can give Brown and Cameron alike something to worry over—and begin to pull the rug from under the Nazis before they have a chance to grow.

Italian lessons

The victory for the coalition around Silvio Berlusconi in Italy is much more serious than the Tory gains in Britain's local elections. It has produced a government in which the hard right have been making the running. The leader of the "post-fascist" National Alliance, Gianfranco Fini, now ensconced as speaker of parliament (replacing Fausto Bertinotti of Rifondazione Comunista), has called a murder committed by Nazi skinheads in Verona less serious than the burning of the Israeli flag by anti-Zionists. Still more seriously, the Northern League, in control of the ministry of the interior after picking up 20 to 30 percent of the votes in parts of the north, has launched a pogrom against immigrants.

The scale of the defeat suffered by the left is such that there is no Communist or socialist representation in parliament for the first time since the Second World War. Umberto Bossi, the Northern League's leader, claims his is "the new party of the workers". By any measure this is a disaster. It does not mean the left cannot fight back. The political defeat is by no means a defeat for the workers' movement as a whole. But the fightback is on more difficult terrain than it needed to be.

The direct responsibility for what has happened lies with the outgoing government of Romano Prodi. It was elected in the aftermath of the great waves of popular protests at the policies of Berlusconi's previous government which began with the Genoa protest against the G8 summit in 2001, and continued with one-day general strikes and huge marches against the Iraq war. Yet in power Prodi's government followed the same economic policies as Berlusconi's, while sending the troops Berlusconi had already begun withdrawing from Iraq to Afghanistan and Lebanon.

At the core of the Prodi government was the Sinistra Democratica (Democratic Left), founded by the majority of what was once the biggest Communist Party outside the Eastern Bloc as a logical follow through to its late 1970s "historic compromise" with the country's Tory Christian Democrats (see below).

By the time April's election took place it had dropped the *sinistra* ("left") to merge with some of the Christian Democrats as the Democratic Party, which now seeks to emulate the US Democrats. In the election campaign it responded to the xenophobia of the Northern League and claims that immigrants were responsible for "insecurity" by insisting it too placed "security" and control of immigration at the top of its agenda. The evening the election result came out it made an offer to Berlusconi "to carry through some of the reforms together".

Unfortunately, however, responsibility for the Italian debacle does not simply lie at the door of this Italian version of Blairism. The mistakes of the biggest organisation of the far left have also played a role.

Learning the lessons the hard way

> The electoral defeat of 13-14 April has historical dimensions. The left is not represented in parliament for the first time since Italy became a republic. The populist right of Berlusconi won a big popular majority, and a xenophobic force within it, the Northern League, doubled its support. This changes the Italian political panorama. We cannot fail to see the negative effects of the inability of our presence in the government to provide an answer to the principal social problems of the country. The victory of 2006 was not about defeating Berlusconi only, but about defeating Berlusconi's policies.
>
> The concrete actions of the government did not respond to this need. On the contrary, they fulfilled the requirements of the powers that be on the principal social questions: the redistribution of income, the fight against precarity, taxation, the laicity [secularism] of the state, just to take some examples. Our political action was ineffectual and the left did not make its presence felt on these issues. That produced a crisis, the depth of which we did not recognise, in our relations with the mass of people and, in particular, with the struggles. It proved impossible to overturn the policies of the last 15 years from within the government and remaining there became as big a problem for us as for the movement.

So run the first few paragraphs of the resolution adopted by the national political committee of Rifondazione Comunista in Italy on 19-20 April. Supported by Paolo Ferrero, who had been the minister of social solidarity in Prodi's government, it was narrowly carried (90 votes to 70) in the face of opposition from Bertinotti's supporters, who will try to reverse the position at the party's congress in late July.

We are tempted to say, "We told you so"—or rather, "Rosa Luxemburg told you so", since the process described in the resolution, of the far left joining a capitalist government and then finding that the movement as a whole suffers as a result, is not something new. It is an experience that goes right back to the beginning of the 20th century, when Rosa Luxemburg polemicised against the French socialist Jean Jaurès' support for entry into a centre-left government. As she put it:

> The government of the modern state is essentially an organisation of class domination, the regular functioning of which is one of the conditions of existence of the class state. With the entry of a socialist into the government, and class domination continuing to exist, the bourgeois government doesn't transform itself into a socialist government, but a socialist transforms himself into a bourgeois minister... The entry of a socialist into a bourgeois government is not, as is thought, a partial conquest of the bourgeois state by the socialists, but a partial conquest of the socialist party by the bourgeois state.[7]

The impact of the far left joining a bourgeois government then, just as now, could only be to remove a pole of opposition to the policies such a government was bound to follow, since the forces at the disposal of capital are much more powerful than the voices of left parliamentarians so long as they are confined to ministerial gatherings. Instead of leading workers in struggles over their conditions or against imperialist intervention, the left end up trying to cool down struggles and curtail movements against imperialism. That was exactly what happened with the government the French socialists entered a century ago. Luxemburg insisted, "The radical cabinet...in a series of equivocal manoeuvres in the course of 19 months... accomplished nothing, absolutely nothing."

Exactly the same can be said of the centre-left government led by former European Union president Prodi and containing Rifondazione

7: See Chris Harman, "The History of an Argument", *International Socialism 105*, www.isj.org.uk/index.php4?id=58

members. While Bertinotti presided over the parliamentary chamber, Rifondazione deputies voted for the budget and to send troops to Lebanon and Afghanistan, with the party expelling the one senator who voted against the Afghanistan war. Now not only the party is paying the price. As the resolution tells:

> We lost votes in all directions—there were those who did not vote because they felt "they are all the same"; there were those who voted for the Democratic Party because they wanted to cast a "useful vote" against Berlusconi; and a section of the proletariat who felt they were not being defended by the left went to the Northern League.

Rifondazione formed an electoral coalition—the Rainbow Left—with the smaller Party of the Italian Communists, and the Greens. Their combined vote in the European elections of four years ago was 11 percent. This time they got only 3.1 percent: "Most of the lost votes went to the centre-left, but about one in five went to Berlusconi's centre-right."

This is not just a disaster for Rifondazione. Seven years ago it was able, with its 80,000 members and its daily paper *Liberazione*, to rally workers and activists across the country to take to the streets in response of the attempt of Berlusconi's newly elected government to brutally crush the movement against capitalist globalisation. Those mobilisations were a check on further moves by the government. Now, much debilitated, discredited by two years of justifying centre-left capitulations and split down the middle, it is unlikely to be able to play a similar positive role in drawing together the many acts of resistance to what the new Berlusconi government and its nefarious allies are setting out to do.

First time tragedy, second time tragedy

For those with a little knowledge of Italian history, there must be a feeling of deja vu. The approach that led the Italian far left to disaster was a repeat performance of that undertaken by the old Italian Communist Party in the mid-1970s. After a huge wave of popular mobilisations, strikes and factory occupations from 1969 to 1975, which derailed all the plans of the Tory Christian Democrat Party and provided the Communists with their biggest vote for 30 years, the Communist leadership opted for a "historic compromise" with that party, entering "the government sphere" (although debarred from the government itself by US opposition) and consciously setting out to bring the popular mobilisation to an end. By the 1980s the Christian Democrats felt strong enough to turn on the

Communists and to oversee an attack on union strength at the key Fiat factory in Turin.

The lesson drawn by the main Communist leaders of the time was, effectively, that they had not gone far enough along the road to conciliation with capitalism. Hence their decision in the early 1990s to dissolve their party into the Democratic Left—an explicitly social democratic formation—and join a succession of governments with the remnants of Christian Democracy.

Rifondazione Comunista (the "Refounded Communist" party) was created by those who rejected the trajectory of the majority of the leadership, and won a considerable proportion of the old Communist Party's activists. But its leaders never carried through an analysis of what had happened in the 1970s, or of the "Eurocommunist" approach to fighting for socialism underlying it. Instead their speeches and writings were an eclectic mish-mash of "Eurocommunist" notions and the autonomist ideas popular in movements based outside the working class.

What was lacking was any real analysis of Italian capitalism and of the continuing key position of the employed working class within it. This in turn meant there was never any clear understanding of the centrality of workers' struggles to achieving the goals of the movements. So long as the other movements were rising, with the big anti-capitalist and anti-war demonstrations, Rifondazione could swing to the left. Once they declined not only the leaders but many of the activists fell for the delusion that joining a centre-left government was the only way to achieve their goals. The way was open to a disaster for the left.

It is not a final catastrophe. The defeat of the left is not the defeat of the workers' movement. This is not Italy in 1922 or Germany in 1933. Italian capitalism is very weakly placed as it faces the backwash of the developing US recession and the impact of soaring global food prices. The bosses' organisation, Confindustria, is pressurising Berlusconi to push through more counter-reforms. And the workers still have enormous potential strength, as was shown at the end of last year when a truck and tanker drivers' strike briefly paralysed the country. Bossi can ignite horrifying witch-hunts against migrants, but he will risk losing his new found votes from some workers if it is his party, rather than Rifondazione, which is part of a government imposing unpopular measures.

There will be workers' fightbacks, as there were with the two previous Berlusconi governments of 1995-6 and 2001-6. The left can only rebuild itself by relating to these struggles, as well as to the defence of migrants', women's and gay rights. There is no alternative to being part of the wider resistance. But there also has to be the ingredient that has been missing before—a serious Marxist analysis of the mistakes of the past, of the nature of present day capitalism and of the capacity of workers to fight back. The weakness of the forces trying to convey such analyses at the height of the far left's influence in 2001-2 is the reason it capitulated so rapidly to the Bertinotti version of the parliamentary road.

Greece: a very different picture

Panos Garganas, editor of the Greek newspaper Workers Solidarity, spoke to International Socialism about the country's strike wave

The social and political trajectory in Greece seems very different to anywhere else in Europe. There have been two years of near continual struggle—the student struggle and successive strikes. Can you explain why this has been so?

The upsurge of the past two years has not been a sudden development. The ground for what has happened was prepared by a wave of struggles under the Pasok centre-left government of the early 2000s. There was a large strike in 2001 against that government's plans for pension reform, there was a wave of sympathy for the 2001 demonstration against the G8 summit in Genoa, Italy, and there was an explosion of anger against the war in 2003. For a couple of months from 15 February 2003, when there were the worldwide demonstrations against the war, until 9 April 2003, when American forces entered Baghdad, there were demonstrations against the war nearly every day. So there is this kind of background to take into account.

At the same time there has been pressure on the Greek government to push ahead with counter-reforms because Greek capitalism is falling behind its rivals. Staging the Olympics in 2004 was a massive effort, and it left behind all sorts of problems. So the Tory government that took office in 2004 had to be much more forceful in pushing its counter-reforms. Where the previous Pasok government would compromise with the unions to avoid confrontation, the Tory New Democracy government was much more aggressive. But at the same time it was faced with young people and workers who had already had experiences of mass mobilisation.

The Tory government did not come in offering right wing policies, did it?
It was not a swing to the right in Greek society—even if that is how it seemed on the surface to many activists, who simply saw a Labour-type government defeated and the Tories victorious at the election.

The reality was that the Greek Tories won the election in 2004 by carefully hiding their real agenda. They presented themselves as being to the left of the Pasok government. When I say "to the left", there were attacks by Tory politicians on Pasok ministers accusing them of being servants of big capital. They used the terminology of the left. There were other token gestures towards the left. Kostas Karamanlis, the Tory leader, went to one of the islands where left wingers were exiled in the post-war decades and declared that this was all in the past, that they would turn it into a museum, and so on. During the election campaign Karamanlis promised to end discrimination against the Muslim minority in Thrace. This was unheard of. The Tories had always been the most racist party. They used all sorts of methods to present themselves as a party that was of the centre and in many ways to the left of Pasok. They knew they had to win votes from people whose opposition to Pasok was from the left.

Then you had the wave of struggle for two years.
There was a lull when the Tories came into office in 2004. For some months the government held back from pressing its agenda, and there was a sense that things would not be so bad. This lasted for about a year. Then in the spring of 2005 the government began to push its measures. There was a reaction from below. New sections of the working class, people who had precarious jobs, started mobilising on the basis of a promise that had been made that they would get permanent jobs.

How did they mobilise?
They had created union sections for themselves. Some unions allowed them to become members, but in many cases they had to start their own sections and, where this happened, they were able to call people out on demonstrations.

Did they just demonstrate, or did they strike as well?
It varied from section to section. They would call demonstrations as a show of anger but then they had the strength to shut down certain areas, like local government, where they won the support of the unions. The strike action was the result of "precarious unions" taking the initiative and then some of the established unions backing them.

Many of those taking a lead were short-term contract workers. The sort of people we are talking about would be employed for, say, eight months, then had to accept a break for three or four months and then got a new contract.

You had these struggles, and then you had the struggles of what might be called more mainstream workers such as those in electricity and telecommunications. The bank workers' action started in the summer of 2005 in the face of attempts to reform the workers' special pension fund. They started with a five-week strike, which got a lot of solidarity. It was impossible for the government to isolate the strikers, so in the end it voted through a new law which it has not been able to implement to this day.

This was the first case showing that people were prepared to take all-out strike action. And to a certain extend they were successful.

The next wave was formed by the students. The government decided that it would be easier to push through educational reform. So it proposed a change in the constitution that would have allowed for private universities, and this provoked a wave of student occupations, which was successful in scuppering the change to the constitution. The government could not get enough votes in parliament to change without support from sections of the opposition—and the opposition had a change of heart under pressure from the student occupations.

So that was the second wave of victorious struggle, and it created the impetus for lots more people to move in the same direction.

The government then called a snap election in September 2007 because it thought a new mandate would help it, but it was wrong. There was a discussion on the left about whether the resistance would decline after the election. But when the re-elected government tried to push through the pension reform a strike by journalists started a new wave of struggle, and the unions were pushed into calling a series of days of action, which were massive. We had days of action on 12 December 2007, 13 February 2008 and 19 March 2008. These days of action came together with all-out strikes by energy workers, local government workers and bank workers, including those at the Bank of Greece.

These were the most advanced strikes we had seen since the 1970s because they were all-out actions. There was mass participation and there were pickets to stop any strike-breaking. They were also in defiance of the law, something which had not happened in a very long time. Not since the 1970s had people ignored court orders and refused to provide staff for "essential" services—the bank workers shut down the stock exchange for a few days, energy workers shut down power stations, and so on.

What was the political feeling during the strikes? What sorts of slogans were people taking up on the demonstrations?

The main slogans were directed against the government. But because of the earlier years of the Pasok government the feeling was different than that during the strikes and demonstrations in the 1990s, when people who were angry at the Tory government looked to Pasok as the alternative. The change in ideas was reflected in the mood of the demonstrations, but it came out most clearly in the opinion polls which showed the left going up for the first time in a long time.

In Greece you have a social democratic party, Pasok, which is now in reality a centre-left party. You have the Communist Party, which used to be the bulwark of the left, but became overshadowed by Pasok in the 1970s, and you have this other formation, Synaspismos, which I suppose you would call left reformist.

The Communist Party in Greece was outlawed for decades—from 1936 and the pre-war dictatorship, through the occupation and the civil war that followed the Second World War, and then the military junta of 1967-74. In 1968 the party split into two wings. One was Stalinist and pro-Moscow and the other was influenced by Eurocommunist ideas.

In the 1970s both were overshadowed by Pasok, which was a new development. There was no tradition of a social democratic party in Greece. Like the socialist parties in Spain or Portugal, Pasok went up like a rocket after the collapse of the junta. So for a long time the two wings of the old Communist Party remained in the shadow of Pasok. They went through a period of cooperation at the end of the 1980s, but by 1989, on top of the broader crisis of the Communist Parties that came with the collapse of the Communist regimes, they had the additional problem of forming a coalition government, first with New Democracy, and then with both the Tories and Pasok (a government of national unity). This period saw new defeats and a new split, and since 1991-2 we have had two wings of the traditional left—the Communist Party, which still defends the Soviet Union, and Synaspismos, which is moving towards a social democratic position. Although it comes from a Eurocommunist background, Synaspismos is a party with a social democratic perspective.

You say they have a social democratic perspective, but don't they also use leftist language?

Yes, they do. The past 15 years have seen ups and downs. There have been three changes of leadership for Synaspismos. The leadership of 1993 now belongs to Pasok—they broke from Synaspismos to become openly social democratic. Then for years there was a leadership that was middle of the road between Pasok and the far left. Now they advocate that Synaspismos should join with Pasok to form a centre-left coalition. But the current leadership of Synaspismos say that there is not enough agreement between Pasok and Synaspismos to form a centre-left government. They say they want a government of the left that would be a break with neoliberal policies. So there has been a shift to the left and an adaptation to the wave of radicalisation we have seen since 2001.

I don't think we can talk about the left in Europe today without talking about the disaster that has taken place in Italy, and the disaster was very much the result of a party that appeared to be far left joining a centre-left government. Are you saying that Synaspismos would like to do this, but under pressure it has not been able to?
Well, Synaspismos is part of the European Party of the Left, which was dominated by the Italian party, Rifondazione. They share the same perspectives, and so forming a centre-left alliance in Greece is on the cards. Pasok cannot get a majority on its own. It needs a partner, and Synaspismos is the obvious partner. That puts a lot of pressure on Synaspismos.

Synaspismos's problem is that this scenario is undermined from two sides. On one side, the Pasok leadership has refused to budge from its neoliberal programme. On the other there is a left—which exists to the left of Synaspismos—that puts a lot of pressure on Pasok. The leadership realise that if they go for an agreement with Pasok at this stage there is the Communist Party, which is to the left of them, and there is the whole radical left milieu, which operates independently both of the Communist Party and of Synaspismos. So they are under pressure from the left.

This seems to me incredibly important. Synaspismos is forced at the moment to act as a focus to the left for fear of the real left.
To get some idea of the radicalisation we have to look at the students. The occupations were run by coordinating committees that were elected in general assemblies in the faculties. In most of the faculties the leadership of these coordinating committees was to the left of the Communist Party and of Synaspismos. That gives an indication of the extent of the radical left milieu that exists in Greece.

There are similar developments within the strikes, although not to the same extent. We have not yet had any strikes that were run from below, with coordinating committees like the students. But the presence of a left milieu among striking workers is fairly visible when it comes to the demonstrations, the picket lines and so on. So, for instance, on the demonstrations, the Communist Party marches separately from the unions. The main bulk of demonstrators may be influenced by the Pasok leadership of the unions, but there are whole sections of the strike rallies where the far left dominates, with its own banners, its own slogans and so on. There are real pressures to the left of Synaspismos.

You mentioned that Synaspismos has gained in the opinion polls during the recent mobilisations.

In the most recent election, in 2007, the Tories lost votes but kept their overall majority. Pasok also lost votes. Those gaining votes were on the left. The Communist Party climbed from 6 percent to 8 percent. Synaspismos climbed from 3 percent to 5 percent. So the combined vote of the left went into double digits for the first time since the 1980s.

After the election people realised that there was a swing to the left, and opinion polls have shown this gathering pace. The main beneficiary has been Synaspismos, which now stands at 15 percent in the polls. The Communist Party has also made further gains. The reason why Synaspismos has been the main beneficiary is that it is not sectarian like the Communist Party—it is much more open. The Communist Party will never work together with Pasok, for instance, and has created barriers for people to move from Pasok to the Communist Party. Synaspismos has also been open to an alliance with groups of the far left. It has created a coalition with the far left, and this is an important factor.

Are people in general aware that there is a political alternative to the left of Synaspismos that is not represented in parliament?

In terms of the parliamentary balance of forces, the far left is barely visible. But this is not the real balance of forces when it comes to influence on various movements, and so on. The existence of a radical left to the left of the parliamentary left is something that can be measured through what happens when the students move, what happens when there are strikes—and this makes an impact on people's consciousness. There are not just two lefts in Greece; there are three lefts. There are the two components of the divided parliamentary left and there is an extra-parliamentary left. People associate this with the 1973 Polytechnic Uprising, which spelt the end of the military dictatorship. The existence of an extra-parliamentary left has been a fact of Greek political life since 1974.

The important thing was that the initiative for the Polytechnic Uprising was taken by the extra-parliamentary left outside the Communist Party. In 1974, 1975 and 1976, when there were massive waves of strikes, people felt that those involved in the Polytechnic Uprising were crucial to all these struggles. So when you talk about an "extra-parliamentary left" in Greece, it is not a derogatory term. It has the aura of the struggle of the 1970s.

What lessons does what is happening in Greece have for the left elsewhere in Europe?

We have to start from Italy to talk about the importance of the Greek experience. What happened in Italy is crucial. The development of radicalisation since the 1999 Seattle protest has gone through different phases. First, people discovered their strength. Then there was the next phase when people discovered the importance of a political expression for the movements, and on the basis of this there was a whole resurgence of the left. The experience of Italy has brought this radicalisation to a new level where the debate over whether we can have a left that is non-sectarian but at the same time does not collapse into centre-left politics is coming to the forefront for the first time in many, many years.

Obviously this debate is raging in Italy, but it is also relevant to France. It may become relevant to Germany as the left advances there—the SPD may try to trap the left into joining a centre-left coalition. This kind of debate is now happening in Greece. And because of the radicalisation and because of the growth of a left very strongly opposed to a centre-left project, what happens now will affect developments in other countries as well. If we have a repeat of the Italian experience in Greece, that will demoralise people throughout Europe. If, however, we are successful in building a radical left that avoids the centre-left trap, it will create a whole new opening for the left everywhere in Europe.

One last question. I have to go back to when the Tories, the New Democracy, won in Greece in 2004. Did that create demoralisation on the Greek left? Because in Britain much of the left is saying since the Tory victory in the local elections and in London, "Everything is moving to the right." Was that the feeling in Greece?

That was the first political debate after the election. People who had been out fighting the previous Pasok government were bewildered by the Tory win. There were two strands of opinion. One was that people were shifting to the right, that the struggles had been transient, without producing any political effect, and that in reality workers were moving to the right. This was used mainly by Pasok as way of explaining their defeat.

Then, on the left, there was another element in the debate. There were illusions that the Tories had changed. There were people on the left who genuinely believed that the Tories had moved to the left of Pasok. We argued against both these currents, saying that there may be a temporary lull in the struggle, but the experience was there, and we would see the radicalisation coming to the fore quite soon. At the same time we said there should be no illusion about the Tories changing their spots.

The first issue of our paper after the election had the headline, "Confrontation With The Tories From Day One!" It looked far-fetched for a few months, but then it became common wisdom and now everyone accepts that was the perspective that should have been in place.

Livingstone pays the price of "triangulation"

Charlie Kimber

In broad terms the story is easily told. Labour's Ken Livingstone was defeated as Mayor of London by Conservative Boris Johnson because the Labour Party is on the slide and the right in British politics has got its act together.

Livingstone's percentage of the first preference votes for mayor was virtually unchanged compared to 2004. But the Conservative share was up more than 14 percent.[1]

The central issue was that Livingstone was a Labour candidate—an unashamed and proud Labour candidate—at a time when Labour was increasingly unpopular. He therefore could not motivate a real movement to beat off the threat from the right. In 2000 hundreds of thousands of Londoners believed the way to hit New Labour was to vote *for* Livingstone, and they voted for him with some enthusiasm. In 2008 they thought the way to hit New Labour was not to vote for Livingstone, and either stayed at home or voted for somebody else.

In 2000 he was widely seen as the radical who Labour had expelled. The result was first preference scores of 39 percent for Livingstone, 27 percent for the Conservative's candidate, 13 percent for Labour and 12 percent for the Liberal Democrats. The final total after transfers was Livingstone 58 percent, the Conservative 42 percent.

By 2004 the warning signs were clear. Having rejoined Labour, and

---

1: Results are available from the London Elects website at http://results.londonelects.org.uk/

with Labour on a sharp decline as Tony Blair waged war on Iraq and pushed through measures that attacked working class people, Livingstone was far less popular. In 2000 the combined first preference vote for Livingstone and the Labour candidate was 52 percent. Four years later Livingstone as Labour candidate could secure only 36 percent. The final totals were 55 percent for Labour and 45 percent for the Conservatives.

So there was always likely to be a close contest in 2008, unless the replacement of Blair by Gordon Brown in June 2007 had decisively revived Labour's fortunes.

For a long period Livingstone was ahead in the polls. In November 2007 he headed Johnson by 45 percent to 39 percent. But even then the detailed results showed that key elements of Livingstone's policies were not shared by his supporters.[2] For instance, a third of those who said they were going to back Livingstone also said that Sir Ian Blair should have resigned as Metropolitan Police commissioner over the police killing of Jean Charles de Menezes. And a majority (46 percent to 41 percent) of Liberal Democrat voters also agreed that Ian Blair should have gone. Yet Livingstone, of course, had steadfastly backed the commissioner.

On 24 January 2008 Livingstone was still ahead in the polls by 44 percent to 40 percent,[3] and he was still four points ahead on 12 February. It was not until 21 February that Johnson moved to the front. After that Livingstone was behind Johnson in almost every poll.

These trends were essentially a delayed version of the general Tory-Labour poll ratings.[4] Hard though it may be to remember, Labour was ahead of the Tories in polls from June 2007 to October 2007 (the "Brown bounce"). At one stage the lead grew to 13 percent. Since 24 October 2007 virtually every poll has put the Tories ahead.

Being associated with Brown and the Labour government became a drain on any candidate, not a benefit. And yet Livingstone never appeared to recognise this. Despite the growing evidence that voters were turning ever more sharply against neoliberal policies as the economic crisis grew, Livingstone stuck doggedly to the very New Labour strategy of "partnership" with big business and the City of London. He also embraced Gordon Brown himself, just as most people were rejecting him. On 20 March the two toured East London, with Brown, who had repeatedly and publicly clashed with him in the past, now describing

---

2: YouGov poll, available from http://tinyurl.com/4mrl9h
3: YouGov poll, available from http://tinyurl.com/5y7lld
4: See http://ukpollingreport.co.uk/blog/voting-intention

Livingstone as "inspirational".[5] Even more disastrously, on 24 April the *Guardian* revealed:

> Tony Blair and Alastair Campbell have both been giving advice to the campaign working for London mayor Ken Livingstone's re-election. The former prime minister and his media strategist had been among Livingstone's most trenchant critics in the past, and he had derided them for being the architects of the New Labour project. However, with the race against the Tory candidate Boris Johnson on a knife-edge, Livingstone's team has sought their expertise, and also the advice of Philip Gould, New Labour's pollster and focus group adviser.[6]

What gems were the war criminal and his court dispensing? Blair's advice was that "Livingstone cannot win solely on the basis of his record, and must be unambiguous that he will continue to attract private sector investment to the capital"—precisely the wrong direction to go in.

The article added, "Livingstone campaign sources said the 10p tax rate abolition was the only national issue intruding into the mayoral campaign and they were relieved that Brown yesterday defused the issue by promising to recompense most losers this year." This delusion—that the issue that had come to focus the anger with Labour's pro-business policies had somehow gone away, or that the backlash against it would not stick to Livingstone—was to be rudely shattered a week later.

Perhaps aware of the danger of aligning too closely with such unpopular duds as Blair and Brown, a Livingstone source, quoted in the article, said:

> Ken welcomes the support he has had in his campaign from anyone including Alastair Campbell, Phillip Gould, and particularly the strong support he has had from government ministers. At the same time, you only have to see the importance his campaign gives to issues like opposing the war in Iraq and bringing the maintenance of the Tube back into public ownership to understand the character of his campaign.

Which might have been interesting, if Iraq or Metronet's fate had featured at all prominently in Livingstone's propaganda.

5: "Livingstone Inspirational Says PM", BBC News, 20 March 2008, http://news.bbc.co.uk/1/hi/uk_politics/7307961.stm

6: "Livingstone's Unlikely Secret Weapons: Tony Blair And Alastair Campbell", *Guardian*, 24 April 2008, www.guardian.co.uk/politics/2008/apr/24/london08.london1

Throughout the campaign Livingstone tried to hold together a creaky alliance which stretched from big business and London's financiers to sections of the "progressive left" and George Galloway MP. This was not a wholly new idea; indeed in 2004 he had already won the endorsement of the *Economist*, a business magazine. And he pursued a pro-business strategy ruthlessly in his second term. In an interview in April 2007 Livingstone said:

> There isn't a great ideological conflict any more. The business community, for example, has been almost depoliticised. One of the first people to lobby me when I became mayor was Judith Mayhew, from the City Corporation. She came and said, "We've all changed; it won't be like the last time; there's so much we can do together." I didn't believe a word of it, but it turned out to be true.[7]

Livingstone's love affair with the City was such that he even found himself to the right of New Labour. When chancellor Alistair Darling proposed some paltry taxation of the "non-doms"—people who live in the capital but are registered abroad and pay no tax—Livingstone waded in to support the billionaires. He "hit out at Darling's plans warning it could drive investment away from London. He argued the government could 'not afford to get it wrong'. In a speech at 'The Global Capital' conference the mayor said concessions made so far by Mr Darling had not gone far enough to satisfy London's financial sector".[8]

So strong was this pitch that a few days before the election the *Financial Times* wrote:

> Livingstone has won the tacit support of the City in his bid to be re-elected as London mayor. Leading business organisations, while stressing their apolitical nature, praise the Labour mayor's "good track record" in running the capital. The City's backing for Mr Livingstone might appear counter-intuitive, given his "loony left" tabloid characterisation during his 1980s leadership of the Greater London Council. However, his actions as mayor have not conformed to this simple stereotyping—witness, for example, Mr Livingstone's recent criticism of the levy imposed by the government on wealthy non-domiciled foreigners in the City. "Big business, big developers, see Ken as a relatively safe bet," said

7: Parker, Goodhart and Travers, 2007.

8: "Axe Non-dom Tax Or City Will Suffer Says Livingstone", *Evening Standard*, 22 February 2008, available from http://tinyurl.com/5f26au

Tony Travers, director of the greater London group at the London School of Economics. "Ken's vision of urban priorities is in its way Thatcherite—he's an über-Blairite who believes in London's rapid development [with] lots of tall buildings, and business success." The CBI employers' organisation praised Mr Livingstone's "good track record" with business.[9]

Bizarrely Livingstone continued to boast of his business support even *after* his defeat. He wrote in the *Guardian*:

> Labour's campaign in London gained major support from business. The *Financial Times* concluded that the majority of big business in London supported my re-election. There is no way to check that, but I know from meetings that very large sections of big business supported my campaign.[10]

Throughout his term of office Livingstone has tried to use a version of New Labour's "triangulation" strategy. This means claiming to stand above and between the "old left-right divisions" and stealing the ideas of your opponents in order to prevent them attacking you. When the left adopts such a strategy it means adopting pro-capitalist policies, to disastrous effect. It also doesn't work. The assumption that working class people will always vote overwhelmingly for the traditional left because they have nowhere else to go ignores the fact that they can always not vote at all, or vote for someone else.

It is, of course, important to look at the right as well as the left. One factor often ignored is that disarray on the right has previously masked Labour's decline. Across Britain Labour lost four million votes between the general elections of 1997 and 2005. Normally that would mean electoral defeat. But Labour hung on to office because the Tory vote also declined by a million from its appalling result in 1997. But now the right is more organised. Tory leader David Cameron has also been lucky that competitors, such as the anti-Europe, anti-immigration United Kingdom Independence Party (UKIP), which took nearly 10 percent of the vote and won two seats in the 2004 London elections, have imploded. Many of their votes will have gone to Boris Johnson.

Johnson also benefited from a slew of articles in the *Evening Standard* attacking Livingstone. Night after night the paper dredged up material which claimed to show corruption in City Hall. Much of this was linked to

9: "Livingstone Wins City Support For Re-election", *Financial Times*, 28 April 2008.
10: Ken Livingstone, "Yes, I Lost. But Still Labour Must Learn From London", *Guardian*, 9 May 2008, www.guardian.co.uk/commentisfree/2008/may/09/livingstone.boris

questions of race. The paper alleged, among other things, that the London Development Agency may have been intimidated by threats of gang violence by black people into giving out grants to organisations in which black people were the majority.

It alleged that an organisation in Brixton, for which London Development Agency funds had been given, was "a vibrant hub for criminals". The mayor's equalities adviser, Lee Jasper, was vilified over corruption allegations, even though he has never been charged with an offence.[11]

And then on 16 April its headline read "Suicide Bomb Backer Runs Ken's Campaign", based on a claim that "an advocate of suicide bombing is among leaders of a group trying to mobilise Muslim voters to back Ken Livingstone... For the past year, the group has been working on a strategy to win an estimated 200,000 Muslim votes in an effort to re-elect the mayor. It is being waged by Muslims 4 Ken, led by 39 year old lecturer Anas Altikriti and Palestinian-born Azzam Tamimi, a supporter of Hamas, the militant group dedicated to the creation of an Islamic state of Palestine".

Elsewhere *New Statesman* editor Martin Bright presented a Channel 4 *Dispatches* television programme, charging the mayor with (as the *Guardian* put it) "financial profligacy, cronyism and links to a Trotskyite faction conspiring to transform London into a 'socialist city state'." Bright wrote a piece in the *Evening Standard* saying, "I now believe Ken is a disgrace to his office. I feel it is my duty to warn the London electorate that a vote for Livingstone is a vote for a bully and a coward who is not worthy to lead this great city of ours."

However nasty these attacks may have been, the left is never going to get very far if it is surprised or demobilised by an assault from such sources. The key question is to rally working people to defeat such assaults. This Livingstone failed to do. This is hardly surprising as his policies have hit working class people across London. Livingstone told union members on the tube to cross picket lines during a strike, failed to use planning controls to promote genuinely affordable housing and presided over a transport system that penalised workers.[12]

Overall there was an average swing of 7.4 percent (on first preferences) across London from Labour to Tory compared with 2004. But this was far from uniform. Most boroughs and wards where the Conservatives are strong in local government recorded big swings to Johnson. In Havering

11: Having resigned from his post on 4 March 2008 after the publication of a sexually explicit email exchange, Lee Jasper has not, at the time of writing, faced any charges.
12: For a full analysis, see Kimber, 2007.

it was 15 percent, in Bexley 14.9 percent, in Bromley 14.2 percent, in Hillingdon 12.1 percent, in Kensington & Chelsea 10.7 percent, and in Wandsworth 10 percent. In general Conservative voters turned out enthusiastically for Johnson and were motivated by his campaign.

This was not true of Labour's efforts. In Brent the swing to Johnson was kept down to 1.3 percent, in Hackney, 1.4 percent, in Waltham Forest, 2.6 percent, in Haringey, 3.6 percent and in Harrow, 4.6 percent. But why were there swings against Livingstone at all in these areas? This was a high profile election where all the focus was on the two major candidates (making it very hard to persuade voters to vote for lesser-known candidates such as those from the Left List). These are working class areas with large numbers of black and Asian voters. Far from the Tories closing the gap, Livingstone should have been able to motivate a swing towards him. His closeness to Labour prevented him doing so. He didn't head up opposition to the removal of the 10p tax rate, he didn't make declining living standards an issue, he didn't champion the six million public sector workers who face Brown's pay cuts, and he didn't visit picket lines or the magnificent demonstration through London on 24 April when teachers, lecturers, and civil service workers struck (indeed I can find absolutely no comment he made about the strikes).

When, as the election campaign began, Labour announced the closure of 171 post offices in London the Tories and the Liberal Democrats took to the streets organising local protests, but Livingstone merely called for a judicial review of the consultation period (and nothing seems to have happened even on that).

There was absolutely no class campaign, no hounding of Johnson for his racist comments in the past, no appeal for activity and action to defend workers' living standards. And that could not happen because Livingstone remained imprisoned in the alliance with the rich and powerful he and Labour had created.

Given Livingstone's defeat, some on the left have reiterated the argument that it was wrong to stand against him. Strong pressure was applied against the Left List for standing Lindsey German as a radical alternative to Labour. George Galloway, for example, argued, "It would be self-indulgence, a luxury the left can no longer afford, to stand a candidate of the left against Livingstone for mayor...a left candidate opposing Livingstone really could aid the Tories and risk handing the keys to City Hall to the rancid reactionaries around Johnson".[13]

13: George Galloway, "Why I Back Red Ken", *Guardian*, 25 January 2008, www.guardian.co.uk/commentisfree/2008/jan/25/whyibackredken

There is no evidence that this happened. Instead the campaign made more people aware of the election and generally raised the left vote. And in any case the London mayor is elected on a supplementary vote system where you can transfer your vote if your first choice is eliminated. The Left List called on its supporters to transfer to Livingstone.

But there is a much bigger question here. Is the left going to be held captive by the threat from Labour that "you will let in the right"? Livingstone spoke out against the Iraq war, and has opposed Islamophobia. He has strongly supported anti-racism. But so have some Liberals. And Livingstone has also backed the City rich, told workers to cross picket lines, backed the killers of Jean Charles de Menezes and called for more police.

The left had to stand in such circumstances.

With Johnson in City Hall and the Tories resurgent, everyone on the left must unite in action to support every workers' struggle, to fight the British National Party, to defend services and jobs, to combat racism and Islamophobia, to keep up and extend anti-war agitation and to strengthen trade unions.

That will mean socialists of all types working together, and left of Labour supporters working with those who supported Livingstone. But while uniting in this way it is also necessary to discuss the lessons of Livingstone's defeat. It is a confirmation of the failure of wooing the right and of trying to appease big business. It is another argument for the need to build a stronger political alternative to Labour.

References

Kimber, Charlie, 2007, "Ken Livingstone: The Last Reformist?", *International Socialism 113* (winter 2007) www.isj.org.uk/index.php4?id=289

Parker, Simon, David Goodhart and Tony Travers, "Interview: Ken Livingstone", *Prospect*, April 2007, www.prospect-magazine.co.uk/article_details.php?id=8636

Behind the world food crisis

Carlo Morelli

The ancestor of modern cheap food policies, the repeal of the Corn Laws in Britain in 1846, was spurred by growing unrest within the swelling cities of an industrialising economy. In 1848 that unrest burst out into a series of revolutions across western Europe.

It is not often that the *Financial Times* warns of impending revolution unless governments act to stem the consequences of untrammelled market capitalism, but that is exactly what Alan Beattie argued in April 2008.[1] The leaders of the World Trade Organisation, the International Monetary Fund and prime minister Gordon Brown, to name but a few, have echoed such comments.

They have much to warn about. By early April 2008 food protests had already occurred in Mexico, Haiti, Ivory Coast, Guinea, Senegal, Mauritania, Morocco, Egypt, Yemen, Mozambique, India and Indonesia. Kazakhstan, Ukraine, Russia, China, India, Vietnam, Cambodia and Argentina had all imposed restrictions upon exports. Even large US retailers had begun to ration rice. Rapidly rising food prices are destabilising much of the developing and developed world, impacting upon billions of the world's population.

1: Alan Beattie, "Governments Can No Longer Ignore The Cries Of The Hungry", *Financial Times*, 5 April 2008.

Figure 1: Food price index
Source: Food and Agriculture Organisation (FAO) of the United Nations

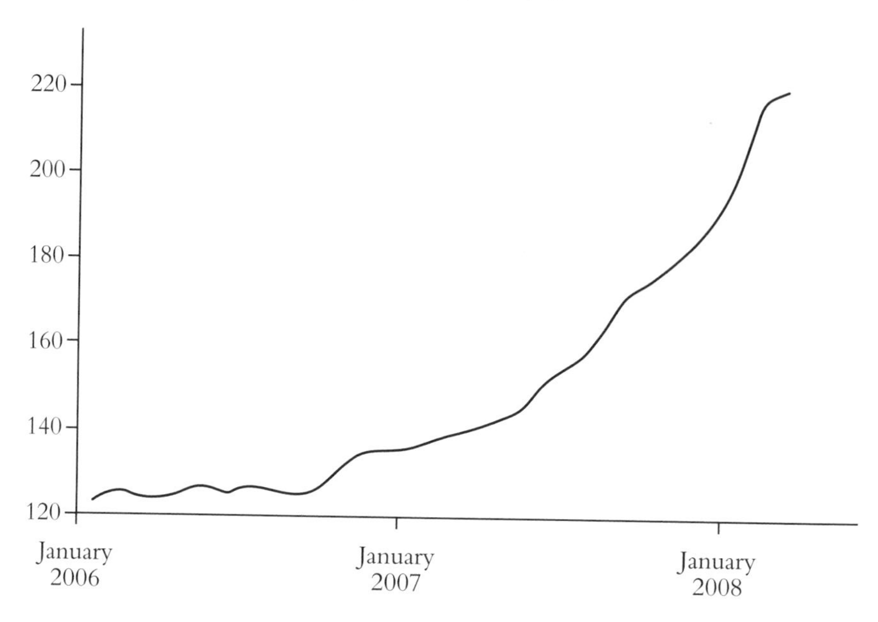

Figure 1 shows just how rapidly food prices have risen. The upward pressure on prices is likely to increase. The UN's Food and Agricultural Organisation says that the world's food import bill will surge above $1 trillion this year for the first time, an increase of about 20 percent from 2007. The food import bill of the poorest countries, those facing a food deficit, is likely to climb to $169 billion—a 40 percent increase. And "prices are unlikely to return to the low levels of previous years".[2]

Short-term causes

Four key factors caused the rising prices of the past 18 months:

- Crop yields have fallen in Australia due to lack of rainfall and in northern Europe because of too much. This suggests that climate change is having an impact on the geography of food production.
- Much of the US's maize crop has been turned over to biofuel production, encouraged by government subsidies. This alone is responsible for about a third of the grain price increase.

2: "Food Outlook", FAO, May 2008, www.fao.org/docrep/010/ai466e/ai466e00.htm

• Increasing meat consumption by the middle classes in China and India is pushing up the demand for grain to feed animals.
• Every increase in oil and gas prices raises farming costs, since it raises the cost of transport, using mechanised equipment, and of pesticides and nitrogen-based fertilisers (which are produced using energy intensive processes).[3]

The failure of the global food production system to respond to these short-term problems points to a deepening long-term fault within it. Throughout the 20th century there was always a surplus of food worldwide; famines resulted from failures of distribution, not production. But today "global grain inventories...are at 40-year lows, equivalent to just 15 to 20 percent of annual demand".[4] This means that harvest failures due to climatic conditions in one region of the world can be disastrous globally. We are now faced with a problem in production, not just distribution. This has its origins in the organisation of capitalist markets.

How the market threatens food security

Rising food prices have been an explicit goal, not an unwelcome by-product, of the wave of agricultural trade "liberalisation" of the past two decades. The IMF estimated in 2006 that import costs for net food importers would increase by "$300 million to $1.25 billion, depending on the degree of liberalisation".[5] The World Bank estimated that full liberalisation would raise international commodity prices by an average of 5.5 percent.[6]

A key element in the Structural Adjustment Programmes (recently renamed Poverty Reduction Programmes) imposed by the IMF across the developing world since the 1980s has been to shift these countries towards the production of high value cash crops to trade on world markets. Developing countries are then supposed to be able to increase living standards by using the earnings of trade to purchase other foodstuffs cheaply.

By the late 1990s over 80 developing countries had implemented such programmes.[7] Coffee was one "high value" crop. By 2000 it had spread to over 80 countries, covering over 100,000 square kilometres

3: See Walker, 2006, for a detailed account of the connections between energy and fertiliser costs.
4: "Financial System Faces Commodity-led Crisis", *Financial Times*, 5 March 2008.
5: IMF, 2006, chapter five, box 5.2.
6: World Bank, 2007, p11.
7: Madeley, 2000, p58.

with an output of 5.7 million tonnes a year.[8] The same approach was pursued with, for example, tobacco in Malawi, oranges in Brazil, cotton in Burkina Faso and cut flowers and fresh vegetables in Kenya. In each case cash crops replaced foodstuffs for the local population on large swathes of land.

The economic theory of "comparative advantage" underpinned this approach. "Gains from trade" were supposed to arise as each economy specialised its production processes. However, the theory ignored how markets really operate. Joseph Stiglitz, former chief economist at the World Bank, criticised the IMF's approach as "based on the outworn presumption that markets, by themselves, lead to efficient outcomes".[9]

Dominant organisations are able to use their market power to capture the "gains from trade" to the detriment of the weaker parties. One way for dominant firms to do this is to ensure that they maintain strict control over supply chains. So just six firms controlled over 84 percent of Kenya's fresh vegetable exports by the end of the 1990s.[10] The five large multiple retailers in the UK dictated stricter delivery schedules, greater product uniformity and higher levels of packaging—conditions that only larger, more capital intensive organisations were capable of satisfying. Those at the bottom of the supply chain—the 15,000 small producers, and below them the hundreds of thousands of subsistence farmers and landless labourers who provided vegetables to the exporters—received barely subsistence levels of payment for their labour. Capitalist accumulation within the global food industry involved the extension and intensification of class-based exploitation strategies across the globe.

Across the developing world the rapid expansion of cash crops led, for many years, to an over-supply of commodities and wildly fluctuating prices, impoverishing medium and small farmers. Coffee output, for example, increased over the two decades up to 2004 twice as fast as consumption. As a result, overall coffee prices fell by two thirds. Similar developments led to the prices for all agricultural exports remaining low into the late 1990s.[11] Agricultural markets in cash crops have been typified by anarchic, uncoordinated boom and bust investment patterns. The beneficiaries have been those firms higher up the supply chain who could begin to dominate commodity markets, determine prices and make big

8: Hellin and Higman, 2003, p36.
9: Stiglitz, 2002, p xii.
10: Dolan and Humprey, 2004.
11: Hellin and Higman, 2003, pp36-37.

profits. Wal-Mart, Tesco, Nestlé, Monsanto and Starbucks all demonstrate how impoverishment was good for business in the 1990s.

Eventually it was recognised that the IMF's approach failed to lead to development, and increased rather than decreased developing countries' dependence upon food aid and borrowing from the World Bank. However, the dominance of big business interests ensured that the blame was placed on continued restrictions within agricultural markets, rather than the liberalisation policies. The outcome was the "Agreement on Agriculture", which emerged, along with the World Trade Organisation (WTO), out of the Uruguay Round of General Agreement on Tariffs and Trade negotiations in 1992. It sought further to intensify the process of market liberalisation by reducing government subsidies. This was intended to permit "market signals" to be transmitted to producers still more strongly.[12]

One response to the current crisis in food prices has been to suggest that yet more market liberalisation is the solution. However, it is not the inability of the developing world's economies to respond to market signals that is the problem; it is the role of the market itself in creating instability.

The capitalist road to greater poverty

As we have seen, trade liberalisation was meant to increase food prices, not decrease them. It also sought to encourage increased investment through increased farm size, as large producers captured the "gains from trade" that comparative advantage was supposed to bring. The liberalisation approach assumes that the numbers of rural poor can only be reduced by transferring most of them out of agriculture and into urban manufacturing. The real goal, then, is the destruction of rural communities and the development of large-scale capitalist agriculture. According to the World Bank's "Agriculture for Development" report:

> Agricultural growth was the precursor to the industrial revolutions that spread across the temperate world from England in the mid-18th century to Japan in the late 19th century. More recently, rapid agricultural growth in China, India and Vietnam was the precursor to the rise of industry.[13]

Just as the Industrial Revolution depended on people being evicted from agriculture, with the enclosure movement in England and the clearances in Scotland, so the developing world's landless labourers,

---

12: Morelli, 2003.
13: World Bank, 2007.

subsistence and small farmers are viewed as a hindrance to economic development in the Global South. Estimates by the UN in 2000 suggested that between 20 and 30 million people had been driven from rural areas as a result of agricultural liberalisation policies.[14] Development is understood within a narrow capitalist framework of the exploitation of wage labour and the impoverishment of the mass of the population, with a tiny minority, the ruling class within the developing world, benefiting from extreme polarisation of wealth. Such a framework ignores sustainability in terms both of the world's resources and the impact on the world's poor.

Even within the logic of capitalism, the primary reason for the failure of developing countries to trade with the developed world does not lie with the structure of agricultural subsidies and support in the developing world. Rather the blame lies with restrictions on exports from the developing world into the developed world. It is in the European Union and the US that subsidies are the most significant. Here subsidies to large-scale producers have existed for over half a century, with direct payments to farmers reaching $235 billion dollars in the developed world by 2003.[15] Agriculture became more, not less, protected in the developed world at exactly the time when the developing world was being forced to become more liberalised.

Even where developing countries have maintained their levels of trade, this has not alleviated poverty. From 1990 to 2005, a period in which world trade increased by a notional 200 percent, low income countries were successful at maintaining their share of world exports in agricultural products at 15 percent. Even Africa, a continent whose economic development has fallen behind much of the developing world, maintained its share of world exports at 3.4 percent. But while exports of meat products, horticulture and oil seeds from developing countries have more than trebled, domestic consumption of cereals and staple crops has remained virtually static since the 1980s, and developing country food imports have risen from 7 to 11 percent of world trade as it became the dumping ground for subsided exports from developed countries.[16]

Food security, the ability of governments to protect domestic production of staple foodstuffs, has been increasingly undermined, leaving the developing world's poor much more vulnerable to the changes in world food prices that we are now seeing.

14: Madeley, 2000, p75.
15: Nash, 2004, p34.
16: World Bank, 2007, figure six.

The problem of the food market is not the failure of market signals to operate on developing countries but the anarchic results than flow from such signals. Market signals provide information about current prices, but do not provide a mechanism for coordinating or planning for the future. When food prices were low there was little possibility of investment by small and medium farmers, who are the key to food security in most of the developing world. Now, with rising prices, things are no better. The spread of marketisation means that farmers are themselves dependent on buying certain foodstuffs. In addition, they are hit by the rising price of agricultural inputs—fertilisers, pesticides and fuel:

> While the price of rice has doubled in recent months, most farmers are benefiting very little—even in Vietnam, the world's second biggest rice exporter. Their revenue has increased, but so too have their input costs—especially fertiliser, closely linked to the price of energy. Fuel, required for pumping water to their rice paddies and transporting their harvest, is another fast-rising cost. In interviews across Vietnam, rice farmers unanimously reported that their costs have nearly doubled since last year, leaving them without any increase in income, despite the surging rice prices in domestic and global markets.[17]

Whether food prices are high or low, the world's hundreds of millions of small farmers are doomed to suffer under the current system.

Speculation and prices

The suffering both of small farmers and of those who depend on the food they produce is being exacerbated by one other factor—speculation.

Commodity markets for food products now operate in exactly the same way as other markets. Each commodity group has exchanges for futures contracts, hedging and price guarantees. Market intermediaries, large-scale buying organisations and dominant firms utilise forecasts to minimise their exposure to risk, and this is supposed to provide a mechanism for stabilising prices. However, commodity exchanges do something else as well. They permit speculation rather than productive investment. Speculation is the ability to recognise market trends and use short-term buying and selling to generate high profits in specific markets at particular points in time. Such speculative investment can have profoundly destabilising effects on markets.

17: "Vietnam's Farmers Face Paradox Of The Paddy", *Globe and Mail*, 1 May 2008. See also "Rice Prices Fail To Benefit Asian farmers", *Financial Times*, 29 April 2008.

The collapse of the dotcom bubble, and then the housing and banking bubbles have left those wishing to invest in short-term markets with fewer opportunities for speculative investments. They have been turning their attention to commodities markets, including food, buying up future supplies on the assumption that their prices will rise and, by doing so, pushing those prices up even more. As the *Financial Times* reports:

> Institutional and retail investment in commodities has shot up, lured by research that shows commodities can diversify a portfolio, and by "performance-chasing"; money pours into sectors that have made money. According to Philip Verleger, a consultant, the amount invested that way has multiplied by five, to about $250 billion, in just three years.[18]

> The volume of speculation has been rising steeply. Since 2003, says Morgan Stanley, open interest in corn futures has risen from 500,000 contracts to almost 2.5 million... More speculative money will go in, and will play a larger role in the rise itself. And, of course, more people in the poorest countries will starve.[19]

The IMF claims that, while speculative investment has not caused prices to rise, once they do, speculative investment rapidly flows into the market.[20] The recent nature of this speculation can be seen in Table 1.

Table 1: Selected commodity prices (2005=100)
Source: IMF

Commodity group	1998	2005	March 2008 (estimated)
All primary commodities	48	100	181
Oil	25	100	191
Food	92	100	170

While average prices for all primary commodities more than doubled between 1998 and 2005, food commodities only rose marginally, from

18: "Classic Films Shed Light On Commodities Boom", *Financial Times*, 9 May 2008.
19: "Speculators Feast On Soaring Commodities", *Financial Times*, 12 May 2008.
20: It should be added that the IMF's research examined historical moves in prices, not those that have taken place so rapidly over the past year, so even their limited rejection of speculation may no longer hold.

92.0 to 100, but since 2005 food prices have risen much more rapidly than previously. However, it is only by understanding the more deep-rooted changes that have occurred in world food systems that we can understand why speculation has had such a devastating impact.

Destructive scenarios

If the cause of the current food crisis is the anarchy brought about by market mechanisms, what are the long-term prospects?

There are two widely held views. The first is that the present upsurge in food prices is similar to the last great upsurge in 1972-3. On that occasion there was a sudden increase in demand for grain imports due to a short-lived world boom, which coincided with the sudden entry of the USSR into the world market for grain following problems with its own harvests. The price upsurge contributed to inflation worldwide, but subsided by the late 1970s, with food prices then falling over the long term compared to other prices. On this view, the present price surge may well be disastrous and lead to up to 100 million people starving due to the inability of world agencies to take counter-measures. But the disaster would only last a few years.

The second view is much more catastrophist.[21] It holds that the present crisis signals the exhaustion of the new ways of increasing crop yields that overcame the food price crisis of the 1970s. These ways, usually referred to as the Green Revolution, were based on the use of new, more productive grain varieties, combined with very big increases in fertiliser and water pump use. Today China uses 40 million tons of fertiliser a year, the US 19 million tons and India 16 million tons. As a result world grain production in 2004, at about two billion tonnes, was three times higher than in 1950s, with worldwide average grain consumption per head up 24 percent, despite population growth. This has banished for more than a generation the spectre of famine and death from starvation (as opposed to widespread malnutrition) from the two Asian giants—India and China.

But the Green Revolution has now reached a "plateau", according to the 2008 report from the International Assessment of Agricultural Knowledge, Science and Technology for Development (IAASTD).[22] World acreage devoted to food is not growing and grain output is only just keeping abreast with population growth in Asia, while the situation in Africa is even worse.

21: For an extreme version of this view, see Pfeiffer, 2004, which was circulated by the student campaign network People and Planet among others.

22: Feldman, Nathan, Raina and Yang, 2008.

The World Bank's latest "World Development Report" admits that conventional strategies of "development" are failing in this respect:

> Many agriculture-based countries still display anaemic per capita agricultural growth and little structural transformation... The same applies to vast areas within countries of all types. Rapid population growth, declining farm size, falling soil fertility, and missed opportunities for income diversification and migration create distress as the powers of agriculture for development remain low. Policies that excessively tax agriculture and underinvest in agriculture are to blame, reflecting a political economy in which urban interests have the upper hand. Compared with successful transforming countries when they still had a high share of agriculture in GDP, the agriculture based countries have very low public spending in agriculture as a share of their agricultural GDP (4 percent in the agriculture-based countries in 2004 compared with 10 percent in 1980 in the transforming countries).[23]

The chances of the catastrophist scenario materialising could be increased enormously by climate change, which is likely to hit food production in the most populated parts of the world. Modelling climate change is fraught with problems, and linking this to agriculture, with assumptions regarding future land use, output growth, consumption patterns and population change, is still more problematic. Nevertheless, modelling suggests that the Intergovernmental Panel on Climate Change's estimates of a 4°C rise could lead to a marginal fall in overall agricultural yields with a dramatic change in the distribution of this output as desertification affects regions closer to the equator. One study suggests that without climate change output of cereal production would double by 2080. With climate change this projected output would be reduced by just 0.6-0.9 percent worldwide. But while in this scenario output in much of the developed world would rise, output in Africa, East Asia and South Asia would fall, by perhaps 20 percent in the case of South Asia.[24] It is possible to dispute the accuracy of the particular models used in studies, but in all likelihood the costs of climate change will fall disproportionately on the poorest of the world's population.[25]

Added to this problem is the diversion of land use for the production

23: World Bank, 2007, p7.

24: Tubiello and Fischer, 2007.

25: This conclusion is confirmed by the IMF's own research. See Cline, 2008.

of biofuel. Nothing highlights the anarchy of market mechanisms more than the movement into biofuels. Across the globe from Brazil to Indonesia rainforests have been destroyed to make way for plantations. Rainforests are burned and the soil depleted in order to produce an oil to reduce carbon use. Similarly, in the US government support for ethanol as a biofuel has ensured major subsidies to the developing industry and with them the shift of maize from food into ethanol production.

Biofuels have been promoted as a way of warding off climate change. Achieving food sustainability requires a recognition that biofuels are not a sufficient solution to climate change. Climate change cannot be addressed simply by the replacement of one form of carbon use with another, but instead by solutions which remove carbon use from society altogether. The motivation of the Bush administration in the US has not been climate change, but to protect energy supplies for US capitalism. Biofuel production is estimated to be profitable against oil-based fuel production once the oil price exceeds $60 per barrel. There is growing belief in capitalist circles that oil production worldwide has peaked and that the price is going to remain above this level. This is leading to an even greater proportion of the US maize crop being devoted to biofuel production this year than last. Overall the use of crops for biofuels is estimated to be responsible for about 30 percent of the increase in grain prices over the past five years.[26]

The high oil price then threatens a double blow to food output and prices. On the one hand, it raises the cost of fuel and fertilisers used in agriculture, in ways which make it much more difficult to increase crop yields using the methods of the past four decades. On the other hand, it diverts land from food to biofuel production. Both shifts aggravate the threat to future food supplies from climate change. Such is the terrible price humanity risks paying because of capitalism's addiction to burning carbon—even though fertiliser production itself only uses about 1 percent of global energy.[27]

The IMF suggest worldwide demand for food may triple by 2080, outstripping supply and creating chronic shortages. Although they do not say it, that would be a catastrophic outcome and an indictment of capitalist accumulation strategies as a means of feeding humanity. By contrast, Tubiello and Fischer suggest that if climate change can be ameliorated, then, even in the "worst case" scenario of a doubling of population pressures on

26: Rosegrant, 2008.
27: Schrock, 2006.

food sources, it would still be possible to provide sufficient food for a population of 12 billion people by 2080.[28]

What seems clear is that the current reliance on large-scale agriculture, the use of monocultures, oil based fertilisers and intensive use of fresh water to expand food output is increasingly unsustainable. The World Bank and IAASTD reports both half accept this. They point to the need for more agricultural investment directed towards the hundreds of millions of small labour-intensive farms, paying attention to water management, organic complements to mineral fertilisers and preventing further degradation of the soil. And they bemoan the failure of governments to direct funds in this direction. Yet their commitment to capitalism means they turn their back on their own insights, and continue to stress high value crops and the growth of large-scale capitalist farming at the expense of the rural poor and the provision of basic foodstuffs.

Under the current model of capitalist food production, the ability to supply enough food to match people's demand is increasingly under threat. If this scenario develops it will not be due to a Malthusian excess of population but to a failure of the capitalist mode of production. Capitalism has proven destructive to food production, giving rise to very specific destructive and unsustainable forms of economic development.

There is no doubt that it lies behind the current price upsurge, with the risk of tens of millions starving in a particularly gruesome version of the slump-boom cycle. And there is a strong likelihood that it is producing a long-term threat to world food security that will persist even if the current crisis eventually passes. The ability to develop a sustainable food production system requires democratic control by the world's population.

28: Cline, 2008; Tubiello and Fischer, 2007.

References

Cline, William R, 2008, "Global Warming and Agriculture", *Finance and Development*, volume 45, number 1, www.imf.org/external/pubs/ft/fandd/2008/03/cline.htm

Dolan, Catherine, and John Humprey, 2004, "Changing Governance Patterns in the Trade in Fresh Vegetables between Africa and the United Kingdom", *Environment and Planning A*, volume 36, number 3.

Feldman, Shelley, Dev Nathan, Rajeswari Raina and Hong Yang, 2008, "International Assessment of Agricultural Knowledge, Science and Technology for Development. East and South Asia and Pacific: Summary for Decision Makers", IAASTD, www.agassessment.org/docs/ESAP_SDM_220408_Final.pdf

Hellin, John, and Sophie Higman, 2003, *Feeding the Market: South American Farmers, Trade and Globalisation* (ITDG).

IMF, 2006, "World Economic Outlook" (September 2006), www.imf.org/external/pubs/ft/weo/2006/02/

Madeley John, 2000, *Hungry for Trade* (Zed).

Morelli, Carlo, 2003, "The Politics of Food", *International Socialism 101* (winter 2003), www.isj.org.uk/index.php4?id=36

Nash, John, 2004, "Agricultural Trade: Reaping a Rich Harvest from Doha", *Finance and Development*, volume 41, number 4, www.imf.org/external/pubs/ft/fandd/2004/12/pdf/picture.pdf

Pfeiffer, Dale Allen, 2004, "Eating fossil fuels", *From the Wilderness*, www.fromthewilderness.com/free/ww3/100303_eating_oil.html

Rosegrant, Mark W, 2008, "Biofuels and Grain Prices: Impacts and Policy Responses", International Food Policy Research Institute, Testimony for the US Senate Committee on Homeland Security and Governmental Affairs, 7 May 2008, www.ifpri.org/pubs/testimony/rosegrant20080507.pdf

Schrock, Richard, 2006, "Nitrogen Fix", *MIT Technology Review* (May 2006), http://www.technologyreview.com/article/16822/

Stiglitz, Joseph, 2002, *Globalisation and its Discontents* (Penguin).

Tubiello, Francesco N, and Günther Fischer, 2007, "Reducing Climate Change Impacts on Agriculture: Global and Regional Effects of Mitigation, 2000–2080", *Technological Forecasting and Social Change* 74, http://pubs.giss.nasa.gov/docs/2007/2007_Tubiello_Fischer.pdf

Walker, David, 2006, "Energy and Fertiliser Costs", *MI Prospects*, volume 8, number 19, www.hgca.com/imprima/miprospects/vol08Issue19/minisite/Vol08Issue19.pdf

World Bank, 2007, "World Development Report 2008: Agriculture for Development", http://go.worldbank.org/ZJIAOSUFU0

More than opium: Marxism and religion

John Molyneux

About 20 years ago I spoke on "Marxism and religion" at the Socialist Workers Party Easter Rally in Skegness. I began, roughly, with the words, "Today, in Britain, religion—fortunately—is not a major political issue." Unfortunately, this is no longer the case. Today religion, or rather one religion in particular, namely Islam, is at the centre of political debate.

Scarcely a day passes without a news item raising the alarm about alleged "hate preaching" imams, or a mosque being taken over by "fundamentalists", or an opinion piece about the deeply flawed nature of Islam, or a radio discussion about whether "moderate" Muslims are doing enough to combat "the extremists" and prevent Muslim youth from being "radicalised", or a TV programme on the plight of Muslim women, or a scare story about some stupidity committed in the name of Islam somewhere in the world. As I start to write this article I see the following report in the *Independent on Sunday*:

> Islamic extremism in Britain is creating communities which are "no-go areas" for non-Muslims, the Bishop of Rochester, the Rt Rev Dr Michael Nazir-Ali, warned yesterday. Bishop Nazir-Ali says non-Muslims face a hostile reception in places dominated by the ideology of Islamic radicals.

Regardless of the merits or accuracy of the individual story or claim, and this is a particularly absurd one, the relentless flow of this kind of comment and coverage has turned Islam into a religion under siege. This incessant problematisation of Islam and demonisation of Muslims have created the phenomenon now widely referred to as Islamophobia.

For readers of this journal, it should be no mystery why this has occurred. It is not an expression of some visceral Christian hostility to Islam stretching back to the Crusades or the conflict with the Ottoman Empire (even though these atavisms are sometimes mobilised ideologically). It is because the majority of the people sitting on the world's most important reserves of oil and natural gas happen to be Muslim and, secondarily, because, since the Iranian Revolution of 1979, much of these peoples' resistance to imperialism has found expression in Islamist form. If the people of the Middle East and central Asia had been predominantly Buddhist or Tibet held oilfields comparable to those of Saudi Arabia or Iraq, we would now be dealing with "Buddhophobia". Seeping out from the White House, the Pentagon, the CIA and Downing Street, coursing through the sewers of Fox News, CNN, the *Sun* and the *Daily Mail* would be the notion that, great religion though it undoubtedly was, there was some underlying and persistent flaw in Buddhism. "Intellectuals" such as Samuel Huntington, Christopher Hitchens and Martin Amis would be on hand to explain that, despite its embrace by naive hippies in the 1960s, Buddhism was an essentially reactionary creed characterised by its deepseated rejection of modernity and Western democratic values, and its fanatical commitment to feudalism, theocracy, misogyny and homophobia.

However, the fact that it has happened—the fact that Islamophobia has been developed, nationally and internationally, as the principal ideological cover and justification for imperialism and war (as straightforward racism was in the 18th and 19th centuries)—has enormously increased the importance of a correct theoretical understanding of, and political orientation towards, religion in its many different forms. Indeed it can be said that a deficient, mechanical or one-sided understanding of the Marxist analysis of religion has been a substantial contributing factor to a number of left individuals and groups completely losing their former political bearings and ending up as left apologists for imperialism.

The most notorious example of this is, of course, Christopher Hitchens, who has written a book on religion, *God is Not Great* (of which more later), and whose trajectory from leftist intellectual and radical critic of the system to "critical" supporter of George Bush has been precipitous and extreme (though in Hitchens' case one cannot help suspecting that material inducements have played a larger role in his race to the right than any mere theoretical error). Other examples include members of the Euston Group, such as Norman Geras, and, among left groups, the

French organisation Lutte Ouvrière, whose hostility to the hijab turned them into temporary allies of the French imperialist state against its most oppressed women citizens,[1] and the sorry case of the semi-Zionist and Islamophobic Alliance for Workers' Liberty.

At the same time, and not by coincidence, in the US and Britain there has arisen a verbally militant anti-religious, pro-atheist campaign, spearheaded by the biologist Richard Dawkins and accompanied by the aforementioned Hitchens, the philosopher Daniel Dennett and others. A critical examination of how these people present their arguments against religion will bring out important features of the Marxist position. But first I want to set out the fundamental principles underlying the Marxist analysis of religion, beginning not with Marx's direct comments on religion but with the basic propositions of Marxist philosophy.

Materialism and religion

Marxist philosophy is materialist. According to Frederick Engels in *Ludwig Feuerbach and the End of Classical German Philosophy*:

> The great basic question of all, especially of latter-day philosophy, is that concerning the relation of thinking and being... The question of the position of thinking in relation to being...in relation to the church was sharpened into this: did God create the world or has the world existed for all time? Answers to this question split the philosophers into two great camps. Those who asserted the primacy of the mind over nature and, therefore, in the last instance, assumed world creation in some form or other...comprised the camp of idealism. The others, who regarded nature as primary, belong to the various schools of materialism.[2]

Marxism, argues Engels, not only stands firmly in the materialist camp but is where "the materialist world outlook was taken really seriously for the first time and was carried through consistently...in all relevant domains of knowledge".[3]

Marxist materialism, reduced to its essentials, involves commitment to the following propositions:

(1) The material world exists independently of human (or any other) consciousness.

1: Boulangé, 2004.

2: Engels, 1989, pp366-367.

3: Engels, 1989, p382.

(2) Real, if not total or absolute, knowledge of the world is possible and has, indeed, been attained.
(3) Human beings are part of nature, but a distinct part.
(4) The material world does not derive, in the first instance, from human thought; human thought derives from the material world.

Propositions (1) and (2) correspond to the presumptions and findings of modern science, and have attained the status of common sense. This is because they are confirmed in practice, millions or billions of times every day, as are most of the findings of science. Proposition (3) also corresponds to the findings of modern science, especially those of Charles Darwin, and modern paleontology and anthropaleontology, but was, as it happens, articulated by Marx before Darwin:

> The first premise of all human history is, of course, the existence of living human individuals. Thus the first fact to be established is the physical organisation of these individuals and their consequent relation to the rest of nature... The writing of history must always set out from these natural bases and their modification in the course of history through the action of men. Men can be distinguished from animals by consciousness, by religion or anything else you like. They themselves begin to distinguish themselves from animals as soon as they begin to produce their means of subsistence, a step which is conditioned by their physical organisation.[4]

Proposition (4) is the most distinctively Marxist and the least widely shared. Many people who take a materialist view of the relationship between humans and nature take an idealist position on the relationship between ideas and material conditions, and on the role of ideas in society, history and politics. Almost without thinking they may accept that "the Cold War was fundamentally a clash of ideologies" or that "capitalism is based on the idea of economic growth". For this reason Proposition (4) is the one Marx and Engels insist on most strongly and often:

> Men are the producers of their conceptions, ideas, etc—real active men, as they are conditioned by a definite development of their productive forces... Consciousness can never be anything else than conscious existence... In direct contrast to German philosophy, which descends from heaven to earth, here we ascend from earth to heaven... We set out from real, active men,

4: Marx and Engels, 1991, p42.

and on the basis of their real life-process we demonstrate the development of the ideological reflexes and echoes of this life-process.[5]

Does it require deep insight to comprehend that people's ideas, opinions and conceptions, in a word, their consciousness changes with every change in their life conditions, their social relations and their social being?[6]

In the social production of their life, men enter into definite relations that are indispensable and independent of their will, relations of production which correspond to a definite stage of development of their material productive forces. The sum total of these relations of production constitutes the economic structure of society, the real foundation, on which rises a legal and political superstructure and to which correspond definite forms of social consciousness. The mode of production of material life conditions the social, political and intellectual life process in general. It is not the consciousness of men that determines their being, but, on the contrary, their social being that determines their consciousness.[7]

Just as Darwin discovered the law of development of organic nature, so Marx discovered the law of development of human history: the simple fact, hitherto concealed by an overgrowth of ideology, that mankind must first of all eat, drink, have shelter and clothing, before it can pursue politics, science, art, religion, etc; that therefore the production of the immediate material means of subsistence, and consequently the degree of economic development attained by a given people or during a given epoch, form the foundation on which the state institutions, the legal conceptions, art, and even the ideas on religion, of the people concerned have been evolved, and in the light of which they must therefore be explained, instead of vice versa, as had hitherto been the case.[8]

Thus it is clear that a definite attitude to religion is present, both implicitly and explicitly, in the most fundamental ideas of Marxism. Moreover it should also be clear that this attitude has a dual character. On the one hand, for the thoroughgoing and consistent Marxist, as for the thoroughgoing and consistent materialist, religious faith, in all its many forms, is

5: Marx and Engels, 1991, p47.
6: Marx and Engels, 1848.
7: Marx, 1977.
8: Engels, 1883.

excluded. Religious ideas, like all other ideas, are social and historical products. They are produced by human beings, and this necessarily precludes religious belief, since religious ideas claim to transcend and take priority over nature, human beings and history. By the same token, philosophical idealism and religion are intimately linked. If mind has priority over matter, whose mind can that be but the mind of god? If ideas are the ultimate driving force in history, where do those ideas come from if not the mind of god? And is not god, as in the terminology of Georg Hegel, "the absolute idea"? As the Bible puts it, "In the beginning was the word, and the word was god." This is why Leon Trotsky, at the very end of his life, wrote that he would die "a Marxist, a dialectical materialist and, *consequently*, an irreconcilable atheist".[9]

On the other hand the same Marxism clearly demands a materialist *explanation* of religion. It is not enough to view either religion as a whole or any particular religion as simply a delusion or folly that happens to have gripped the minds of millions for centuries. A common habit of less thoughtful religious believers (especially religious believers in imperialist countries) is to mock or dismiss as superstition the religious beliefs of others (especially so-called "natives") on the grounds that they are obviously irrational or contrary to well known laws of nature, without realising that exactly the same applies to their own beliefs—in the virgin birth, the resurrection, the feeding of the 5,000 or whatever.

But Marxism does not just generalise this mistake by pointing to the equal stupidity of the cargo cultist and the Catholic, the Rastafarian and the Anglican. It requires an analysis of the social roots of religion in general and of specific religious beliefs; an understanding of the real human needs, social and psychological, and the real historical conditions, to which such beliefs and doctrines correspond. A Marxist needs to be able to understand why a belief in the divinity and immortality of Haile Selassie could inspire a musician of the calibre of Bob Marley in Trenchtown, Jamaica, in the 1960s, or why the belief in the divinity and immortality of Jesus inspired an artist (and mathematician) of the calibre of Piero della Francesca in 15th century Florence.

If we now turn to Marx's most important statement directly on religion, the first couple of pages of *The Introduction to a Contribution to the Critique of Hegel's Philosophy of Right*,[10] we find it to be a condensed expression of all these elements. It begins with the assertion, "For Germany, the

9: Trotsky, 1964, p361 (my emphasis).
10: Marx, 1970.

criticism of religion has been essentially completed, and the criticism of religion is the prerequisite of all criticism."

By this Marx means that the combined work of the scientific revolution, the Enlightenment (especially the French encyclopaedists) and the Bible criticism of German secular left Hegelians has demolished the claims of Christianity and the Bible to offer a factually true account of nature or history, or even an internally coherent theology. Moreover this work was necessary and progressive because a genuinely critical analysis of the world was not possible until human thought was liberated from the fetters of religious dogma. But this single sentence is all Marx says on this aspect of the question. Taking the factual refutation of religion as given, he proceeds rapidly to his main point, the analysis of the social basis of religion: "The foundation of irreligious criticism is: *man makes religion, religion does not make man.*" This is the starting point. What follows is a paragraph of exceptional density, typical of Marx, in which a PhD's worth of insights are compressed into a few sentences:

> Religion is, indeed, the self-consciousness and self-esteem of man who has either not yet won through to himself, or has already lost himself again. But man is no abstract being squatting outside the world. Man is the world of man—state, society. This state and this society produce religion, which is an inverted consciousness of the world, because they are an inverted world. Religion is the general theory of this world, its encyclopaedic compendium, its logic in popular form, its spiritual *point d'honneur*, its enthusiasm, its moral sanction, its solemn complement, and its universal basis of consolation and justification. It is the fantastic realisation of the human essence since the human essence has not acquired any true reality. The struggle against religion is, therefore, indirectly the struggle against that world whose spiritual aroma is religion.

Thus religion is a response to human alienation—man who has "lost himself". But this is not an abstract or ahistorical condition; rather it is a product of certain specific social conditions. This society produces religion, an inverted view of the world in which humans bow to an imaginary god of their own making, because it is an inverted world in which people are dominated by the products of their own labour. But religion is not just a random collection of superstitions or false beliefs; it is the "general theory" of this alienated world, the way in which alienated people try to make sense of their alienated lives and alien society. Therefore it performs the rich array of diverse functions listed by Marx: "encyclopaedic compendium", "logic in popular form", etc. And *therefore* to struggle against religion is to struggle against that world "whose spiritual aroma is religion"—this world of alienation in which people *need* religion.

Two points need to be made about this passage. The first is that it is almost universally ignored by commentators offering summaries or explanations of Marx's views on religion. This may be because they have not read it (unlikely) or have not understood it (more likely), or (most likely) because it is radically incompatible with the attempt to reduce the Marxist theory of religion to a simple one-dimensional analysis such as, "Marx argues that religion is a tool of the ruling class" or "according to Marx religion functions to pacify the toiling masses". Of course, Marx does say this kind of thing about religion but he says much else besides. To reduce the complex totality of his theory to just one of its strands is effectively to falsify it. The second point is that Marx is so keen on its conclusion that he repeats it again and again in a veritable storm of metaphors and aphorisms.[11]

However, before concluding his argument on religion, Marx inserts one more highly significant paragraph:

> Religious suffering is, at one and the same time, the *expression* of real suffering and a *protest* against real suffering. Religion is the sigh of the oppressed creature, the heart of a heartless world, and the soul of soulless circumstances. It is the opium of the people.[12]

This passage is much better known than the previous one, but that is largely because of its much quoted final phrase (often presented as the essence or the totality of Marx's analysis). In fact it is the first sentence that is probably the most interesting and most important for understanding the political role of religion. Marx's insistence that religion is both an expression of suffering and a protest against it is the key point, giving the lie to any analysis which focuses only on religion's narcotic and soporific effects. It also points in the direction of the important historical fact (to which I shall return) that there have been many progressive, radical and even

11: "The abolition of religion as the illusory happiness of the people is the demand for their real happiness"; "The criticism of religion is...the criticism of that vale of tears of which religion is the halo"; "Criticism has plucked the imaginary flowers on the chain not in order that man shall continue to bear that chain without fantasy or consolation, but so that he shall throw off the chain and pluck the living flower"; "The criticism of heaven turns into the criticism of earth"; etc.

12: Marx's emphasis.

revolutionary movements that have either taken a religious form, had a religious coloration or been led by people of religious faith.

In the course of their work Marx and Engels made numerous references to and analyses of religion. In particular the young Marx wrote *On the Jewish Question*, a polemic in favour of Jewish emancipation;[13] Engels contributed a number of interesting studies of the historical development and role of Christianity, particularly in *The Peasant War in Germany*, *Anti-Dühring*, the introduction to the English edition of *Socialism: Utopian and Scientific*, *Bruno Bauer and Early Christianity*, and *The History of Early Christianity*.[14] However, all these comments have one thing in common: they never take religious doctrines, sects, churches, movements and conflicts at face value, nor treat them as simple follies or deceptions practised by the priests, but regard them always as distorted reflections and expressions of real social needs and interests. A few extracts will illustrate the point.

From *The Peasant War in Germany*:

> In the so-called religious wars of the 16th century, very positive material class interests were at play, and those wars were class wars just as were the later collisions in England and France. If the class struggles of that time appear to bear religious earmarks, if the interests, requirements and demands of the various classes hid themselves behind a religious screen, it little changes the actual situation, and is to be explained by conditions of the time in Germany. The revolutionary opposition to feudalism was alive throughout all the Middle Ages. According to conditions of the time, it appeared either in the form of mysticism, as open heresy, or of armed insurrection.

From the introduction to *Socialism: Utopian and Scientific*:

> Calvin's creed was one fit for the boldest of the bourgeoisie of his time. His predestination doctrine was the religious expression of the fact that in the commercial world of competition success or failure does not depend upon a man's activity or cleverness, but upon circumstances uncontrollable by him.

From *The History of Early Christianity*:

13: This rather obscure text has been particularly controversial because it has been cited as evidence of Marx's anti-Semitism. John Rose discusses this in detail in his article in this issue of *International Socialism*. See also Draper, 1977; Bhattacharyya, 2006.

14: All available in Marx and Engels, 1957.

> Christianity was originally a movement of oppressed people: it first appeared as the religion of slaves and emancipated slaves, of poor people deprived of all rights, of peoples subjugated or dispersed by Rome...
>
> [The risings of peasants and plebeians in the Middle Ages], like all mass movements of the Middle Ages, were bound to wear the mask of religion and appeared as the restoration of early Christianity from spreading degeneration... But behind the religious exaltation there was every time a very tangible worldly interest.

And, incidentally, from the same work, a footnote on Islam:

> Islam is a religion adapted to Orientals, especially Arabs, ie, on one hand, to townsmen engaged in trade and industry, on the other, to nomadic Bedouins. Therein lies, however, the embryo of a periodically recurring collision. The townspeople grow rich, luxurious and lax in the observation of the "law". The Bedouins, poor and hence of strict morals, contemplate with envy and covetousness these riches and pleasures. Then they unite under a prophet, a Mahdi, to chastise the apostates and restore the observation of the ritual and the true faith and to appropriate in recompense the treasures of the renegades. In a hundred years they are naturally in the same position as the renegades were: a new purge of the faith is required, a new Mahdi arises and the game starts again from the beginning. That is what happened from the conquest campaigns of the African Almoravids and Almohads in Spain to the last Mahdi of Khartoum who so successfully thwarted the English... All these movements are clothed in religion but they have their source in economic causes.

The point here is not the historical truth or falsity of all or any of these specific observations, but the consistent methodology underlying them.

Dawkins, Hitchens and Eagleton

Richard Dawkins is an evolutionary biologist who first came to prominence with his book *The Selfish Gene*, and thereafter built himself a considerable reputation and career as a populariser of science. In 2006 he published *The God Delusion*, a full frontal assault on religion and defence of atheism, which became an international bestseller, generated huge controversy, especially in the United States, and attracted plaudits from sources as diverse as Ian McEwan, Michael Frayn, the *Spectator*, the *Daily Mail* and Stephen Pinker.

I should say at the outset that I do not at all share the apparently widespread admiration of Dawkins's style and intellect. Reading Dawkins after Marx is like going from Leo Tolstoy or James Joyce to Kingsley Amis or Agatha Christie. Where Marx packs a book into a paragraph, Dawkins expands a short essay into a large book. In fact all 460 odd pages of *The God Delusion* do not take us intellectually beyond what Marx summed up in the first sentence of his analysis in 1843, namely that the criticism of religion is essentially complete. What Dawkins offers is an "Enlightenment", empiricist, rationalist refutation of religion—a "scientific", ie positivist, demonstration that there is a complete lack of factual evidence to support what he calls "the god hypothesis" and that, on the contrary, the evidence makes it almost (if not absolutely) certain that god does not exist. This is supplemented by logical refutations of the various arguments advanced for god's existence ranging from the venerable "proofs" of Thomas Aquinas and "Pascal's Wager" to the bizarre recent speculations of one Stephen Unwin, and numerous examples of the follies and crimes perpetrated in the name of religion. I suppose there are some people for whom this will be revelatory and others who may enjoy it because it makes them feel smarter than the ignorant masses who swallow these superstitions, but theoretically there is nothing new here, indeed very little that is not at least 200 years old.

The only real exception to this lies in Dawkins's attempt to explain why religion is so widespread in human society, but this attempt is a rather miserable failure. Being a committed evolutionary biologist he feels obliged to frame his explanation in terms of genetic advantage in the process of natural selection, but his blanket hostility to religion also obliges him to deny that religion can be advantageous for individual or societal survival. He tries to wriggle out of this contradiction by suggesting that religion is a side-effect of a characteristic that he claims is advantageous in the struggle for survival, namely a propensity for children to believe what they are told by their elders. Clearly this does not withstand criticism. First, the extent to which youthful suggestibility outweighs youthful scepticism, especially into adolescence, is debatable. Second, it is equally debatable whether such suggestibility is, on balance, advantageous. Third, it seems highly likely that both the extent and advantageousness of suggestibility are massively socially conditioned and very different in different societies. Finally, like any theory that explains the behaviour or beliefs of children by the behaviour or beliefs of their parents, it is left with the problem of explaining the parents' disposition in the first place if it is to avoid being caught in an infinite regress.

As Marx pointed out, "The educators themselves must be educated".[15] In other words Dawkins's explanation turns out to be no

15: Marx, 1845.

explanation at all. Moreover it is symptomatic of his whole approach that neither in this section nor any anywhere else in *The God Delusion* does the author find time seriously to consider the Marxist theory of religion.

However, intellectual unoriginality and mediocrity are by no means the main objection to this book. (It would be churlish to cavil so over a work that was second rate but reasonably sound.) The main objection is to the reactionary political conclusions that flow from the weak methodology. As Marx argued in relation to the German philosopher Feuerbach, mechanical materialism invariably leaves the door open to idealism, and Dawkins is a particularly clear example of this. Without noticing it, he flip flops from a vulgar materialist genetic determinism in his view of human nature and behaviour in the abstract, to a rampant idealism in his view of the role of religion in concrete historical circumstances. Again and again he makes the mistake of assuming that when people do something in the name of religion it really is religion that is determining their behaviour. The following passage from his essay "The Improbability of God" epitomises his approach:

> Much of what people do is done in the name of god. Irishmen blow each other up in his name. Arabs blow themselves up in his name. Imams and ayatollahs oppress women in his name. Celibate popes and priests mess up people's sex lives in his name. Jewish shohets cut live animals' throats in his name. The achievements of religion in past history—bloody crusades, torturing inquisitions, mass-murdering conquistadors, culture-destroying missionaries, legally enforced resistance to each new piece of scientific truth until the last possible moment—are even more impressive. And what has it all been in aid of? I believe it is becoming increasingly clear that the answer is absolutely nothing at all. There is no reason for believing that any sort of gods exist and quite good reason for believing that they do not exist and never have. It has all been a gigantic waste of time and a waste of life. It would be a joke of cosmic proportions if it weren't so tragic.[16]

In fact this is no more than a souped up version of the familiar nostrum that lots of wars are caused by religion. It will not stand a moment's critical scrutiny. Let us take the example of Ireland. The view that the conflict in Ireland was essentially or primarily about religion is both manifestly false and plainly reactionary. It is false even in terms of the declared statements and consciousness of the principal protagonists. If many, though by no means all, Republicans were Catholics, no Republican would have said

16: Dawkins, 1998.

(or believed) that they were fighting for Catholicism; they fought for an independent, united Ireland. Things were less clear on the Unionist side where religious bigotry played a much larger role; nevertheless the principal declared goal was a "national" one, namely remaining "British". Moreover, it is abundantly clear that behind these conflicting national aspirations lay not religious differences about the doctrine of transubstantiation or the fallibility of the pope but real economic, social and political issues of exploitation, poverty, discrimination and oppression. To see the conflict as basically about religion was reactionary because it fitted with the racist stereotype of the Irish as primitive and stupid (after all "we" gave up fighting about religion centuries ago) and helped to legitimise British rule as a neutral arbiter between warring religious factions.

To his credit, Dawkins opposed the Iraq war, and politically he is no friend of George Bush, but, in the context of the "war on terror", his approach to religion becomes, even if unintentionally, even more reactionary. For it is central to the ideology of the neocons, Bush, Cheney, Blair and Brown that Muslim hostility to "the West" is unprovoked and unjustified. It is not seen as a reaction or response to Western imperialism, exploitation and domination, but rather an offensive religion-based campaign aimed at destroying, conquering or perhaps converting the non-Muslim world.

Some see these aims as inherent in mainstream Islam,[17] while for Bush, Blair and Co it derives from an "evil" misinterpretation or perversion of Islam, but in both cases the motivation is religious. It is an interpretation which flies in the face of the declared statements of both Al Qaida, who made explicit political demands such as the removal of US troops from Saudi Arabia, and the 7/7 bombers in London, who said they were motivated by what was being done to Iraq, and defies reason. The notion that America, Britain or any big Western nation could be destroyed, conquered or, indeed, converted by planting bombs on the underground or flying planes into buildings is so utterly absurd that it cannot be the real motive for any sustained campaign. The idea that the US could be induced by a terrorist campaign to stop supporting Israel or to get out of Afghanistan is also mistaken but it is not completely implausible. For Bush, Blair and Co, however, the "religious" interpretation is mandatory, as without it they would be forced to concede the culpability of imperialism and of their own policies—and the Dawkins approach dovetails with this and reinforces it:

17: Dawkins himself seems to hold this view or something like it—see Dawkins, 2007, pp346-347.

> "Mindless" may be a suitable word for the vandalising of a telephone box. It is not helpful for understanding what hit New York on 11 September... It came from religion. Religion is also, of course, the underlying source of the divisiveness in the Middle East which motivated the use of this deadly weapon in the first place. But that is another story and not my concern here. My concern here is with the weapon itself. To fill a world with religion, or religions of the Abrahamic kind, is like littering the streets with loaded guns.[18]

Similar to Dawkins, but worse, is Christopher Hitchens. His book, *God is Not Great*, is on an even lower intellectual level than *The God Delusion*, with a more arbitrary combination of self-serving personal anecdote and rambling journalistic polemic. Its adaptation of the atheist case to Islamophobia is embodied in the title (a mocking reference to the Muslim cry, "God is Great!") and blatant throughout. I suppose out of deference to his radical past he actually quotes, approvingly, a couple of the key paragraphs of Marx on religion. He then proceeds to ignore their meaning completely. In the key section, "Religion Kills", he takes us on a whistlestop tour of six strife-torn cities—Belfast, Beirut, Bombay, Belgrade, Bethlehem and Baghdad—in each case offering a swift summation of the conflict exclusively in terms of religious hatreds, without any reference to history, imperialism, oppression or class. It is a travesty of socio-political analysis. The "analysis" of Palestine is especially striking:

> I once heard the late Abba Eban, one of Israel's more polished and thoughtful diplomats and statesmen, give a talk in New York. The first thing to strike the eye about the Israeli-Palestinian dispute, he said, was the ease of its solubility... Two peoples of roughly equivalent size had a claim to the same land. The solution was, obviously, to create two states side by side. Surely something so self-evident was within the wit of man to encompass? And so it would have been, decades ago, if the messianic rabbis and mullahs and priests could have been kept out of it. But the exclusive claims to god-given authority, made by hysterical clerics on both sides and further stoked by Armageddon-minded Christians who hope to bring on the Apocalypse (preceded by the death or conversion of all Jews), have made the situation insufferable, and put the whole of humanity in the position of hostage to a quarrel that now features the threat of nuclear war. Religion poisons everything.

This is risible, but when Hitchens says, and I quote verbatim from

18: Richard Dawkins, "Religion's Misguided Missiles", *Guardian*, 15 September 2001.

YouTube, "I am absolutely convinced that the main source of hatred in the world is religion",[19] he is also saying the cause is not the material facts of capitalism, imperialism, inequality, exploitation or class conflict, just a mistaken idea people have lodged in their heads.

Vigorously opposing the arguments of Dawkins and Hitchens does not, however, involve diluting in any way the classical Marxist critique of religion or opening the door to some kind of theoretical compromise with religious ideas. At this point we need to leave the odious Hitchens for the far more congenial Terry Eagleton, who provides an example of what should be avoided. Eagleton is an eminent cultural and literary theorist, friendly to Marxism, who, in the past, attacked the racism and other bigotries of Philip Larkin. He recently distinguished himself by denouncing the Islamophobia of his academic colleague Martin Amis. In 2006 he wrote a highly critical review of *The God Delusion* for the *London Review of Books*. Although Eagleton's review advances some of the same arguments as this article, for example in relation to Ireland, the general terms of his critique are not Marxist. His principal argument is that Dawkins has attacked fundamentalist religion, Christian and Islamic, as if it represents all religion, while ignoring more sophisticated "liberal" theology of which Dawkins is largely ignorant:

> What, one wonders, are Dawkins' views on the epistemological differences between Aquinas and Duns Scotus? Has he read Eriugena on subjectivity, Rahner on grace or Moltmann on hope? Has he even heard of them? Or does he imagine like a bumptious young barrister that you can defeat the opposition while being complacently ignorant of its toughest case?[20]

As a criticism of Dawkins's book this has some validity, but there are also serious problems here. First, it is not reasonable to argue that it is necessary to master all the ins and outs of Christian (or Buddhist, or Zoroastrian) theology before one can make an intellectually sound case for atheism and for rejecting theology as such. Second, in demonstrating his understanding of the liberal theologians' concept of an immaterial, impersonal god of love and tolerance, in contrast to the Old Testament god of vengeance, Eagleton

19: It is not easy to grasp how far Hitchens has gone. Again I quote from him on YouTube, debating with Reverend Al Sharpton: "You see, I don't love our enemies, and I don't love people who do love them. I hate our enemies and think they should be killed... And I'm absolutely sure there should be no other country that has a budget that threatens ours, and I'm not sentimental about it." And by "our enemies" and "our budget" he means the enemies and budget of US imperialism.

20: Eagleton, 2006.

leaves decidedly open the possibility that this liberal god may actually exist, or be worthy of worship. He does the same when he offers his picture of Jesus as proto anti-imperialist revolutionary:

> Jesus did not die because he was mad or masochistic, but because the Roman state and its assorted local lackeys and running dogs took fright at his message of love, mercy and justice, as well as at his enormous popularity with the poor, and did away with him to forestall a mass uprising in a highly volatile political situation.[21]

For a Marxist the loving, caring, impersonal god of Dietrich Bonhoeffer and the radical Jesus of Terry Eagleton are both just as much human creations, illusory projections, as the unpleasant bigoted gods of Ian Paisley or Osama Bin Laden.

Religion and socialist politics

To conclude this article I shall outline a brief and rather schematic summary of the principal political conclusions that flow, and have flowed historically, from the foregoing analysis.

First, and contrary to widespread opinion (fostered by widespread misrepresentation), Marxist socialists are absolutely opposed to any idea of banning religion. This is not some new position but was explicitly stated by Engels as far back as 1874 in response to a proposal by followers of the French socialist Louis Blanqui. The reasons given by Engels remain valid to this day:

> In order to prove that they are the most radical of all they abolish god by decree as was done in 1793:
>
> > "Let the Commune free mankind for ever from the ghost of past misery" (god), "from that cause" (non-existing god a cause!) "of their present misery. There is no room for priests in the Commune; every religious manifestation, every religious organisation must be prohibited."
>
> And this demand that men should be changed into atheists *par ordre du mufti* is signed by two members of the Commune who have really had opportunity enough to find out that, first, a vast amount of things can be ordered on

21: Eagleton, 2006.

paper without necessarily being carried out; and, second, that persecution is the best means of promoting undesirable convictions![22]

Far from banning religion, Marxists argue that religion should be a private matter in relation to the state, and complete freedom of religion should prevail under both capitalism and socialism. Lenin spelt this out unambiguously in an article from 1905:

> Religion must be of no concern to the state, and religious societies must have no connection with governmental authority. Everyone must be absolutely free to profess any religion he pleases, or no religion whatever, ie, to be an atheist, which every socialist is, as a rule. Discrimination among citizens on account of their religious convictions is wholly intolerable. Even the bare mention of a citizen's religion in official documents should unquestionably be eliminated.[23]

The only sense in which Marxists contemplate the elimination of religion is through its gradual withering away as a result of the disappearance of its underlying social causes—alienation, exploitation, oppression, etc. Marxist socialists are, however, opposed to any state privileges for religion and call for the disestablishment of any or all official state churches (such as the Church of England).

Inevitably the general perception of the Marxist attitude to religion is considerably influenced by the experience of the Stalinist regimes in Russia, Eastern Europe, China, Cuba, North Korea, etc. A systematic investigation of this experience is impossible in this brief article and, hopefully, readers of this journal are well aware that the policies of these regimes were in no way representative of genuine socialism or Marxism. Nevertheless, certain observations are worth making. Stalinist repression of religion is often both exaggerated and misunderstood. It is exaggerated in that, in general, the Stalinist regimes did not repress the main religions or churches but tolerated them and even formed alliances with them, on condition that these churches were politically compliant (which they mainly were). It is misunderstood in that, where religious groups or individuals were persecuted, it was primarily because they were politically troublesome, rather than because of their faith as such. But then these were societies in which all political opposition was suppressed. A broad overview of the "Communist" states' treatment of the

---

22: Marx and Engels, 1957.
23: Lenin, 1965.

religious can be found in the last chapter of Paul Siegel's *The Meek and the Militant*,[24] and an especially useful case study of the Russian Revolution's dealings with its Muslim minority is provided by Dave Crouch in an earlier issue of this journal.[25] Crouch shows how in the early years of the revolution the Bolsheviks adhered strictly to the Leninist principles outlined above and thus met with considerable success in winning Muslims over, whereas the rise of Stalin led to the adoption of increasingly top-down authoritarian policies, including an assault on the veil, which proved disastrous.

In determining their attitude to popular movements with a religious coloration, which are many and varied, Marxists take as their point of departure not the religious beliefs of the movement's leaders or of its supporters, or the doctrines and theology of the religion concerned, but the political role of the movement, based on the social forces and interests which it represents.

To put this in perspective consider the respective historical roles of Catholicism and Protestantism. In the Middle Ages and the Early Modern period Catholicism was essentially the religion of the feudal aristocracy and therefore almost universally reactionary. By contrast radical Protestantism tended to represent either the rising bourgeoisie or the plebeian elements below and to the left of it. The great rebels and revolutionaries of those times, the Thomas Muenzers, John Lilburnes and Gerard Winstanleys, were passionate Protestants—extremists and fundamentalists in the language of today. But the moment these bourgeois rebels came to power, in the Netherlands and England, they became participants in what Marx called "the primitive accumulation of capital" and thus vicious colonists and slavers. Oliver Cromwell, the revolutionary and regicide in England, became Cromwell the oppressor in Ireland (where his name still lives in infamy), and specifically the oppressor of the Catholic peasantry. Dutch protestant burghers could be the heroes of Europe in the Dutch Revolt but villains in Africa with apartheid. The strongly reactionary role of the Catholic church continued in Europe, especially southern Europe, and saw it give active support for Franco in Spain and strike deals with Mussolini and Hitler. It still continues in attenuated form in the main conservative parties in Italy, Spain and southern Germany today. But the countries in Europe where Catholicism and religion in general remained strongest were Ireland and Poland where the church was able, very moderately but powerfully, to identify itself with opposition to national oppression.

24: Siegel, 1986.

25: Crouch, 2006.

Any socialist looking back to the 17th century will identify immediately with the Protestant rebels and against the Catholic monarchs and emperors. Any socialist looking at Ireland in 1916 or Belfast in the 1970s will identify with the "Catholic" Nationalists, not the "Protestant" Unionists. Any socialist who saw the rise of Solidarnosc in Poland as a conflict between the "backward" Catholics of Gdansk and the "progressive" atheist Communists of the Soviet state ended up on the side of the imperialist oppressor. The same applies today to the Tibet/China conflict and, above all, to the "war on terror" and the struggles in the Middle East.

Many other cases can be adduced to reinforce this argument. Where would a socialist be who decided their political attitude to Malcolm X on the basis of his reactionary religious beliefs as a member of the Nation of Islam, to Bob Marley on the basis of his belief in the divinity of that old tyrant Haile Selassie or even to Hugo Chavez on the basis of his self-proclaimed Catholicism and admiration of the pope? Unfortunately some would-be socialists who have no difficulty grasping this in relation to Chavez or Marley, under the pressure of intense bourgeois propaganda are unable to apply the same approach when the religion in question is Islam. To put the matter as starkly as possible: from the standpoint of Marxism and international socialism an illiterate, conservative, superstitious Muslim Palestinian peasant who supports Hamas is more progressive than an educated liberal atheist Israeli who supports Zionism (even critically).

It also follows that Marxist socialists do not accept the idea that any of the major religions is inherently, or in terms of its doctrines, more or less progressive than any of the others. For a religion to become "major", that is to survive over centuries in many locations and different social orders, it is a precondition that its doctrines be capable of almost infinite selection, interpretation and adaptation. Once again, what is decisive is not doctrine but social base in the specific social situation. Thus in the US we find a right wing racist imperialist Christianity in the Moral Majority or the Mormons and a left wing anti-racist anti-war Christian tradition in Martin Luther King. In South Africa there was a pro-apartheid Christianity and an anti-apartheid Christianity; in Latin America there has been a right wing, pro-oligarchy, pro-dictator Catholicism and a leftist "theology of liberation" Catholicism; and, of course, there are a multitude of different, often sharply conflicting, versions of Islam.

The main argument used to justify the notion of Islam as an especially backward religion is, of course, the attitudes to women and homosexuality prevalent in Muslim countries. Those who put this argument need to be reminded that much the same attitudes were prevalent in Western societies until very recently and are still present in the teachings of many Christian churches. But the fundamental flaw in this argument takes us back to the basics of Marxist materialism—the secret of the Muslim Holy Family lies in the earthly Muslim family. It is not Muslim religious consciousness that determines the position of women in Muslim society, but the real position of women that shapes Muslim religious beliefs. Islam was born in the Arabian peninsular, spreading west across North Africa and east across Central Asia. For centuries this great belt has been largely poor, underdeveloped and rural, and to a considerable extent remains so today. Other societies, from Ireland to China, with similar levels of development and similar social structures but different religions, exhibit similar oppression of women and gays.

Finally, there is the question of the relationship of the revolutionary party to religious workers. Any such party operating in a country where religion remains strong among the mass of the population, which is much of the world, must reckon with, indeed count on, the fact that the revolution will be made by workers of whom many will still be religious. The vast mass of workers will be liberated from their religious illusions not by arguments, pamphlets or books, but by participation in the revolutionary struggle, and beyond, in the building of socialism. In such a situation it is incumbent on the party to ensure that religious differences, or differences between the religious and the non-religious, do not obstruct the unity of working class struggle. Moreover, insofar as the party becomes a truly mass party, leading the class in its workplaces and communities, it will inevitably find in its ranks a layer of workers who remain religious or semi-religious. To reject such workers because of their religious illusions would be sectarian and non-materialist. It would be to share the religious/idealist mistake of regarding religion as the most important element in consciousness and consciousness as more important than practice. At the same time, the party must not become a religious party, or party whose policy, strategy or tactics are shaped by religious considerations. Revolutionary victory requires that the party should be guided by the theory that expresses the collective interests and struggle of the working class, namely Marxism. Therefore the party must ensure that on this matter it educates and influences its religious members rather than vice versa.

One revolutionary party working in such a situation was the Bolshevik Party, and its leading theorist, Lenin, wrote on these matters with insight and clarity in his 1909 article "The Attitude of the Workers' Party to Religion". Here are a few extracts:

Marxism is materialism. As such, it is as relentlessly hostile to religion as was the materialism of the 18th century Encyclopaedists or the materialism of Feuerbach... But the dialectical materialism of Marx and Engels goes further...for it applies the materialist philosophy to the domain of history..... It says: We must know how to combat religion, and in order to do so we must explain the source of faith and religion among the masses in a materialist way. The combating of religion cannot be confined to abstract ideological preaching, and it must not be reduced to such preaching. It must be linked up with the concrete practice of the class movement, which aims at eliminating the social roots of religion.

Why does religion retain its hold?... Because of the ignorance of the people, replies the bourgeois progressivist, the radical or the bourgeois materialist. And so: "Down with religion and long live atheism; the dissemination of atheist views is our chief task!" The Marxist says that this is not true, that it is a superficial view... It does not explain the roots of religion profoundly enough; it explains them, not in a materialist but in an idealist way... The deepest root of religion today is the socially downtrodden condition of the working masses and their apparently complete helplessness in face of the blind forces of capitalism

Does this mean that educational books against religion are harmful or unnecessary? No, nothing of the kind. It means that Social Democracy's atheist propaganda must be subordinated to its basic task—the development of the class struggle of the exploited masses against the exploiters.

The proletariat in a particular region...is divided, let us assume, into an advanced section of fairly class conscious Social Democrats [the name used by socialist groups in Russia], who are of course atheists, and rather backward workers...who believe in god, go to church, or are even under the direct influence of the local priest... Let us assume furthermore that the economic struggle in this locality has resulted in a strike. It is the duty of a Marxist to place the success of the strike movement above everything else, vigorously to counteract the division of the workers in this struggle into atheists and Christians, vigorously to oppose any such division. Atheist propaganda in such circumstances may be both unnecessary and harmful—not from the philistine fear of scaring away the backward sections, of losing a seat in the elections, and so on, but out of consideration for the real progress of the class struggle, which in the conditions of modern capitalist society will convert

> Christian workers to Social Democracy and to atheism a hundred times better than bald atheist propaganda.
>
> We must not only admit workers who preserve their belief in God into the Social Democratic party, but must deliberately set out to recruit them; we are absolutely opposed to giving the slightest offence to their religious convictions, but we recruit them in order to educate them in the spirit of our programme, and not in order to permit an active struggle against it.[26]

What these extracts confirm is what this whole article has argued, namely that handling correctly the issue of religion, so vital in the present political situation, is not just a matter of ad hoc judgments or tactics, still less of electoral opportunism, but of understanding the most basic ideas of Marxist dialectical materialism.

26: Lenin, 1973.

References

Bhattacharyya, Anindya, 2006, "Marx and Religion", *Socialist Worker*, 4 March 2006, www.socialistworker.co.uk/art.php?id=8373

Boulangé, Antoine, 2004, "The Hijab, Racism and the State", *International Socialism 102* (spring 2004), www.isj.org.uk/index.php4?id=45

Crouch, Dave, 2006, "The Bolsheviks and Islam", *International Socialism 110* (spring 2006), www.isj.org.uk/index.php4?id=181

Dawkins, Richard, 1998, "The Improbability of God", *Free Inquiry*, volume 18, number 4 (autumn 1998), available from: www.positiveatheism.org/writ/dawkins3.htm

Dawkins, Richard, 2007, *The God Delusion* (Black Swan).

Draper, Hal, 1977, "Marx and the Economic-Jew Stereotype", in *Karl Marx's Theory of Revolution, volume one: State and Bureaucracy* (Monthly Review), www.marxists.de/religion/draper/marxjewq.htm

Eagleton, Terry, 2006, "Lunging, Flailing, Mispunching", *London Review of Books*, 19 October 2006, www.lrb.co.uk/v28/n20/eagl01_.html

Engels, Frederick, 1883, speech at Marx's graveside, from *Der Sozialdemokrat*, 22 March 1883, www.marxists.org/archive/marx/works/1883/death/dersoz1.htm

Engels, Frederick, 1989 [1886], *Ludwig Feuerbach and the End of Classical German Philosophy*, in Karl Marx and Frederick Engels, *Selected Works*, volume three (Progress), alternative version online: www.marxists.org/archive/marx/works/1886/ludwig-feuerbach/

Lenin, Vladimir, 1965 [1905], "Socialism and Religion", in *Collected Works*, volume ten (Progress), www.marxists.org/archive/lenin/works/1905/dec/03.htm

Lenin, Vladimir, 1973 [1909], "The Attitude of the Workers' Party to Religion", in *Collected Works*, volume 15 (Progress), www.marxists.org/archive/lenin/works/1909/may/13.htm

Marx, Karl, 1845, *Theses on Feuerbach*, translation online: www.marxists.org/archive/marx/works/1845/theses/

Marx, Karl, 1970 [1844], *Introduction to a Contribution to the Critique of Hegel's Philosophy of Right* (Cambridge University) www.marxists.org/archive/marx/works/1843/critique-hpr/intro.htm

Marx, Karl, 1977 [1859], *Preface to a Contribution to the Critique of Political Economy* (Progress), www.marxists.org/archive/marx/works/1859/critique-pol-economy/preface.htm

Marx, Karl, and Frederick Engels, 1957, *On Religion* (Progress), www.marxists.org/archive/marx/works/subject/religion/

Marx, Karl, and Frederick Engels, 1848, *Manifesto of the Communist Party*, alternative translation online: http://ebooks.adelaide.edu.au/m/marx/karl/m39c/

Marx, Karl, and Frederick Engels, 1991 [1845], *The German Ideology* (Lawrence & Wishart), www.marxists.org/archive/marx/works/1845-gi/

Siegel, Paul, 1986, *The Meek and the Militant—Religion and Power Across the World* (Zed), sections available online: www.marxists.de/religion/siegel-en/

Trotsky, Leon, 1964, *The Age of Permanent Revolution* (New York).

China, Tibet and the left

Charlie Hore

The riots and protests in Tibet earlier this year were the most significant since China's takeover in the 1950s. Together with the protests that have accompanied the Olympic torch relay around the world, they have shown that Tibetan nationalism remains a potent force and that opposition to the Chinese occupation is still widespread. But the international left has been divided on whether to support the Tibetan protesters, with some openly backing the Chinese occupation, while others have raised important questions about the leadership of the Tibetan nationalist movement and about US support for Tibetan nationalism.

In this short article, I aim firstly to look at the extent of the 2008 protests and then give a sketch of Tibet's history since 1949. I will then look at some controversial arguments over China and Tibet, and finally consider how Tibet fits into the wider analysis of China today. Crucially, I want to argue that the protests cannot simply be seen as part of the general pattern of protest in China. There is a particular dimension of national oppression (as there is in the Muslim-majority province of Xinjiang), which both the Chinese and the international left have to pay attention to.

The extent of the protests

The protests in Lhasa, Tibet's capital, began on 10 March 2008, the anniversary of a failed uprising in 1959 (see below). Small numbers of Buddhist monks and nuns staged peaceful demonstrations, which were met with heavy handed policing. As larger numbers protested against mass arrests, Chinese security forces first used tear-gas and cattle-prods, and then live

China and Tibet

ammunition.[1] By the end of the week thousands of people were fighting back with stones against a massive police and army presence, and rioters controlled substantial parts of Lhasa. The Chinese government claimed that 19 people had died, while Tibetan exile sources put the figure at over 80.

There were also protests in other parts of Tibet and, more importantly, in Tibetan areas of other Chinese provinces, in what is often called "greater" or "historic" Tibet. The government admitted opening fire and killing demonstrators in the towns of Luhuo and Aba in Sichuan province.[2] In Gansu province the BBC reported major unrest in the town of Hezuo, led by nomads on horseback[3] and the *Guardian*'s website showed video footage of several thousands of people demonstrating in the town of Xiahe, where protesters were tear-gassed.[4] The Free Tibet campaign reported large numbers of similar protests in Sichuan, Gansu and Qinghai provinces, including further large-scale shootings in Kardze, Sichuan province, in April.[5]

One Tibetan expert from the London School of Economics argued that "in terms of the scale of the protests and the subsequent troop deployment, there has not been anything like this since the 1950s".[6] There were major riots in Lhasa in October 1987 and March 1989, but they were not substantially echoed outside Lhasa.[7] The geographic spread of these protests is unprecedented and poses a distinctly new problem for China's rulers. While it is impossible to know the relative numbers involved, Tibetan specialist Robert Barnett argued in the *New York Review of Books* that "roughly 80 percent of the protests came from the eastern areas of the Tibetan plateau—within Qinghai, Sichuan and Gansu provinces—which China does not recognise as Tibet".[8]

The Chinese government has made much of attacks on Han Chinese

1: "Gunfire On The Streets Of Lhasa As Rallies Turn Violent", *Guardian*, 15 March 2008, www.guardian.co.uk/world/2008/mar/15/tibet.china1

2: "Paramilitaries Open Fire On Hundreds Of Monks And Nuns At Tibet Rally", *Times*, 25 March 2008, www.timesonline.co.uk/tol/news/world/asia/article3612661.ece

3: "Tibetan Monk Speaks Out", *BBC News*, 21 March 2008, http://news.bbc.co.uk/1/hi/world/asia-pacific/7308890.stm

4: "Tibet Protests Spread To Neighbouring Provinces", *Guardian* online, 16 March 2008, www.guardian.co.uk/world/video/2008/mar/16/xiahe.gansu

5: "Tibet Demonstrations 2008", Free Tibet, www.freetibet.org/newsmedia/2008-protests-summary

6: Cited in "Tibet Untamed: Why Growth Is Not Enough At China's Restive Frontier", *Financial Times*, 31 March 2008.

7: Barnett, 2006, includes eyewitness reports of the rioting in 1987.

8: Barnett, 2008. This is one of the best accounts of the spread of the protests that I have seen, and I quote details from it elsewhere in this section.

and Hui Muslims by Lhasa protesters,[9] arguing that the riots were essentially a racist pogrom. While this is untrue—rioters' targets also included symbols of Chinese occupation such as the Bank of China and government buildings—there certainly were numerous attacks on Chinese businesses and some attacks on individual Chinese in the streets.

These attacks are a product of the rapid economic development of Tibet since the 1980s, which has seen Lhasa in particular grow very fast, but without benefiting the majority of Tibetans. Most of the new jobs and economic opportunities have been taken by Han and Hui Chinese migrants, who are at best indifferent to and at worst racist towards Tibetans.[10] In addition, the growing numbers of Chinese tourists (over two million last year) exacerbate the sense that Tibetans are being squeezed out of Lhasa, except as "exotic" tourist attractions. In these circumstances, it is not surprising that Tibetans should take out their frustrations on Chinese businesses, in much the same way as African-American rioters in the US in the 1960s targeted white-owned businesses that symbolised their oppression. But much greater violence was meted out by the Chinese police and army in "restoring order".

The stunning resurgence of Tibetan nationalism poses huge problems for the Chinese government. Huge numbers of Tibetans still look to Tibetan nationalism or independence as preferable to the present situation, and they see the Dalai Lama as embodying their aspirations (even though the Dalai Lama does not call for independence). The history of Tibet since 1949 explains why this is so.

Tibet since 1949

"Historic" Tibet in 1949 was a cultural/linguistic entity, but not a political one. The two central provinces (which are now the Chinese province of Tibet) were ruled by an independent Tibetan government of senior landowners and lamas (Buddhist spiritual leaders). Amdo province to the north (now incorporated into Qinghai and Gansu provinces) and Kham to the east (now incorporated into Sichuan and Yunnan provinces) were in theory parts of China but in practice were ruled by local monarchs. Tibetan society as a whole accepted the Dalai Lama as the spiritual leader of Tibetan

9: According to China's official listing of nationalities, the Han are the majority ethnic group, making up over 90 percent of the population, while the Hui are ethnic Chinese who are Muslims.

10: For Western accounts of how Tibetans are excluded from economic growth, see "Tibet: Death by Consumerism", *New Statesman*, 30 August 2007, www.newstatesman.com/200708300022 as well as Barnett, 2006, and French, 2003.

Buddhism, but there were major political differences inside the Tibetan ruling class, which the Chinese Communist Party initially exploited intelligently.

The Communist Party's armies took control of Amdo and Kham by early 1950. In 1951, following the entry of Communist troops into Lhasa, the Dalai Lama signed a peace agreement that essentially preserved the existing Tibetan ruling class on condition they accepted Chinese rule. As one Tibetan Communist remembered:

> Our immediate priorities were to establish cordial relations with the Tibetan government and the elite... All thoughts of socialist reform, therefore, were put on the back burner, and we did not pay any attention to propagandising the masses about them, let alone issues of class struggle and exploitation.[11]

Although there was some initial resistance—one history notes that "Chinese who were in Lhasa in the early 1950 reported that the streets were unsafe for the Han"[12]—at this stage the Chinese occupation was passively accepted, not least because of the active collaboration of the Panchen Lama, the second figure in Tibet's religious/political hierarchy. In 1955 the Dalai Lama, the Panchen Lama and numbers of leading monks and landlords joined the committee set up to prepare Tibet's full integration into China.

But the honeymoon could not last, partly because of the "great Han chauvinist" attitudes of many leading Communist Party officials within Tibet,[13] but also because of the very different strategy in Kham and Amdo. In 1955 the government began enforcing land collectivisation, in the process forcing nomads to settle. From late 1955 fighting was general in Kham and Amdo, and in early 1956 a major rebellion erupted in Kham, centred on Yunnan province.[14] This was put down with great ferocity—the army at one point bombing a monastery that peasants had taken refuge in—and thousands of refugees fled to Lhasa.

The Guomindang government in Taiwan and the CIA gave a limited amount of support to the rising, but it was in no sense inspired by them, and the limited arms they delivered made no substantial difference.[15]

11: Goldstein, Sherap and Siebenschuh, 2004, pp160-161.
12: Grunfeld, 1987, p111.
13: For graphic examples, see Goldstein, Sherap and Siebenschuh, 2004, pp169-184.
14: For an account of the revolt, see Shakya, 1999, pp136-144.
15: Grunfeld, 1987, pp144-160, and Shakya, 1999, pp 170-180, give very different accounts of the extent of CIA and Guomindang involvement. What comes out from both accounts is the tiny numbers of people involved and the lack of impact on the outcome.

However, the limited US involvement no doubt contributed to the Chinese Communist Party's determination to strengthen its control, and both this and the presence of refugees from Kham increased the tensions. In March 1959 Lhasa rose in rebellion, and following its defeat the Dalai Lama and thousands of refugees fled to India.

In Chinese accounts the rising is represented as directly organised by the CIA. Tsering Shakya argues convincingly that this is not the case, and that the original demonstrations were "not only expressing their anger against the Chinese but their resentment against the Tibetan ruling elite who, they believed, had betrayed their leader". He also points to the leading role played by the artisans' guilds and mutual aid societies of the (very small) Tibetan working class.[16] Following the revolt the Communist Party's cautious policy was abandoned, and full Chinese control was established over central Tibet.

Central Tibet was at least spared the Great Leap Forward,[17] but the effects in Kham and Amdo were among the worst anywhere in China. In one county in Qinghai province half the population may have starved to death. Across Gansu, Qinghai and Sichuan provinces, the death toll was far higher than in the rest of China. The rural economy in these areas was desperately fragile and on the edge of subsistence at the best of times. On top of the repression of the revolts and the forced collectivisation, Chinese officials now forced peasants to give up their traditional crop of barley and instead grow wheat, which would not thrive at such high altitudes.[18] The Panchen Lama, who was born in Qinghai, wrote a long letter to Mao pleading for a change of policy, and was first criticised and then jailed for it.

Living standards fell across Tibet because of the Great Leap Forward, and the demands of the huge Chinese government and military presence. But far worse was to come in the Cultural Revolution from 1966 onwards.[19] In Tibet the campaign turned into a wholesale assault on Tibetan culture. As one generally pro-Chinese writer described it:

16: Shakya, 1999, pp192, 199.

17: The Great Leap Forward (1958-1960) was a disastrous attempt to accelerate industrial development by collectivising agriculture, and massively increasing exploitation in both the cities and the countryside.

18: Becker, 1996, is the best general source on the Great Leap Forward. He describes its impact on Amdo and Kham on pp166-182.

19: The "Great Proletarian Cultural Revolution", which began in 1966, was launched by Mao and his close supporters to maintain his grip on power and remove their political rivals. It involved unleashing "Red Guards" (mainly young people) to attack "authority figures" and symbols of pre-revolutionary China in a movement that from early 1968 onwards spun increasingly out of Mao's control.

> The damage caused by the wanton destruction and the fighting was awesome... Even if we discount stories of thousands of Tibetans killed... verifiable activities of the Red Guards are horrifying enough. There were killings and people hounded into suicide. People were physically attacked in the streets for wearing Tibetan dress or having non-Han hairstyles. An attempt was made to destroy every single religious item.[20]

In 1969 a millenarian revolt broke out, marked by the wholesale slaughter of Chinese and Tibetan officials, which at its high point covered 18 counties.[21] The rebels were hunted down by the Chinese army and their leaders publicly executed in Lhasa, but the revolt showed the scale of China's failure in Tibet.

Mao's death in 1976 allowed senior leaders around Deng Xiaoping to abandon his failed economic strategy in favour of "market socialism" and an opening to the world economy. Part of the new strategy involved allowing a greater degree of personal freedom in order to win back popular support. In Tibet this led to the government admitting that most Tibetans had become worse off. Hu Yaobang, a leading associate of Deng, pushed through a radical change of policy which delivered emergency relief, reopened monasteries and rapidly promoted Tibetan officials. Almost half of all the Chinese in Tibet left between 1980 and 1985.[22]

These partial reforms undoubtedly increased living standards and removed some of the worst restrictions on Tibetans' everyday lives. However, they also whetted appetites for much greater change. The demonstrations in September 1987 were staged to coincide with the Dalai Lama's visit to the US for maximum Chinese embarrassment and were followed by a series of smaller protests. Hu Yaobang was sacked at the start of 1987 because of arguments over national economic strategy, but this was widely seen as a repudiation of "liberalisation" in Tibet.

At the beginning of 1989 the Panchen Lama died, removing the most senior supporter of the Chinese occupation, a month after the Communist Party secretary in Tibet had been sacked for being too soft on Tibetan nationalism. Funeral marches for the Panchen Lama turned into scuffles with the police, and on 5 March 1989 the police opened fire on a demonstration, killing at least ten people. The riots that followed were the largest since 1959, and the centre of Lhasa was taken over for three days. Hundreds

---

20: Grunfeld, 1987, pp180-181.

21: See Shakya, 1999, pp344-347.

22: See Wang Lixiong, 2002.

of people were killed and thousands jailed in the subsequent repression.[23] The protests were overshadowed by the much larger movement that began in Beijing in May 1989, and the "Tiananmen Square" massacre on 4 June 1989. Tibet was further hit by the nationwide crackdown that followed.

Finding the replacement for the Panchen Lama led to China losing yet more supporters in the Tibetan religious hierarchy. In Tibetan political theology the two most senior figures are the Panchen Lama and the Dalai Lama.[24] When one dies, their spirit is supposedly reincarnated in a young boy born at the time of their death. Monks from their monastery then find the reincarnation and bring them to the other person for their final approval. So the Dalai Lama chooses who will be the next Panchen Lama, and—crucially—the Panchen Lama chooses who will be the next Dalai Lama.

So the government organised a group of pro-Chinese monks to conduct a search for the reincarnation at the same time as the Dalai Lama was also searching. In 1995 both announced that they had found the reincarnation. The Dalai Lama's choice is presumed to be under arrest, together with his family. But the government's attempt to impose their choice on the Panchen Lama's home monastery in Shigatse, Tibet's second city, produced open revolt in what had been the base of religious support for Chinese rule.[25] That revolt was followed by the departure into exile of numerous leading monks who had previously supported Chinese rule. The attempt to capture the religious succession only succeeded in confirming the Dalai Lama as the pre-eminent religious and political figure in Tibetan society.

Economic development has similarly done little to reconcile Tibetans to Chinese rule. After 1978 Deng Xiaoping's economic reforms led to very fast economic growth, firstly through the privatisation of agriculture and the consequent growth of rural industry and then, after 1991, through the growth of export-orientated small-scale industry in south eastern coastal provinces.[26] Tibet experienced neither of these, and has become more marginal to the Chinese economy as the gap between western and eastern China has widened. Although the Tibetan economy has grown at a faster rate than China as a whole, this is from an incredibly low starting point. And growth has been almost entirely in the state sector, either in

23: There are very few detailed descriptions of the riots in print. The details here are taken from Hilton, 2000, pp197-198. For the death toll, see "Chinese Said To Kill 450 Tibetans In 1989", *New York Times*, 14 August 1990.

24: For the origins of these institutions see Grunfeld, 1987, pp37-45. Siegel, 1986, pp137-162, gives a good materialist account of Buddhism.

25: Hilton, 2000, pp262-274.

26: For more on this see Hore, 2004.

government or construction work from which Tibetans are increasingly excluded.[27] The recent growth of tourism has similarly opened up far more job opportunities for migrant workers and traders from the rest of China than for Tibetans.

As the Tibetan Environmental Network pointed out:

> What is exceptional about the growth since 2000 is that it has been fuelled by a sudden increase in government spending by about 75 percent in 2001 alone. As a result the provincial government deficit in 2001 was worth over 70 percent of the provincial GDP.[28]

But almost none of this went into agriculture, still the major source of employment for most Tibetans. However, "development" is fast coming to rural Tibet in the shape of a very ambitious forced rehousing programme. The government claims to have rehoused 10 percent of Tibet's population in 2006 alone. In March 2008 the Channel 4 *Dispatches* programme "Undercover in Tibet" showed the reality of these settlements, which seem aimed at forcing the nomadic population to settle in reservations like those imposed on Native Americans or indigenous Australians: rural ghettos without employment ravaged by alcoholism. Partly motivated by security concerns, the government's rehousing plan may also be the precursor to a greater exploitation of Tibet's presumed mineral riches, which will further exclude Tibetans from economic development.

Marx famously argued:

> Religious suffering is at one and the same time, the *expression* of real suffering and a *protest* against real suffering. Religion is the sigh of the oppressed creature, the heart of a heartless world, and the soul of soulless circumstances. It is the opium of the people.[29]

That fits very closely with Tibet's experience since the 1950s. Tibetan Buddhism has been, if anything, strengthened as it has become the primary vehicle through which Tibetans express their opposition to Chinese rule. It may seem paradoxical that violent protests against occupation should be powered by a religion which stresses the acceptance of fate and suffering,

---

27: The government's figures are analysed by a Tibetan exile at: www.phayul.com/news/tools/print.aspx?id=8864

28: "Deciphering Economic Growth in the Tibet Autonomous Region", www.tew.org/development/eco.growth.tar.html

29: Marx, 1844 (emphasis in original).

but this is to miss the real significance of religious ideas, which is that they express both alienation and the belief that change is possible. Fifty years of Chinese rule in Tibet have seemingly done nothing to change the belief that the return of the Dalai Lama would be preferable to what currently exists.

Tibet and the left

The Tibetan protests provoked a wide range of responses on the left, not all of them expected. It was no surprise that the Cuban Communist Party should fully support China,[30] nor that almost all Communist Parties should echo them. The Sino-Soviet split is long over and for most Stalinists China at least still has a ruling Communist Party. Venezuelan president Hugo Chavez's declaration—"We are strongly with the people of China. We fully support the People's Republic of China on the Tibet issue. It has our complete and unrestricted solidarity"—could equally be explained by the growing oil trade between Venezuela and China.[31]

But it was a genuine shock to see anti-capitalists such as Michael Parenti[32] or Slavoj Zizek[33] defending Chinese rule as being good for ordinary Tibetans. Their arguments echoed a more widespread unease among some left wingers about any criticism of China, exemplified by this comment on a *New Statesman* article:

> Until we no longer interfere or invade other countries and leave them in peace to live as they choose, we have no right to criticise China. The first step would be to leave Iraq and Afghanistan.[34]

Of course, much of the press coverage of the protests was hypocritical, and it is important to expose such hypocrisy. But at the same time there is no contradiction between opposing the occupations of Iraq, Afghanistan and Palestine, and opposing the occupation of Tibet. The old slogan "Neither Washington nor Moscow"—that we can refuse to support either ruling class—needs dusting off in a context in which much of the left seems to see China as a bulwark against an all-powerful US imperialism.

30: See, for example, Aida Calviac Mora, "Cinco Preguntas Sobre Tíbet", *Granma*, 11 April 2008, English translation available online at: www.walterlippmann.com/docs1875.html

31: "We Fully Support China On Tibet: Hugo Chavez", the *Hindu*, 11 April 2008; "Venezuela And China Boost Ties With Refinery Deal", Reuters, 10 May 2008.

32: Parenti, 2007.

33: Slavoj Žižek, "No Shangri-La", letter to *London Review of Books*, 24 April 2008, www.lrb.co.uk/v30/n08/letters.html

34: Comment on *New Statesman* website, www.newstatesman.com/200805010020

The substance of left opposition to the Tibetan protests, the Tibetan independence movement and the idea of Tibetan independence (which are three quite different things) comes down to three main strands: China has a right to be in Tibet; the Dalai Lama is a reactionary who would restore feudalism; and the Tibetan movement is a pawn of US imperialism.

"Tibet is Chinese"

In one sense, we can deal with this summarily, since the unspoken second half of this sentence is, "whatever the Tibetan people may think about it". Even if the Chinese account of Tibetan-Chinese relations since the 8th century was correct in every respect, the Tibetan people would still have the right to self-determination. China's claims to Tibet are based on imperial conquest, rather than any impulse on the part of ordinary Tibetans to become Chinese.

Yet this account of history is not correct. While it is true that Tibet was conquered on several occasions, it is also true that Chinese rule was either driven out or notional for much of Tibet's past. Around 1250 Tibet was incorporated into the Mongol Empire, which had been established by Chinggis Khan (better known in English as Genghis Khan), and which later formed the Chinese Yuan dynasty. The overthrow of Mongol rule by the ethnic Chinese Ming dynasty in 1368 loosened these ties, and, according to one generally pro-Chinese historian, "from 1566 to the fall of the Ming in 1644, political relations between Beijing and Lhasa were apparently non-existent".[35]

The Ming were in turn overthrown by the far more expansionist Qing dynasty, who reasserted Chinese control over Tibet, but this was essentially a diplomatic fiction. As Wang Lixiong argues:

> Between 1727 and 1911, the principal symbol of Chinese sovereignty over Tibet was the office of the Residential Commissioner... The imperial presence in Lhasa, however, consisted "solely of the commissioner himself and a few logistical and military personnel"... Speaking no Tibetan, they had to reply on interpreters and spent most of their time in Lhasa, making only a few inspection tours a year outside the city.[36]

In 1904 British troops briefly invaded Tibet in what was essentially a prolonged looting expedition unsanctioned by London. The invaders

35: Grunfeld, 1987, p37.
36: Wang Lixiong, 2002.

withdrew after less than a month, leaving both the Tibetan government and Chinese control severely weakened. When the Qing empire collapsed in 1911 the Tibetan government saw its opportunity and expelled all Chinese troops and officials. From then until 1949 Tibet was effectively independent. In some histories this is seen as a period of British domination, but while the Tibetan government took some arms from Britain and increasingly traded with India, they refused to allow the British to open a permanent mission. On one count, by the late 1940s there were just six Europeans in Tibet, three of whom were employed by the Tibetan government.[37]

What the history shows is that, given the opportunity, the tendency was for Tibetans to reject Chinese rule or influence. Defending China's "right" to rule Tibet means, in effect, defending China's right to impose its control over the population by force. This is often justified by the awfulness of the old Tibetan regime, and it is true that Tibet before 1959 was a desperately poor, disease-ridden society ruled by feudal slave-owners. But the same defence could be made of British, French, Spanish or Dutch colonial conquests in Africa, Asia and Latin America. Almost none of the societies they colonised were ones we would want to see re-established. And the claims of "historical progress" in Tibet seem increasingly difficult to justify.

The Dalai Lama

Many people who would agree so far are nevertheless put off by the pro-Tibetan movement. This is even true of some who have previously been active in that movement.[38] It is certainly true that the Dalai Lama's government in exile doesn't look anything like "classic" national liberation movements such as the Algerian or Vietnamese National Liberation Fronts or the Sandinistas. It is also very easy to mock the celebrity hangers-on that the Dalai Lama attracts. And, as I noted above, the movement doesn't even demand independence, just greater autonomy inside China.

However, for socialists, the judgement about whether to support a nationalist struggle is separate from whether we support any particular organisation that claims to represent that struggle. So over the past 60 years Kurdish organisations in Iraq and Iran have variously allied with their own ruling classes, neighbouring countries' ruling classes, or Russian and American imperialism, depending on the shifting politics of the region.[39] None of this makes the national oppression of the Kurdish people any less real.

37: Grunfeld, 1987, pp72-78.
38: See, for example, French, 2003.
39: See, for example, Chaliand, 1980.

Similarly, supporting a particular liberation movement is not conditional on the politics of its leadership, still less on the likely nature of the society that might result from its victory (although it would be practically impossible to restore the pre-1959 society, and there is no evidence that the Dalai Lama or those around him wish to do so). What the Dalai Lama does think or want is curiously difficult to pin down. In one much quoted interview he argued:

> Of all the modern economic theories, the economic system of Marxism is founded on moral principles, while capitalism is concerned only with gain and profitability. Marxism is concerned with the distribution of wealth on an equal basis and the equitable utilisation of the means of production. It is also concerned with the fate of the working classes—that is, the majority—as well as with the fate of those who are underprivileged and in need, and Marxism cares about the victims of minority-imposed exploitation. For those reasons the system appeals to me, and it seems fair.[40]

However, it seems unlikely that this is what he focuses on when meeting George Bush or Angela Merkel. In reality the leadership of the Tibetan movement is composed of people who want to form a government and are willing to do whatever deals necessary with other world powers in order to achieve that aim. They may look very different from, say, the leadership of the Irish Republican movement or the African National Congress, but politically they are remarkably similar.

A tool of US imperialism?

This is the most serious of the arguments against supporting the Tibetan struggle, and it is a widely held view. The Israeli peace activist Uri Avnery summed this attitude up in a widely reprinted article:

> I support the Tibetans in spite of it being obvious that the Americans are exploiting the struggle for their own purposes. Clearly, the CIA has planned and organised the riots, and the American media are leading the worldwide campaign.[41]

Tsering Shakya has rebutted this specific argument in the *Far Eastern*

40: "Tibet and China, Marxism, Nonviolence", http://hhdl.dharmakara.net/hhdlquotes1.html
41: Avnery, 2008.

Economic Review,[42] but the wider point about US involvement cannot be so easily dismissed. It is certainly true that the CIA was involved in training and arming some Tibetans who took part in the risings in 1956-8, though the numbers were small and their impact minuscule. The CIA also helped with the Dalai Lama's escape in 1959, though this involved just two agents.[43] But compared to what the US would later spend in Vietnam, Laos and Cambodia, the amounts of money and equipment involved were tiny, and diminished throughout the 1960s as the Vietnam War escalated. Grunfeld notes that by 1970 "CIA money had totally dried up" and concludes that "American involvement did not alter the situation in Tibet in any discernable manner after 1959".[44]

Following Richard Nixon's visit to China in 1971, China and the US entered into a mutually convenient alliance against the USSR—something often forgotten by those who see China as a constant target for US imperialism. Part of the price for that alliance was ending all US support for Tibetan emigre groups. As tensions between China and the US have again risen, with US strategists becoming worried about Chinese economic, political and military competition, US support for some Tibetan organisations has started up again. The National Endowment for Democracy, which was heavily involved in the "colour revolutions" in Ukraine and Georgia, seems to be one of the major conduits for this.

In 2006, the last year for which they have published figures, they admitted giving just under $300,000 (about £150,000) to 11 organisations in Tibet—peanuts, essentially. More money was given to just one trade union project in Pakistan.[45] And both sums pale into insignificance when compared to the tens of millions given to the various Afghan mujahideen groups. In practical terms Tibet is irrelevant to the overall nexus of Chinese-American relations. As two supporters of the neoconservative Project for a New American Century wrote:

> Americans need to recognise that, for better or worse, we have no practical alternative to Chinese sovereignty in Tibet... It would be pointless to

42: Shakya, 2008. The article also traces some of the dividing lines between the various Tibetan exile movements and the resistance inside Tibet.

43: Norbu, 1994, p195.

44: Grunfeld, 1987, pp157, 158.

45: Figures available online, www.ned.org/grants/06programs/grants-asia06.html

make independence a goal when there is no chance that such a goal can be reached.[46]

This could change. If the US seriously targeted China as a military opponent, it would undoubtedly try to use the Tibetan movement as allies in that strategy (the "Kosovo option"), and undoubtedly parts of the Tibetan movement would go along with that. But that is a very big if. While the possibility cannot be ruled out, it is a long way from the current reality. The Chinese and American economies are deeply enmeshed and each reliant on the other for future growth (or in the US's case a shallower recession), as well as being competitors. Politically, too, China and the US are both rivals and allies. China supports the continuing occupation of Afghanistan and the wider "global war on terror" (not least because it provides a cover for repression in both Tibet and Xinjiang), while the US relies on Chinese help in dealing with North Korea's nuclear weapons.[47] While that situation lasts, the US may give some token support to Tibetan organisations for their nuisance value towards China, but Tibet will remain peripheral.

And while the various Tibetan organisations will undoubtedly accept whatever they are offered, Tibetan nationalism, and the various organisations that represent it, cannot be reduced to a tool of American imperialism. Instead it draws its support from the harsh realities of Chinese rule in Tibet and from the fact that most Tibetans continue to refuse to accept it. Recognition of that national oppression and resistance is mostly missing from discussions of whether the Western left should support the Tibetan independence movement.

But this recognition should be the starting point. Whether we support particular Tibetan organisations, whether Tibetan independence is feasible, what the borders of an independent Tibet might be: these are secondary questions. What is important about the riots and protests of 2008 is that they have conclusively demonstrated the vitality of Tibetan resistance to Chinese rule and an awakening of Tibetan national consciousness in Tibetan areas of Qinghai, Gansu and Sichuan. Just as socialists welcome other challenges to the power of the Chinese state, so we should welcome these.

46: Bernstein and Munro, 1998, p214.

47: For fuller details, see Hore, 2004.

Conclusion

The immediate reaction inside China to the Tibetan riots was a rise in Chinese nationalist sentiment,[48] and a flurry of demonstrations against the disruption of the Olympic torch relay (directed, for some reason, particularly against the French supermarket chain Carrefour).[49] An upsurge of nationalism inside China is not all good news for the government, however, as such demonstrations can easily turn against them. Even if they do not, the publicity they attract makes it easier to stage protests on other subjects.

The past ten years have seen a major escalation of public protests, strikes and riots in both urban and rural China, with the government effectively conceding both the right to demonstrate and the right to strike.[50] However, there are fundamental differences between these outbreaks and the Tibetan protests. One aspect of this is shown in the respective death tolls. Across China as a whole in the first half of 2005 about 100 people were killed in mass protests[51]—fewer deaths than in Tibet's March protests. This reflects a fundamental difference in the "rules of engagement" for both the army and police. In Han-majority areas they very rarely open fire on demonstrators; indeed most protests do not attract the army or armed police at all. In Tibetan and Muslim areas an armed response is the norm.

In China localised protests are often aimed at local officials or managers, in the belief that the central government will put things right once they know the truth. As one peasant activist put it, "Some wicked officials have sealed off the centre from reality. If peasants do not lodge complaints, the emperor will never know what is going on. If I tell the emperor, he should thank me and take care of me".[52] By contrast, protests in Tibetan (and Muslim) areas are directed, or seen to be directed, against the central state, and there is no ambiguity about whether local officials are carrying out the centre's policy.

There has been some support for the Tibetan protests inside China. A group of dissident intellectuals circulated a petition accusing the government of fanning racism and calling for talks with the Dalai Lama.[53] And

48: See, for example, "Sympathy On The Streets, But Not For The Tibetans", *New York Times*, 18 April 2008, www.nytimes.com/2008/04/18/world/asia/18china.html

49: "Protests In China Target French Stores, Embassy", *Washington Post*, 20 April 2008.

50: I recently reviewed a number of books on these topics in Hore, 2008.

51: Leonard, 2008, p73.

52: Quoted in O'Brien and Li, 2006, p45. The reference is not to any particular emperor, but rather to traditional views about the social contract between rulers and subjects.

53: "Chinese Intellectuals Condemn Tibet Crackdown", *International Herald Tribute*, 24 March 2008, www.iht.com/articles/2008/03/24/asia/chinasub.php

there are significant numbers of Chinese, often influenced by Buddhism, who respect Tibetan culture and will be to some extent sympathetic to the protests, if not to Tibetan independence. But for the moment these are very definitely minority views. A recent Muslim protest in Xinjiang may have been inspired by the Tibetan riots,[54] but this will, if anything, harden the Chinese response to future protests.

China's rule in these areas is primarily motivated by strategic reasons and nationalist pride, rather than for the direct economic benefit they bring—indeed it is probable that the occupation of Tibet costs more than the profit Tibet produces, though this is less likely to be true of Xinjiang. Between them Muslim-majority and Tibetan-majority areas of China account for one third of China's total surface area, though less than 2 percent of the population. The Chinese state will not concede control of these areas short of a major upheaval across China as a whole. At the same time the spread of economic development which marginalises the majority of the population is likely to deepen the resentment of Chinese rule. The potential for further clashes is huge, and so it matters that the left understands which side it should be on.

54: "Muslim 'Separatists' Protest As Unrest Spreads In China", *Guardian*, 2 April 2008, www.guardian.co.uk/world/2008/apr/02/china

References

Avnery, Uri, 2008, "Tibet and Palestine", *Counterpunch*, 7 April 2008, www.counterpunch.org/avnery04072008.html

Barnett, Robert, 2006, *Lhasa: Streets with Memories* (Columbia University).

Barnett, Robert, 2008, "Thunder from Tibet", *The New York Review of Books*, volume 55, number 9, 29 May 2008, www.nybooks.com/articles/21391

Becker, Jasper, 1996, *Hungry Ghosts* (John Murray).

Bernstein, Richard, and Ross H Munro, 1998, *The Coming Conflict with China* (Vintage Books).

Chaliand, Gerard (ed), 1980, *People Without a Country—the Kurds and Kurdistan* (Zed).

French, Patrick, 2003, *Tibet, Tibet* (HarperCollins).

Goldstein, Melvyn C, Dawai Sherap and William R Siebenschuh, 2004, *A Tibetan revolutionary: The Political Life and Times of Bapa Phuntso Wangye* (University of California).

Grunfeld, A Tom, 1987, *The Making of Modern Tibet* (Zed).

Hilton, Isabel, 2000, *The Search for the Panchen Lama* (Penguin).

Hore, Charlie, 2004, "China's Century?", *International Socialism 103* (summer 2004), www.isj.org.uk/index.php4?id=50

Hore, Charlie, 2008, "China's Growth Pains", *International Socialism 118* (spring 2008), www.isj.org.uk/index.php4?id=430

Leonard, Mark, 2008, *What does China Think?* (HarperCollins).

Marx, Karl, 1844, *Introduction to a Contribution to the Critique of Hegel's Philosophy of Right*, www.marxists.org/archive/marx/works/1843/critique-hpr/intro.htm

Norbu, Jamyang, 1994, "The Tibetan Resistance Movement and the Role of the CIA", in Robert Barnett and Shirin Akiner (eds), 1994, *Resistance and Reform in Tibet* (Christopher Hirst).

O'Brien, Kevin J, and Lianjiang Li, 2006, *Rightful Resistance in Rural China* (Cambridge University).

Parenti, Michael, 2007, "Friendly Feudalism: The Tibet Myth", www.michaelparenti.org/Tibet.html

Shakya, Tsering, 1999, *The Dragon in the Land of Snows* (Pimlico).

Shakya, Tsering, 2008, "The Gulf between Tibet and its Exiles", *Far Eastern Economic Review*, volume 171, number 4, www.feer.com/essays/2008/may/the-gulf-between-tibet-and-its-exiles

Siegel, Paul, 1986, *The Meek and the Militant* (Zed).

Wang Lixiong, 2002, "Reflections on Tibet", *New Left Review 14* (March-April 2002), www.newleftreview.org/?view=2380

Zimbabwe: imperialism, hypocrisy and fake nationalism

Leo Zeilig

One striking feature of Zimbabwe's crisis has been the vocal support of the British and US governments for "democratic change". In April, George Bush's assistant secretary of state for African affairs, Jendayi Frazer, undertook a whirlwind "democracy tour" to support Zimbabwe's opposition. Global media outlets seemed to be counselling the opposition to organise a mass uprising in defence of the results, while the International Monetary Fund and World Bank promised to provide funds to an opposition-led government. Gordon Brown, who has, for years, been part of a government deporting Movement for Democratic Change (MDC) activists back to Zimbabwe, added his voice to calls for election results to be respected.

By contrast, South African president Thabo Mbeki divided his own African National Congress by standing beside Robert Mugabe's regime and declaring that there was "no crisis in Zimbabwe". The violence perpetrated by Mugabe's ruling Zanu-PF party, once it registered the scale of its defeat in March's elections, was proclaimed to be in the name of anti-imperialism and independence. So on 18 April, the anniversary of independence, Mugabe called on Zimbabweans "to maintain utmost vigilance in the face of vicious British machinations and the machinations of our other detractors, who are allies of Britain".[1] By the middle of May at least 22 people had been killed, thousands made homeless and scores of activists beaten.[2]

1: "Mugabe Attacks Opposition And UK", *BBC News*, 18 April 2008.

2: "Zimbabwe Violence Reaches Crisis Levels", Amnesty International, 16 May 2008.

Faced with the hypocrisy of Western governments, many have believed Zanu-PF's claim to be defending the country's sovereignty against imperialism. For others, the regime is the incarnation of evil, personified by the president. Amazon lists seven biographies of Mugabe written in the past six years; each promises to get to the "man behind the monster".[3] Neither of these explanations helps us understand what is happening in Zimbabwe.

The country's crisis is tied inextricably to the nature of global power. As Brian Raftopoulos, a Zimbabwean activists and academic, has explained:

> On the one hand there is a global superpower, espousing liberal democratic values, but policing a global economic agenda producing widespread global improverishment; on the other hand this system of global inequalities is breeding an authoritarian nationalism in countries like Zimbabwe.[4]

It is vital that socialists steer a path between the authoritarian nationalism of Mugabe's Zanu-PF and Western imperialism that is seeking to pull Zimbabwe back into its orbit.

South Africa, SADC and the West

South Africa's President Mbeki has been an important, if malevolent, factor in the crisis. His government has deported thousands of Zimbabweans. Those who remain, living in desperate poverty, are demonised by politicians and the media, and face violent attacks.[5] Mbeki has sought to shield Zanu-PF from regional and international criticism, and refused to engage with the opposition.

Following a violent assault on the leadership of the opposition MDC on 11 March 2007 and a wave of repression across the country, the Southern African Development Community (SADC) launched a new initiative for a mediated solution to the Zimbabwean crisis. Mbeki was the official facilitator of this process. At emergency SADC meetings in April this year the regional organisation showed its true colours by congratulating "the SADC facilitator, President Mbeki...for the role they had played in helping to contribute to the successful holding of elections [in Zimbabwe]".[6] Still,

3: The most recent biography is Heidi Holland's *Dinner with Mugabe: The Man Behind the Monster*.

4: Cited in Kibble, 2003.

5: In May 2008 there were attacks against "foreigners" in some of South Africa's poorest townships.

6: "2008 First Extraordinary Summit of Heads of State and Government", SADC communique, 13 April 2008. Negotiations in the run-up to the elections did see the MDC accept the principle of a "transitional government", which would include an honourable departure for Mugabe and power sharing with members of Zanu-PF.

the negotiations had certain important consequences for the elections. Parliamentary seats were increased from 120 to 210 and the results from individual polling stations were posted outside the station.

Many commentators argue that Mbeki's support for Mugabe has been driven by his solidarity for another national liberation movement, but his principal motivations are quite different. Mbeki sees the crisis in Zimbabwe as a lever Western powers can use to reassert themselves over a former colony to the detriment of South Africa's companies, which are already active in Zimbabwe, and its political influence.

Zimbabwe is a hive for regional and international capital. No sector illustrates this more than minerals. The country is home to the second greatest platinum reserves in the world—a centre of activity for the South African mining giant Impala. There are new mines developing in the Midlands province, and the London-based mining company Rio Tinto extracts diamonds among other minerals. But the development of this sector has been hindered by the economic crisis. Rio Tinto has seen the quantity of diamonds mined drop from 240,000 carats in 2006 to 145,000 in 2007. The company blames the erratic power supply and has recently started to import power directly from Mozambique. The mining corporations are desperate to see stability in Zimbabwe to secure their investments from both the possibility of nationalisation (recently threatened by Mugabe) and the current economic chaos.[7]

A new Zimbabwean government might be less dependent on Mbeki's patronage, but South African companies would still benefit from a post-Mugabe settlement. Undoubtedly a new MDC-led government could see the intervention of the British and American governments, quickly followed by the IMF and the World Bank. This would certainly not signal the end of hardship for millions of Zimbabweans, though it would temporarily alleviate the crisis and open political debate for activists and their organisations.

The election

The election on 29 March 2008 seemed to open the way to a new future in Zimbabwe. Several days before the election some could see that the tables had turned on Mugabe and Zanu-PF. One opposition candidate wrote the day before the poll, "Everyone now seems to be happy to say to me 'Mugabe must go', which last elections everyone felt, but no one dared to say." The

7: "Output At Rio Tinto Zimbabwe Diamond Mine Down Forty Percent", Reuters, 27 February 2008.

opposition MDC, led by Morgan Tsvangirai, was going to make massive gains. Even before votes had been cast it was clear that "support is huge and varied. From everything I have seen...the MDC (Tsvangirai) is massively popular. The Mugabe Zanu-PF is massively unpopular".[8]

However, even with the changes to the elections following the SADC mediation, few believed that an MDC victory was possible. Zanu-PF had turned Zimbabwe into a very uneven playing field. New repressive legislation was introduced in 2007 and Zimbabwe's *Socialist Worker* commented in February this year that "the entire state machinery, including the media, is being mobilised to ensure a Zanu-PF victory... War veterans and chiefs will ensure that rural areas...remain no-go areas for the opposition. Thousands of rural families are receiving ploughs, carts, harrows".[9]

Despite all this, parliamentary elections gave Tsvangirai's MDC 99 seats compared to Zanu-PF's 97; a breakaway faction of the MDC led by Arthur Mutambara won ten seats and the former Zanu-PF minister Simba Makoni's organisation won eight. This gave the combined opposition a majority in the 210-seat parliament.[10] On 2 May the electoral commission was finally forced to concede that Tsvangirai had beaten Mugabe in the presidential poll, even if, it claimed, he had not broken through the 50 percent needed to avoid a run-off.[11] But these results conceal a far more remarkable truth.

There was an electoral revolution in rural areas. Ruling party strongholds, held since the first multiracial elections in 1980, fell for the first time to the opposition. While Tsvangirai's MDC won most of their seats in urban areas, the party also triumphed in the previously Zanu-PF provinces of Manicaland and Masvingo. Mutambara, who broke with Tsvangirai, was routed by Tsvangarai's MDC in the urban constituency of Zeneza in Harare. Welshman Ncube and Gibson Sibanda, respectively Mutambara's secretary general and vice-president, were also defeated in the urban consistencies they contested in the southern city of Bulawayo.

Unlike in previous elections, there was no widespread campaign of violence and the MDC operated in a degree of openness. Rural areas had also been politicised by the crisis. Policies pursued by Zanu-PF backfired

8: Michael Laban, "Zim Fight On", 28 March 2008.

9: "Crisis In Zimbabwe: No To Fake Elections! Jambanja Ndizvo!", *Socialist Worker* (Zimbabwe), February 2008.

10: The combined opposition also triumphed in the senate elections, where Tsvangirai's MDC won 24 seats, Zanu-PF 30 and Mutambara's faction of the MDC six.

11: The official results gave Tsvangirai 47.9 percent of the popular vote and Mugabe 43.2 percent.

dramatically. The urban slum clearances of 2005, known as Operation Murambatsvina, had driven out informal traders and market stalls, often run by workers retrenched in the recent crisis who formed the backbone of MDC support. This helped create a rural class that campaigned for and supported the MDC in the countryside. New areas opened up to the opposition.

In the past the regime was able to shore up support with its land redistribution programme; now this was no longer possible. People who had been given plots of land lacked the resources to cultivate them, while the ruling party's big supporters benefited from handouts. Willias Mudzimure, an MDC MP, explained that in rural areas Mugabe's "pro-poor" bribes and "anti-imperialism" fell on deaf ears:

> Mugabe's land reform has been a catastrophe, so he couldn't talk about that. Moreover, when he tried to win votes by giving out tractors and farm implements these just went to the fat-cats who now have the land... So he fell back into talking about the 1970s war against Ian Smith. This meant nothing at all to young people.

Mugabe then attempted to blame the British, but again no one was fooled:

> People would say, 'You've said that before but what are you doing about it?' They were in no mood for more excuses.[12]

When the parliamentary results were announced the shock was palpable across the country. When the regime recovered, it refused to allow results from the presidential poll to be released. First, the government relocated the electoral commission office to a secret place; then, absurdly, the regime accused the commission of manipulating results in favour of the MDC. The ruling party even demanded a run-off for the presidential elections before the results of the first round were known. As Nelson Chamisa, a spokesperson for the MDC, explained, this was the equivalent of a student requesting a re-mark of an examination when the results had not been announced.[13]

Zimbabwe's ruling class was more divided then at any time in recent

---

12: Cited in Johnson, 2008.

13: "Velvet-glove Inaction Will Have Dire Results", *Sunday Independent* (South Africa), 13 April 2008.

years. However, once Zanu-PF had recovered from the surprise defeat, repression against the opposition quickened. Many of the political forces that the government had developed to defend the regime were resuscitated. War veterans—a category of ex-fighters in the 1970s guerrilla war, though frequently with few actual former combatants—were used to make high-profile seizures of some of the remaining white farms. But the worst attacks were not carried out against the dwindling class of white landowners—MDC activists and supporters were the main targets. Zanu-PF youth were also used in the wave of repression.

The Zimbabwe Peace Project reported beatings and torture against suspected opposition supporters in Mashonaland East and Mashonaland West, previously ruling party strongholds. "Bases of torture" were established in one constituency. Elsewhere war veterans drew up "lists of MDC activists who are then systematically targeted for abuse".[14] In Mashonaland West one MDC election agent and three activists were forced in flee into the mountains after receiving threats of violence. Members of the Zimbabwe Association of Doctors for Human Rights recorded 157 cases of injury from organised violence and torture between the election on 29 March and 14 April, a fraction of the actual level. Violence increased again in the following weeks. The UN reported at the end of April that politically motivated arson had destroyed almost 300 homes. Both electoral commission and MDC polling officers were arrested.[15] The opposition reported that at least ten of their supporters had been killed.[16] The political space that had opened temporarily during the election campaign in March was shut down.

The MDC's response

The MDC had added up the results posted outside polling stations and declared itself victor of both the parliamentary and presidential vote. According to the party's own calculations, Tsvangirai won 50.3 percent in the presidential poll compared to Mugabe's 43.8 percent. This was enough to avoid a second round run-off. Tsvangirai and his secretary general, Tendai Beti, gave dozens of interviews to the international media declaring that the MDC was now the constitutionally elected government. But still the party advised caution. Tsvangirai urged "people to remain calm...we would rather caution against opportunistic reaction...at the end of the day

14: "Zimbabwe Behind New Wave Of Human Rights Abuses", Human Rights Watch, 30 April 2008.

15: "UN Experts Concerned About Deteriorating Human Rights Situation In Zimbabwe", United Nations Office at Geneva, 29 April 2008.

16: "Zimbabwe Police Raid Opposition Elections Office", Associated Press, 25 April 2008.

they should wait...until the results are known".[17] If decisive leadership was needed then the "people" would be sorely disappointed.

The MDC contested the results in the high court, which rejected the opposition's bid to compel the release of the presidential election results. The MDC delegation to the extraordinary SADC summit of heads of state, called to discuss the Zimbabwe crisis, was also frustrated. The cautious MDC strategy also infected their call for a stayaway (a one-day general strike) on 15 April, which was issued with scant regard to grassroots mobilisation and organisation. The stayaway was news internationally but not in Zimbabwe. MDC supporters in the capital, Harare, were reported on the BBC as saying they "did not even know about this stayaway".[18]

As the regime tightened repression the MDC called for international intervention: "Outsiders should come and intervene to try to persuade this regime it has no legitimacy".[19] The window, when action could have been escalated, had now been closed. Zanu-PF unleashed its repressive apparatus in a bid to hang on to power.

The role of the military

There was a moment when Zanu-PF seemed to have accepted defeat. According to Tsvangirai, the day after the elections the ruling party sent an emissary to see him. The emissary explained that they had been trying to persuade Mugabe to go. "Mugabe has accepted," Tsvangirai was told. "Now the question is how you can accommodate us." But the hawks in the military refused to accept a transfer of power.[20]

Zimbabwe's military has played a vital role in the country since independence. In the recent crisis leading military figures have maintained an iron grip on power. The six commanders of the security services—chiefs of the defence force, the army, the air force, the commissioners of police and prison services, and the head of the national intelligence organisation—are members of the joint operational command (JOC) and are widely recognised as the prime movers behind Mugabe's throne.

Take two figures from the JOC. Air Marshal Perence Shiri commanded the North Korean trained Fifth Brigade in the 1980s, which stands accused of the massacre of 20,000 so-called "terrorists" in Matabeleland. The defence force chief, Constantine Chiwenga, has been a major player

17: "Hot Seat Interview", SW Radio Africa transcript, 11 April 2008.
18: "Zimbabwe Opposition Strike Fails", BBC News, 15 April 2008.
19: "Hot Seat Interview", SW Radio Africa transcript, 11 April 2008.
20: "Outrage And Consequence In The Twilight Of A Tyrant", *Business Day* (South Africa), 30 April 2008.

in recent repression, rolling out youth militias, soldiers and war veterans to terrorise opposition supporters after the last election. Chiwenga has also enriched himself on the recent economic collapse, amassing a personal fortune and ensuring that his wife secures defence force supply contracts.

After the elections Mugabe offers less to the heads of the military. Previously he had been able to guarantee a degree of political support in the country. The elections showed that this has substantially evaporated. While Mugabe no longer ensures political cohesion, the JOC still have their guns. Mugabe now leads a divided ruling party and has been humbled in front of his commanders. Jonathan Moyo, a former Zanu-PF loyalist, explained on 29 April that the generals "can see that the political ship is sinking...because it no longer has a captain".[21]

Economic meltdown

Zimbabwe was once regarded as an exception in a continent of so-called failed states and bad governance. But since the late 1990s Zimbabwe's economy has been in free fall. From 2000 to 2005 the economy contracted by more than 40 percent. Today GDP per capita is estimated to be the same as it was in 1953. The country has the highest inflation rate in the world, soaring to 165,000 percent in February. There are regular shortages of basic goods, from food to fuel. At the beginning of 2007 the IMF calculated that 80 percent of the population lived below the poverty line. The Consumer Council of Zimbabwe stated in September 2007 that the people needed a minimum of Z$22 million a month to survive, far above the income of most Zimbabweans. Schools have collapsed, major hospitals suffer from basic shortages and unemployment is estimated at about 80 percent.[22]

While much of the economic crisis has been triggered by the land seizures, this explanation, favoured by the media commentators and IMF economists, gives only a fraction of the picture. Zimbabwe has been squeezed by the implementation of direct and indirect sanctions by Western countries. An international legislative structure has forced the pace of this strangulation; this has included the US Zimbabwe Democracy and Economic Recovery Act, which immediately cut access to international credit for the state and Zimbabwean companies. The reduction in aid and investment means that the country is now the recipient of less than $10 for every HIV-infected person, compared to the regional average of $100. As international funds have dried up the state has been largely incapacitated,

21: "Consistency Is The Virtue Of A Donkey", *City Press* (South Africa), 26 April 2008.
22: Chagonda, 2007.

with welfare provision now, often in the form of food aid, being provided by international agencies and NGOs.[23]

In the face of the economic collapse, the regime has been unable to sustain its attempts to capture support through a limited programme of reforms. Early this decade Zanu-PF introduced price controls on basic commodities but was forced to suspend them as massive shortages hit most shops. Since then the regime has swung wildly backwards and forwards between price controls and leaving the market to decide. Though land reform in Zimbabwe was celebrated across much of Africa as a historical retribution for colonial inequality, this was also a failure. In 2002 Zanu-PF stated that it intended to seize 8.5 million hectares of land before the presidential elections that year, the majority of land owned by white farmers. They succeeded in doing this only by 2003, as the pace of land seizures and occupations came to an end.[24] Although the regime could provoke high-profile land seizures, most of the large farms went to the fat-cats. Even for those Zimbabweans who were granted small parcels of the seized land, the regime did not have the resources to provide them with the training and equipment so that they could profitably cultivate their holdings.

Caught in a global economic vice, the regime resorted to what it had always done. Land and business contracts were distributed to cronies while Mugabe mouthed platitudes about "foreign powers". Zanu-PF relied increasingly on violence, as each reform was snatched back under pressure from the economic crisis. The regime's authoritarian neoliberalism has continued unabated, albeit chaotically, for years. For the past four years Zimbabwe's reserve bank governor Gideon Gono has pursued a haphazard programme of cuts in subsidies, privatisation and debt repayment.[25]

But even the regime's capacity to maintain its repression has suffered in the economic meltdown. Though politicians and security chiefs have remained insulated from hardship, with access to foreign currency and subsidised fuel, ordinary forces have not. While there is a statutory requirement to provide food rations to defence personnel, agricultural collapse has restricted the state's ability to do this. Soldiers lack new uniforms and the police are unable to carry out their routine patrols because of the lack of transport.

Zimbabweans have managed to survive hyperinflation largely through

23: "Only Mass Mobilisation Can Defeat The Dictatorship And Stop A Neoliberal Elitist Deal", *Socialist Worker* (Zimbabwe), April-May 2008.

24: Zeilig, 2007, p299.

25: He made a large payment to the IMF in 2005 and recently repayed some of the country's loan to the African Development Bank. "Zimbabwe Settles $700 ADB Loan", *Business Daily* (South Africa), 17 May 2008.

remittances from those who have managed to flee. The International Organisation for Migration reported in 2007 that approximately 3.4 million Zimbabweans have left the country since 2000. Most eke out an existence in the informal sector across the border in South Africa, but hundreds of thousands live in Britain. Some 74 percent of these Zimbabweans send money back home. These remittances come in direct transfers, through Western Union type cash-transfer companies. But more innovative methods have developed in recent years using websites and text messaging to turn cash into a tank of petrol, medication or a sack of mealie-meal (the staple food). One study in 2008 claimed that half of all households received overseas payments to pay for essential goods. Those households that did not were unable to cope.[26]

Zimbabwe's intifada

Zimbabwe's *biennio rosso* of 1996-8 saw a two-year revolt by students and workers. Strikes by nurses, teachers, civil servants and builders rippled across the country. In January 1998 housewives orchestrated a "bread riot" that became an uprising of the poor living in Harare's township.

The protests, strikes and campaigns were often explicitly against the government's programmes of structural adjustment. The first of these was introduced in 1991 as the Economic and Structural Adjustment Programme (ESAP), and was sponsored and advocated by the World Bank and IMF. The second, nicknamed ESAP II, was introduced in 1996. Factories closed, workers were laid off, and state funding to the national university and students was slashed.

Inspired by the largely urban movement the rural poor, veterans of the war for independence, started to invade white-owned farms. Initially the regime evicted the "squatters" and arrested the movement's leaders. In June 1998 the University of Zimbabwe in Harare was closed for five months and students started to demand that the opposition forces be organised into a national political party—a workers' party. Students organised protests, marching with workers. The revolt in Indonesia in 1998 against Suharto inspired those protesting in the streets.

These years of popular mobilisation and political debate were described by one activist as a "sort of revolution". Eventually the revolt gave way to the formation of the Movement for Democratic Change in September 1999. The new party was formed by the Zimbabwe Congress of Trade Unions. At this point the MDC was resolutely pro-poor, formed

26: Bracking and Sachikonye, 2008.

by the working class and for them. As Job Sikhala, a founding member, explained, "It was basically a party of the poor with a few middle class".[27] For many of those who had been involved in the exuberant protests that had rocked Zimbabwe, and who saw a parallel between the revolution in Indonesia and the protests in Zimbabwe, the new party would bring about a radical—and even socialist—transformation.

As the opposition movement grew, Zanu-PF started to worry. From being a government lauded by Western leaders, its leader dined by the queen, the regime made a "left turn" in an attempt to outflank the new party. War veterans, excluded for years from the independence settlement, were encouraged to invade white farms and were famously paid off through the War Veteran Levy in 1997. Now Mugabe talked of the third Chimerenga (anti-colonial uprising) and boasted about correcting a historical wrong by redistributing the land to the poor. Outside Zimbabwe thousands believed his claims and saw him as a genuine pan-Africanist. Inside the country the regime continued to arrest and torture workers and students.

Other political forces began to flock to the MDC. It was now seen by respectable NGOs, some white farmers and the middle classes as a force that could appease foreign interests and replace Zanu-PF with a government that respected property rights and business interests. So, under the influence of these groups, the MDC did not attack the hypocrisy of the regime but instead allied itself to those whose farms had been seized and who saw a continuation of structural adjustment as the solution to Zimbabwe's woes.

Zanu-PF strikes back

Zanu-PF's hallmark is violence. There has always been a remarkable degree of continuity, with pre- and post-election violence since independence in Zimbabwe. This violence, often justified as legitimate punishment against those audacious enough to vote for the opposition, has also frequently been instigated by a politicised Zanu-PF youth movement.[28]

The period from February 2000, when the government lost a vote on a new constitution, and the first elections contested by the MDC in June that year was marked by a rapid escalation of violence. The MDC almost won the election in 2000, gaining 57 seats against a backdrop of escalating violence. The regime maintained its pressure on the opposition in subsequent years. Along with the regime's politicisation of the war veterans it launched the National Youth Service (NYS). In 2001 the first NYS

27: Interview, Harare, 31 July 2003.

28: Kriger, 2005, p2.

camp was opened, named after the government minister who initiated the training, Border Gezi. One graduate described the courses as "a combination of things but mainly Marxism, socialism and business management".[29] This expressed Zanu-PF's schizophrenic mix of state capitalism and neoliberalism, set against a background of economic crisis. By 2006 the NYS had opened eight training centres. In the first five years of the NYS more than 40,000 youths had completed training programmes.[30]

By 2003 the regime seemed to have gained the upper hand. Zanu-PF increasingly sold itself internationally and at home as the true inheritors of the liberation movement. The MDC, by contrast, seemed cowed and unable to mount a serious resistance, either politically or on the streets. One decisive moment was in June 2003. The so-called "final push" on 2 June was launched by the MDC and meant to turn the tables on the regime with a week-long stayaway and a march on Mugabe's State House. No serious efforts were made to mobilise the available forces, leaving only students in Harare to organise a protest that was violently crushed. The week gave the MDC neither its international media coup nor mass action. The government scored another victory against the opposition and emerged stronger.

Zimbabwean activism began to suffer from "donor syndrome", as foreign funded NGOs increasingly filled the political vacuum that had been left by the failure of the opposition and the collapse in the economy. Zimbabwean-based organisations saw a massive inflow of funds. This distorted grassroots activism, leading to what has been described as the "commodification of resistance" as mobilisation is increasingly "paid for" from NGO funds.[31]

In parliamentary elections in 2005, also widely believed to have been rigged, the MDC lost 16 seats to Zanu-PF, which secured the necessary two-thirds majority needed to unilaterally change the constitution. Though the opposition had faced years of violent intimidation, the MDC was also by this stage hopelessly divided by a regime that had succeeded in outmanoeuvring it. The MDC became a contested space, with voices and groups criticising the direction of the leadership.

Munyaradzi Gwisai, a Zimbabwean socialist who was at one time inside the MDC, drew attention to the mistakes being made by the party's leadership, criticising the "hijacking of the party by the bourgeoisie, marginalisation of workers, adoption of neoliberal positions and cowardly failure to physically confront the Mugabe regime and bosses. It is...

29: Interview, Chegutu, 10-12 June 2003.

30: Shumba, 2006, p1.

31: Interview, Bulawayo, 22 May 2003.

imperative that the party moves much more leftward...in order to realign to its base".[32] But it was not only socialists who criticised the opposition. In 2003 one loyal MP, Job Sikhala, explained how the party core had become "really fat and thick...it is almost a party of the rich. You cannot look at a person who was with you during the foundation of the MDC as the person who is there now".[33] The disarray in the MDC eventually led to the party splitting in 2005, with one faction now being led by Arthur Mutambara, who had been an important student activist in the late 1980s and returned to Zimbabwe after an academic career in the US.[34]

Although important efforts were made to mount opposition to the ruling party after 2003, increasingly these did not come from the MDC. New organisations attempted to fill the vacuum. Women of Zimbabwe Arise is an activist organisation that led some of the most important protests in recent years, often on issues of violence against women. The Zimbabwe Social Forum, formed in 2002, became an alternative space for political discussion and a forum that attempted to group together those who sought to resist the regime. These organisations never became alternatives to the MDC, or attracted mass support; rather they can be seen as occupying a space that emerged only after the real movements of workers and students that had led to the formation of the MDC at the end of the last century were in retreat.

Contradictions of the MDC

The MDC has long been a curious paradox. As the election results proved, the party maintained, and has even increased, mass support among poor and working class Zimbabweans in conditions of astonishing hardship. But the MDC has also flirted with the organisations of imperialism and has been avowedly neoliberal in its policies. The party is advised by the International Republican Institute and Cato Institute. In April international media reported that an MDC government would immediately access $2 billion each year in "aid and development", which Patrick Bond describes as "top-heavy with foreign debt and chock-full of conditions".[35]

32: Cited in Zeilig, 2007, p160.
33: Interview, Harare, 31 July 2003.
34: Readers of *International Socialism* will also be interested to note that Mutambara was briefly in Britain and considered himself a fellow traveller of the Socialist Workers Party, speaking at the Marxism event in the early 1990s and attending Oxford meetings of the party. An important layer of the current leaders of the MDC, especially many ex-students, have their roots in the International Socialist Organisation of Zimbabwe. This includes secretary general Tendai Beti and spokesperson Nelson Chamisa.
35: Bond, 2008.

The party emerged out of the great upheavals that shook the country in the late 1990s. These protests were themselves a product of the failures of independence and the government's implementation of two structural adjustment programmes.

But these protest movements took place in the aftermath of the collapse of the regimes in Eastern Europe and Russia, and with them the ideological moorings for a generation of trade union bureaucrats and activists. To many it seemed that the set of ideas that championed economic adjustment and so-called democratisation—the Washington Consensus—had triumphed.

Paradoxically the continent was exploding in protests. Across Africa in the early 1990s protest movements developed at an astonishing speed. While there had been approximately 20 annual incidents of political unrest in the 1980s, in 1991 there were 86 protest movements across 30 countries. Between 1990 and 1994 a total of 35 regimes had been overthrown by protest movements, often led by opposition coalitions with a trade union leadership. In many cases free elections were held for the first time in a generation.[36]

As in Zimbabwe, protests and new political parties were born out of anger at structural adjustment programmes. By the early 1990s this economic devastation had left Africans consuming 25 percent less, and their governments spending a much smaller amount on education and social services than at any time since independence.[37]

However, the emerging opposition movements lacked a political alternative to structural adjustment. They often used the word "change" as their slogan—it became the rallying cry in Zimbabwe and Senegal ("chinja" and "sopi" respectively). But once the new movements became new governments, economic adjustment resumed. Many of those who had been active in the movements that swept the continent became disillusioned as governments that had emerged from the "transition" committed themselves to IMF and World Bank programmes.

The MDC is an expression of the revolt against structural adjustment programmes carried out by Zanu-PF. It was formed directly by the labour movement and supported by students who appealed openly to the trade union bureaucracy for a party to confront Zanu-PF. The MDC's core support came from the urban working class in the main cities of Harare, Chitungwiza and Bulawayo. But the MDC also attracted a middle class

36: Seddon and Zeilig, 2005, p14.

37: See the *Africa Research Bulletin*, volume 37, number 9.

group representing local and international business interests, who quickly gathered round the leadership of the party. As early as the parliamentary elections in 2000 workers made up only 15 percent of candidates.

Why did the party become politically dominated by groups of the middle class that gathered in its ranks? Some of the answer lies in the weakness of an alternative vision that could have argued inside the new party against the reorientation towards neoliberalism. Socialists were active in the MDC, as they were in similar organisations in other countries across the continent, but their voices were marginal. Though the mass struggles of 1996-8 showed the potential power of the working class, the protests, strikes and movements remained controlled by the trade union bureaucracy.

The MDC was an important step forward. After all, here was an organisation that was the product of the mass struggles of the Zimbabwean working class. Although the party, now openly "Brownite" in its politics, has travelled a long way from its founding purpose, it is still the crucial repository of the hopes of millions of Zimbabweans battered by the crisis. In fact the party's support base has actually grown substantially in recent years. This article has suggested that there are two important factors that have contributed to this. The first is the mass expulsions following the "slum clearances" in 2005, uprooting MDC supporters to rural areas and swelling the party's rural constituency. Second, the assault on the MDC leadership of 11 March 2007 meant that they again became the symbol of resistance to Zanu-PF.

But we should not minimise the problems the MDC has posed for activists in Zimbabwe. In the early years of this century the MDC seemed unwilling to take on Mugabe. The organisation has been characterised by confusing vacillations, calling mass action then retreating from it, seeking to align itself with right wing policies and accepting shoddy compromises with the regime. In the political space that was left, new forces played a temporary role. These groups often sought to substitute themselves for the failures of the MDC and even replace the organisation entirely. In February a national People's Convention was held in Harare with over 3,000 delegates. Some saw the initiative as a first step in building an alternative to the MDC in a "united and democratic front of all movements of the commons".[38]

But despite the compromises and vacillations the MDC rose again. It proved impossible to appeal for another "united front", purer and less compromised, when millions were looking in increasing numbers to the

38: "Crisis In Zimbabwe: No To Fake Elections! Jambanja Ndizvo!", *Socialist Worker* (Zimbabwe), February 2008.

MDC as the only alternative capable of removing Zanu-PF. Radical forces in Zimbabwe must relate to and organise with the MDC, but this does not mean quiet acquiescence to the politics of the organisation. A vocal and powerful radical minority in Zimbabwe has impressively kept alive the hope for genuine political transformation in the face of state repression and a disorientated opposition. This minority needs to work to recruit MDC activists to its ideas and organisations, while bringing pressure to bear on the party's leadership.

The election results expressed clearly the role the MDC has always played: both expressing and holding back mass struggles against the Mugabe regime. The party continues to act as a beacon to the poor.

The MDC is not an instrument of Western imperialism, even if it is funded by groups that are sympathetic to that power. But we can confidently predict that there will be attempts by Washington and London to co-opt it, should it come to power. The MDC is not a homogenous and wholly neoliberal organisation. The party's very contradictions make it porous and responsive to both struggle and critical debate. This presents socialists with potentially exciting possibilities.[39] Though real transformation will not come with an MDC government, the political alternative that the MDC momentarily became in 1999 can only be built within the mass ranks of MDC supporters and voters.

Conclusions

A run-off presidential election had been promised, against a background of increased repressions and rumours of deals and coups, as this journal went to press. Defeat of Mugabe and Zanu-PF would be reason for great celebration, not only by those who have suffered so much at the hands of the regime, but also by activists and socialists across the world. The political space created by a new government would give radical forces new opportunities to resist the encroachment of Western governments, international corporations and the IMF and World Bank. This political alternative, already with important advocates in Zimbabwe, must be based on the extraordinary power of the region's working class.

On 16 April news spread around the world of the arrival of a Chinese cargo ship, the *An Yue Jiang*, owned by China's state shipping company, in the major container port in Durban, South Africa. The ship included three million rounds of ammunition and 1,500 rockets bound for Zimbabwe,

---

39: One recent and graphic example of this was the invitation the International Socialists Organisation received to train MDC activists in the party's cadre school!

two days drive from the port. The South African government explained to the world that there was nothing they could do: this was a legal transfer of cargo that had already been paid for by a neighbouring sovereign state. The problem was that the sovereign state of Zimbabwe was busy stealing an election and crushing the opposition. The South Africa Transport and Allied Workers Union (Satawu) refused to be browbeaten by claims of legality. The union refused to unload the ship, while Satawu truckers said that they would not transport the cargo by road. The ship was paralysed in "outer anchorage" in "off-port limits".[40] Within a few days trade unions with members in ports near Zimbabwe followed suit: Mozambique and Namibia also refused to unload the weapons. The ship was forced to sail to Angola, where dock workers "maintained a watch" to ensure that the 77 tons of weapons were not unloaded.[41]

Socialists argued that Zimbabwe's opposition had to turn away from the sham talks led by South Africa's President Mbeki. These yielded nothing but a breathing space for the regime in Harare. The solidarity shown by the trade union movement in Southern Africa tantalises us with the prospect of an alternative in the mass action of the regional working class. These are not abstract dreams, but real and pressing possibilities. If socialists were able to forge a link between these ideas and the Southern African working class, both the Mugabe dictatorship, and the agenda of structural adjustment and neoliberalism across the region could be flushed away. But this politics needs to be organised, argued and built for. In Zimbabwe this must take place among those who have voted massively for the MDC.

40: "Union Refuses To Unload Arms Ship", *Sapa*, 17 April 2008.

41: "Arms Ship Leaves Angola", iafrica.com, 7 May 2008.

References

Bond, Patrick, 2008, "Vultures Circle Zimbabwe", *Counterpunch*, 5-6 April 2008, www.counterpunch.org/bond04052008.html

Bracking, Sarah, and Lloyd Sachikonya, 2008, "Remittances, Poverty Reduction and Informalisation in Zimbabwe 2005-6: A Political Economy of Dispossession?", Brooks World Poverty Institute (University of Manchester).

Chagonda, Tapiwa, 2007, "The Response of the Working Class in Harare to the Economic Crisis, 1997-2007", University of Johannesburg, department of sociology, seminar series, www.uj.ac.za/2007series/tabid/5994/Default.aspx

Johnson, R W, 2008, "Where Do We Go from Here?", *London Review of Books*, 8 May 2008, www.lrb.co.uk/v30/n09/john01_.html

Kibble, Steve, 2003, "Zimbabwe: Identity, Security and Conflict", Catholic Institute for International Relations, www.brandonhamber.com/documents/docs-zimkibble1.htm

Kriger, Norma, 2005, "Zanu-PF Strategies in the General Elections, 1980-2000: Discourse and Coercion", *African Affairs,* volume 104, number 414.

Seddon, David, and Leo Zeilig, 2005, "Class and Protest in Africa: New Waves", *Review of African Political Economy*, volume 31, number 103.

Shumba, R, 2006, "Constructing a Social Identity: The National Youth Service of Zimbabwe", MA dissertation, University of Johannesburg.

Zeilig, Leo, 2007, *Revolt and Protest: Student Politics and Activism in sub-Saharan Africa* (Tauris).

Benjamin's emergency Marxism

Chris Nineham

A review of Esther Leslie, ***Walter Benjamin*** *(Reaktion, 2007), £10.95*

There has been a growing fascination on the left for the often cryptic work of Walter Benjamin ever since he was first widely published in the 1960s. In 1968 Berlin students took him up as a libertarian Marxist likely to have approved of direct action, non-violent or otherwise. In the 1970s and 1980s he became a muse for Marxists taking the "cultural turn". In the 21st century his philosophy of history has been proposed as a guide for revolutionaries in a world facing environmental catastrophe.

When Benjamin died in 1940 his essays, notebooks and reviews lay scattered across Europe in the homes of correspondents and comrades. Now every scrap of his huge output has been lovingly deciphered, pieced together and reproduced, and there is an accompanying boom in Benjamin commentary.

His post-1968 popularity is partly a result of the New Left's interest in all things cultural. Benjamin was, among many other things, a brilliant cultural critic, one of the pioneers of the Marxist study of mass culture and the avant garde. Unlike some who have followed in his wake he was driven by a sense of emergency—"that things are status quo *is* the catastrophe".[1] Benjamin was, in Esther Leslie's words, an "active symptom" of desperate times. Even the manner of his death symbolised the tragedy of his generation. He almost certainly committed suicide after being captured by

1: Benjamin, 2002, p184.

General Franco's police at Portbou on the French-Spanish border while he was fleeing the Gestapo.

But he also has lasting appeal because of the way he responded to the failure of both the social democrats and stalinised Communist Parties to confront the Nazis in the 1930s. He publicly defied the Nazis even as he fled across Europe days ahead of their secret police. At the same time he berated the left for its passivity and tried to develop a Marxist method that could guarantee against the fatalism that crippled the socialist movement of his time.

Esther Leslie has done a great service in this and her previous book on Benjamin[2] in helping to rescue his life and work from the obscurity of academic writers of many different stripes. Here, using new archive material, she presents for the first time a coherent narrative of Benjamin's extraordinary life and his complicated intellectual development. She explains how and why this supremely cultured German Jewish intellectual came to denounce European bourgeois culture from the top of the "crumbling mast" of the shipwreck "from where he can at least signal and have a chance of being rescued".[3] She gives a careful account of his move towards Marxism in the mid-1920s and of his strange, disjointed life afterwards, in penury and on the hoof, collaborating with leading left wing intellectuals trying to fight an "intellectual civil war" against the gathering forces of reaction.

In the process she rescues Benjamin the revolutionary from many who have tried to tame him. The liberal critic George Steiner claimed Benjamin knew that humane and critical intelligence "resides in the always threatened keeping of the few".[4] Sociologist Jurgen Habermas insisted his criticism was aimed at personal redemption rather than "consciousness raising".[5] Lifelong friend and religious scholar Gershom Sholem described him as "a theologian marooned in the realm of the profane".[6] Leslie shows that despite his many influences the direction of his development was from idealism and romanticism towards a close engagement with historical materialism.

Benjamin is best known for a few essays about the opportunities and the threats to human culture presented by technological change. His most famous essay, "Art in the Age of Mechanical Reproduction", celebrates the democratic possibilities of new technologies. In it he argues that mass reproduction strips artworks of the aura of originality and uniqueness that mystify them.

2: Leslie, 2000.

3: Leslie, 2007, p117.

4: Cited in Eagleton, 1981, p ii.

5: Cited in Leslie, 2007, p228.

6: Cited in Leslie, 2007, p229.

> To an ever greater degree the work of art reproduced becomes the work of art designed for reproducibility...the instant the criterion of authenticity ceases to be applicable to artistic production, the total function of art is reversed. Instead of being based on ritual, it begins to be based on another practice—politics.[7]

Put more bluntly, "mechanical reproduction of art changes the reaction of the masses towards art. The reactionary attitude towards a Picasso painting changes into a progressive attitude towards a Chaplin movie".[8]

In the era of *Celebrity Big Brother* such enthusiasm sounds naive. In his defence, in this essay Benjamin was characteristically investigating different possibilities or choices available for humanity. In the essay's epilogue he argues that in the hands of the Nazis the same technology which can politicise culture can also glamorise and corrupt mass politics. "Self alienation has reached such a degree that it (mankind) can experience its own destruction as a pleasure of the first order".[9]

Benjamin was preoccupied with these two possibilities for the rest of his life. He argued that all culture carries a promise of liberation, but also the scars of the suffering that was part and parcel of its making:

> It owes its existence not just to the efforts of the great geniuses who fashioned it, but also in greater or lesser degree to the anonymous drudgery of their contemporaries. There is no cultural document that is not at the same time a record of barbarism.[10]

It was the job of the critic to uncover these contradictions and all they tell us about human possibilities. Any cultural history that discussed the development of art separate from the tensions of the wider world was reactionary: "It may well increase the burden of the treasures that are piled up on humanity's back. But it does not give mankind the strength to shake them off, so as to get its hands on them".[11]

Many of Benjamin's colleagues in the 1930s and 1940s were overawed by the conformist aspects of mass culture and developed a very pessimistic approach. In the 1970s and 1980s left wing intellectuals often stressed the positive value of "working class culture" without putting it in

---

7: Benjamin, 1978, p224.
8: Benjamin, 1978, p223.
9: Benjamin, 1978, p242.
10: Benjamin, 2000, p359.
11: Benjamin, 2000, p361.

context or exploring the ambiguities of the concept. Benjamin's activist approach to culture, his concern for context and the contradictory nature of all culture under capitalism are vital legacies for anyone trying to develop a Marxist understanding of culture.

Benjamin also tried to overcome the elitism implied by the idea of liberation through cultural practice. He championed the work of his close friend the communist playwright Bertolt Brecht because Brecht was trying to bridge the gap between the stage and the auditorium, between art and politics. Brecht's theatre did not just expose the contradictions of modern life but sought to turn the audience into active participants. Brecht was looking for the moment:

> when the mass begins to differentiate itself in discussion and responsible decisions or in attempts to discover well founded attitudes of its own, the moment when the false and deceptive totality called "audience" begins to disintegrate and there is new space for the formation of separate parties within it—separate parties corresponding to conditions as they really are.[12]

Benjamin's championing of radical modernism was complemented by a preoccupation with history, memory and time. He argued that capitalism had transformed time in a way that turned quality into quantity, smoothing historical development into a linear process that encouraged forgetting. Partly this was a product of the process of commodification itself. In his unfinished "Arcades Project", Benjamin used a montage of documents and observations of 19th century French life to try to uncover hidden history. According to Leslie's description:

> An awakened consciousness scrutinised "the residues of a dream world" in the form of arcades and interiors, exhibition halls and panoramas... The utopias of a classless society, traces of which were stored in the unconscious of the collective, in memories of a primal past, left deposits "in a thousand configurations of life, from enduring edifices to passing fashions".[13]

But the reduction of history to a linear process was also reinforced by the dominant theoretical approaches to history. Benjamin attacked the way social democrats had accepted bourgeois ideas of inevitable human progress based on the gradual development of technology. In different

12: Benjamin, 1977, p10.
13: Leslie, 2007, p155.

ways, Benjamin argued, the Communist Party during its "third period" and its "popular front period" also put its faith in the forward march of history.[14] For him, this "historicism" was the secret of the left's passivity in the face of the rise of the fascists.

He wrote his "Theses on the Concept of History", one of his last works, to challenge this approach:

> The conformism which has been part and parcel of social democracy from the beginning attaches not only to its political tactics but to its economic views as well. It is one reason for its later breakdown. Nothing has corrupted the German working class so much as the notion that it was moving with the current.[15]

Real, effective class consciousness meant understanding the need to break the flow of history:

> The awareness that they are about to make the continuum of history explode is characteristic of the revolutionary classes at the moment of their action. The great revolution introduced a new calendar.[16]

Elsewhere he wrote, "What our generation has learnt is that capitalism will not die a natural death".[17]

The final chapter of Leslie's biography celebrates the breadth and insight of Benjamin's criticism. He predicted earlier than many the rise of authoritarianism in Germany, partly because he saw the growing militarisation of culture. He warned of the suspension of "militant communism" in Stalin's Soviet Union, and in general "he detailed the flaws and contradictions of cultural life in ways that demanded awareness of the conditions of the dispossessed".[18] And, despite living in the "midnight of the century", he consistently pointed up hidden human potential that could be unlocked by breaking open conformism.

Benjamin's sense of urgency, and his insistence on the need for the

---

14: The "third period" (1928-1934), saw Communist Parties adopt an "ultra-left" rhetoric, equating social democratic parties with rising fascist organisations, rather than uniting workers against the far right. Subsequently, the "popular front" period saw the same Communist Parties seeking to form very broad alliances including even mainstream pro-capitalist parties.

15: Benjamin, 1978, p258.

16: Benjamin, 1978, p260.

17: Cited in Löwy, 2007, p86.

18: Leslie, 2007, p231.

left to actively break with capitalist logic are important in today's unstable world in which so much of the mainstream left has capitulated to neoliberalism. But in her enthusiasm to place Benjamin in the Marxist tradition and to promote the subversive strengths of his approach to culture, Leslie tends to overlook the problems in his method. This matters because some on the far left are making great claims for Benjamin. Michael Löwy has recently said that Benjamin's work is central to enabling us to "conceive of a revolutionary project with a general mission to emancipate".[19]

As the American critic Hannah Arendt pointed out in the 1960s, Benjamin's intellectual method echoes the techniques of contemporary artists. In the "Arcades Project" in particular, as Leslie notes, he "followed the surrealist procedure to the letter, montaging disparate industrially produced fragments, trash and parodies of natural form".[20] The problem is that montage is an artistic method. It can be effective in the hands of someone sensitive to the half hidden, symbolic significance of appearances, but it does not add up to a method of analysing how society works or the role of culture within it. At times this was the weight Benjamin tried to place on it.

The Marxist writer Theodor Adorno made this point in a criticism of an essay that was to introduce the "Arcades Project". Leslie summarises Adorno's analysis:

> Motifs were assembled not developed. There was no theoretical interpretation of various motifs of trace, flaneur, panorama, arcades, modernity and the ever-same. Ideas were "blockaded" behind "impenetrable walls of material". It lacked "mediation".[21]

Benjamin never wrote the great theoretical work that Adorno and others were hoping for, so theoretical assessments are difficult. But when he does discuss his own method he sometimes suggests that being open to raw experience is the key to breaking through illusion. At other times he comes close to a technological determinism which became more and more pessimistic as reaction gained ground:

> The questions which mankind asks of nature are determined amongst other things by its level of production. This is the point where positivism breaks

19: Löwy, 2007, p112.
20: Leslie, 2007, p85.
21: Leslie, 2007, p191.

> down. In the development of technology it saw only the progress of science, not the retrogression of society. It overlooked the fact that capitalism has decisively conditioned that development. It also escaped the positivists among the theoreticians of social democracy that the development of technology made it more and more difficult for the proletariat to take possession of it—an act that was seen to be more and more necessary.

Benjamin explains the drive to war as a product of uncontrollable technology:

> The energies that technology develops beyond their threshold are destructive. They serve primarily to foster the technology of warfare, and the means to prepare public opinion for war.[22]

The danger of letting determinism in via the back door is averted by Benjamin's consistent commitment to class struggle. But the problem is that he never grounds class struggle in any social or historical process. It is longed for, it is remembered, but remembrance itself is suggested as the most likely source of renewal: "The materialist presentation of history leads the past to bring the present into a critical state".[23]

Of course, consciousness of history is an important factor in current struggles. An unresolved history of repression and resistance can help stir up contemporary battles. All kinds of resistance movements call on historical precedents to lend weight to their cause. And, of course, it is also true, in the words of historian Eduardo Galleano, that "the past says things that concern the future".[24] A correct interpretation of the past is a key element in shaping consciousness and is crucial in orientating workers in present and future struggles. One of the most important roles of the revolutionary party is to keep alive the memory of past struggles that the ruling class want to suppress, and to fight for their revolutionary interpretation.

When Benjamin says, in his sixth thesis on the concept of history, "Every age must strive anew to wrest tradition away from the conformism that threatens to overpower it," we can follow him.[25] But, when he suggests that "historical materialism wishes to hold fast that image of the past which unexpectedly appears to the subject of history in a moment of danger", he

22: Benjamin, 2000, pp357, 358.
23: Cited in Leslie, 2007, p213.
24: Cited in Löwy, 2007, p57.
25: Benjamin, 1978, p255.

is asking too much of history.[26] By itself, or even with the help of the best historians, history cannot make people struggle.

The truth is that Benjamin never completely solved the problem that haunted him. He correctly warned against blind faith in progress. He knew the potential of the explosive struggles capitalism stores up, but he never arrived at a rounded explanation of how those struggles could develop and mature. Sometimes he fell back on a catastrophe theory of consciousness: "Marx says that revolutions are the locomotives of history. But perhaps it is quite otherwise, perhaps revolutions are an attempt by passengers on this train—namely the human race—to activate the brake." This is characteristically thought provoking, but it is also voluntaristic. It is not clear where the action arises from. Revolutions are always partly a response to a sense of emergency, but Benjamin's own epoch demonstrates all too vividly that impending catastrophe does not automatically mean the brake will be applied. Revolutionary consciousness is made possible by the everyday contradictions of capitalism, and active intervention in them, not just a sense of the horror of its ultimate destination. What we do now to develop it will effect what happens when the train approaches the bumpers.

No doubt anticipating such criticisms, Leslie complains that "few take the time to reconstruct the political context or possibility, or to carefully set Benjamin's action and thinking within the realm of such context and possibility".[27] This is fair comment. Benjamin's lingering mysticism is not just a result of eclectic thinking. His approach was shaped by the desperate situation that emerged soon after he became a Marxist. It would have been very hard to elaborate a coherent theory of developing class consciousness faced with the growing threat of fascism and a demoralised, misled working class. George Lukács, the Marxist who most successfully theorised the way revolutionary working class consciousness can develop, formed his politics during the high point of revolutionary struggle in the years immediately after the Russian Revolution. He was an active participant in the Hungarian Soviet Republic in 1918. By the time Benjamin came to Marxism, the great wave of struggle was declining. In the years that followed, the policy of Communist Parties became more and more distorted by the priorities of the Stalinist bureaucracy. In the mid-1920s Lukács himself succumbed to Stalinist revisionism.

In these circumstances it is easy to see how history itself must have seemed the only source of revolutionary inspiration at a time when such

26: Benjamin, 1978, p255.
27: Leslie, 2007, p231.

inspiration was needed more than ever. Benjamin's defiance is exemplary. His cultural criticism remains a rich resource. But we owe it to his memory, and the memory of the millions of others who died alongside him, to analyse both the strengths and weaknesses of his theoretical legacy. If the moment of danger proves to be the moment at which the authentic image and understanding of the past emerge it will probably be too late.

References

Benjamin, Walter, 1978, *Illuminations: Essays and Reflections* (Schocken).
Benjamin, Walter, 1977, *Understanding Brecht* (Verso).
Benjamin, Walter, 2000, *One Way Street and Other Writings* (Verso).
Benjamin, Walter, 2002, *Selected Writings*, volume four (Belknap).
Eagleton, Terry, 1981, *Walter Benjamin, or Towards a Revolutionary Criticism* (Verso).
Leslie, Esther, 2000, *Walter Benjamin: Overpowering Conformism* (Pluto).
Leslie, Esther, 2007, *Walter Benjamin* (Reaktion).
Löwy, Michael, 2007, *Fire Alarm* (Verso).

Karl Marx, Abram Leon and the Jewish Question—a reappraisal

John Rose

At some point, quite early on in our revolutionary "careers", Jewish students of the 1968 generation had to confront *On the Jewish Question* written by the young Karl Marx in 1843. This was an awesome moment—almost a political virility test. (If you passed it, you were at least ready for the intellectual barricades, if not the physical ones.) Struggling through the high levels of Hegelian abstractions was unnerving enough. True, once you have more or less understood it, this is a robust, even outstanding, defence of Jewish emancipation. Yet, at the same time, Marx seems hostile to the Jews, with Jewish history reduced to crude economics:

> Let us look for the secret of the Jew not in his religion, but rather the secret of the religion in the actual [meaning "economic"] Jew.[1]

Worse, the essay is peppered by what seems to be anti-Semitic polemic. And the reason for all this is that Marx appears to be equating the values of the new economic system that he would come to describe as capitalism with Jewish "trading", "market" or money values.

1: From one of the best translations, Marx, 1967, p243. I thank Martin Tomkinson, former 1968 student leader at the London School of Economics, for what became the permanent loan of this book. I would also like to thank Alex Callinicos, Moshe Machover, Sabby Sagall and Brian Klug for commenting on the first draft of this article, and Neil Davidson and Mark Thomas of the *International Socialism* editorial board for comments on the second draft.

It was a peculiar and very unsatisfactory experience. But there was an antidote.[2] Not least of the great innovations of 1968 was the publication of Abram Leon's *The Jewish Question.* This was an astonishing document, buried away since the war. Now Marxist academic Maxime Rodinson had brought it to the light of day, symbolically enough, at the Sorbonne university in Paris.

Abram Leon had been the leader of a tiny (Trotskyist) revolutionary socialist group in Nazi-occupied Belgium. He was hunted down, captured and perished in Auschwitz. Remarkably, he had written the manuscript under these wartime conditions. There was a touch of genius about Abram Leon to match Karl Marx and a touch of heroism to match Che Guevara any day.

Leon took the "economistic" passage quoted above from Marx's original essay and developed a dynamic and ambitious study of Jewish history, locating the growing historical trading function of Jewish communities from antiquity to modernity at the root of Jewish survival. In fact Leon's argument owes far more to the "mature" Marx's theoretical structure in *Capital* than to the essay written in Marx's young Hegelian days. Marx hardly mentions Jews in the three volumes of *Capital.* A rare and famous reference, discussed below, suggests the Jews are marginalised. This is the starting point for Leon. He argues that the Jewish trading communities were excluded from the rise of capitalism from the 13th century onwards, despite their long history—or, in fact, because of this history, since they were seen as potential competitors by new Christian traders in the developing national market economies.

I believe these positions require some amendment. Leon never discussed Marx's original essay. Marx himself never explained why he abandoned the arguments he used in it. Moreover, new research stimulated in recent years by the emergence of academic Jewish studies suggests that there was indeed a modest Jewish contribution to the rise of capitalism. Indeed, treated with care and in a critical spirit, there is a case to be made that the connection the young Marx makes, between the Jewish economic role and modernity arriving so late in Germany, can complement and refine the position adopted by Leon.[3]

2: Meeting Tony Cliff, the Jewish leader of the International Socialism group, forerunner of today's Socialist Workers Party, was the second antidote. He prepared you for the physical barricades as well...

3: An early warning about reading Marx's young Hegelian works is in order. His fondness for plays upon meaning makes this task very difficult. See Leopold, 2007, p8, an excellent book on the young Marx. Frederick Engels put it nice and simply: the young Marx wrote "very

But because of the intense ideological, and often prejudicial, claims and counter-claims about this subject, analysis of the evidence requires caution in the context of a sophisticated theoretical framework. Consider what follows to be no more than work in progress to achieve this objective. First, let us look more closely at the original writings of Marx and Leon.

Marx's Jewish Question

At the risk of considerable over-simplification, Marx's essay can be reduced to the analysis of his two key concepts—emancipation and Jewishness—and how they are linked. However, a complication that cannot be avoided is that both of these concepts are open to two distinctively different meanings.

Marx's idea of emancipation for Jews roughly corresponds to the granting of citizenship rights to them and the removal of discrimination by the French Revolution of 1789. A similar revolution is anticipated in the crumbling semi-feudal principalities that constituted Germany at the time. Marx defends emancipation in these terms from attacks from the leading "left" Hegelian, Bruno Bauer:

> The disintegration of man into Jew and citizen, Protestant and citizen, religious man and citizen, does not belie citizenship or circumvent political emancipation. [On the contrary,] it is political emancipation...from religion.[4]

Religion becomes a private matter for the individual. Hence in the public sphere the individual, whatever his religion, ought to be entitled to equal rights.

But Marx sees this political emancipation as producing a very restricted form of freedom. It is an advance, but "political emancipation is a reduction of man to a member of civil society, to an *egoistic independent* individual on the one hand and to a citizen, a moral person, on the other".[5] This is an alienated emancipation—a split in a person's human nature and a loss of potentially expanding human powers.

badly" just like a "German philosopher". Heinrich Heine (like Engels himself) was a more positive stylistic influence. See below.

4: Marx, 1967, p227.

5: Marx, 1967, p241. Emphasis in the original.

Thus man was not freed from religion; he received religious freedom... He received freedom of property. He was not freed from the egoism of trade but received freedom to trade.[6]

This passage leads directly to the two meanings of Jewishness. The first meaning is the one with which we are familiar—religious consciousness. (Marx never considered Jewishness as ethnic identity or as a nationalism—see footnote 11.) The second meaning is as a trading activity, and this does pose difficulties for us today. In other words, the last sentence in the passage above would be reinterpreted by Marx to state that man "received freedom to trade", ie to become Jewish.

At the time Marx was writing, it was taken for granted that "Jewishness", *Judentum* in German, was associated with money and trade. As the source book used on many of today's university Jewish Studies courses puts it, "The German word Judentum had, in the language of the time, the secondary meaning of commerce".[7] This makes us very uneasy when we read it now. It is virtually impossible to shake out anti-Semitic connotations. But Marx really did mean that Judentum now dominated society to such an extent that "the Jewish practical spirit has become the practical spirit of Christian nations...and Christians have become Jews".[8]

Judentum then appears as a symbol of an erupting and misunderstood modernity dominated by trade and money. In one of many notorious passages in the essay, Marx wrote:

What is the secular basis of Judaism? Practical need, self-interest. What is the worldly cult of the Jew? Bargaining. What is his worldly god? Money. Very well! Emancipation from bargaining and money, and thus from practical and real Judaism would be the self-emancipation of our era... The emancipation of the Jews, in the final analysis, is the emancipation of mankind from Judaism.[9]

This passage can legitimately be read as a clarion call for Jewish emancipation.

Market trading would no longer dominate Jewish everyday life. Everyone, Jews and non-Jews alike, would be emancipated from an

6: Marx, 1967, p240.
7: Mendes-Flohr and Reinharz, 1995, p327.
8: Marx, 1967, p244.
9: Marx, 1967, pp243-244.

economic system dominated by the market. Even Julius Carlebach, the most recent and most sophisticated in the long line of scholars to insist that Marx was anti-Semitic, was forced to conclude that:

> Marx was determined to elevate Judaism into an abstract element like labour and that he no more intended personal harm to individual Jews by calling for the dissolution of Judaism than he would have wanted workers to be attacked when he called for the abolition of labour.[10]

Carlebach is acknowledging that Marx is dealing with an economic and social system that needs to be abolished. Abolish a particular labour system and workers will be free. Indeed, they will no longer be workers, in the capitalist sense—they will be free producers. Abolish a Jewish dominated economic system and Jews will be free. Indeed they will no longer need to be Jewish. Marx took it for granted that in a free society there would be freedom of worship, but he also took for granted that in a free society religion would wither away.

Incidentally, Marx did acknowledge his own Jewish origins, despite claims to the contrary. David Leopold has unearthed a particularly fascinating example where Marx additionally points to his Jewish heritage as a factor in his own intellectual creativity. Bauer had accused the Jews of being an "eyesore". Marx responds in *The Holy Family*:

> Something which has been an eyesore to me from birth, as the Jews have been to the Christian world, and which persists and develops with the eye is not an ordinary sore, but a wonderful one, one that really belongs to my eye and must even contribute to a highly original development of my eyesight.[11]

The youthful Marx was not the first to use Judentum to mean the domination of money:

> Marx's earlier demonisation of the money economy was stimulated...by none other than Moses Hess... Marx would seem to have drawn inspiration for his

10: Carlebach, 1978, p178.

11: Leopold, 2007, p172. Yes, it could be argued that Marx is ignoring a distinctive Jewish cultural identity in both secular and religious societies. Nevertheless, there is an equally robust defence of Marx as a product of the Jewish assimilationist movement at that particular time in the Germany of the 1840s—Traverso, 1994, pp21-22; Draper, 1977, pp591-608. Fifty years later Marx's daughter, Eleanor, would insist on restoring her father's Jewish identity, see Rose, 2005.

most "anti-Semitic" invectives from the work of the subsequent "father of Zionist socialism" and precursor of Jewish nationalism more generally.[12]

Carlebach provides a flavour of Hess's writing at the time. It was the Jews' "world historic mission to turn mankind into predators, and they have completed their mission"; "Money is social blood, but alienated, spilt blood". Such "alienated, spilt blood" of man is symbolically consumed (in communion), so that in "the modern Jewish Christian pedlar world, the symbolism becomes actuality".[13]

But to what degree did Judentum dominate the expanding "money" economy? This remains the subject of intense controversy.

Both supporters and critics of Marx are agreed that Jewish economic activities made a public impact. It is not necessarily anti-Semitic to assume they accelerated the way the market was bursting the ramshackle network of semi-feudal principalities that constituted the German economy prior to unification.

Thus Hal Draper, one of the most prominent defenders of Marx's essay, explains how Jews had been "forced into a lopsided economic structure by Christendom's prohibition on their entrance into agriculture, guild occupations and the professions".[14] This, in turn, affected the socio-economic structure of the Jewish community in three ways. First, "the upper stratum of Jews did play an important role in the development of post-feudal society, especially considering the tiny proportion of the population that they constituted". Second, there was a "great tilt in the economic structure of Jewry toward middleman and financial occupations, including the bulk of poor Jews in huckstering occupations, for example, peddlers, petty merchants". And third, there was the "relatively high visibility of the Jews' economic role—as, for example, when Junkers (German feudal princes and officials) employed Jews as loan collectors and

---

12: Fischer, 2007, pp41-42. Fischer is a young Jewish studies scholar, not at all sympathetic to Marx or the Marxist method. Nevertheless he seeks to rescue Marx and his essay from its constant misuse by revisionist German social democrats in the 19th and early 20th centuries. Fischer has also translated and published in the same book a previously unknown defence of Marx's essay by Rosa Luxemburg. See also Isaiah Berlin, the 20th century Jewish liberal philosopher, who denounces Marx's alleged "anti-Semitism", praises Hess for "virtually inventing" Zionism but remains silent about Hess's "anti-Semitism"—Berlin, 1959, pp17-18.

13: Carlebach, 1978, pp117, 123. Zionist socialism can be regarded as at least in part a Jewish revolt against the high visibility of Jews in the European "money" economy in the 19th century. See the study of Nachman Syrkin, Hess's Zionist socialist pioneering disciple, in Frankel, 1981, pp288-328.

14: Draper, 1977, p597.

mortgage foreclosers, thus gaining the profits while the Jews gained the onus as 'bloodsuckers' ".[15]

Similarly Carlebach recognises:

> Like other medieval groups Jews were seen as a single socio-economic unit... The Prussian monarchs defined the value of the Jews as entrepreneurs and industrial innovators. They wanted economic expansion, foreign trade and currency and industrial investments. To this end, Prussia's Great Elector permitted...Jews..to settle in Prussia and...would have regarded the protests of Christian traders—that Jews use innovatory, aggressive trading methods as opposed to their own sedate settled methods—as a full vindication of his intentions.[16]

In other words, the Prussian ruler used the Jewish elite simultaneously as a device to help develop capitalism and as a buffer against the nascent Prussian capitalists or bourgeoisie. This arrangement might have been very convenient for the Prussian aristocracy but it was extremely dangerous for the Jews. Without doubt, it helped fuel the way modern anti-Semitism would develop, and link Jews with money and power. The ferocious anti-Semitic riots against Jewish traders in Frankfurt in 1614-5 anticipated exactly these dangers.[17]

Most Jews remained very poor in the 18th century. But a standard text in contemporary Jewish studies notes the rapidity of the "rags to riches" story of the German Jewish middle class in the 19th century: "Ironically, it was the ignominious occupations of itinerant peddling, used clothes dealing and small scale usury, that had prepared the ground for this change".[18]

A surprising source has inadvertently reinforced the argument being pursued here. Niall Ferguson, the neo-conservative historian and writer, was given first time access to the private documents of the Rothschild family—arguably the world's most famous international banking family, with its roots in the Jewish ghetto of Frankfurt, Germany, in the 18th century. He has

---

15: Draper, 1977, p598. Draper's note on Marx and the Economic Jew Stereotype, pp591-609, argues that Marx was merely reflecting the deeply entrenched view about Jews and the expanding money economy at the time. But Draper does not properly address just how this view became so entrenched.

16: Carlebach, 1978, pp13, 53. See also Penslar, 2001, pp46-47, who shows how convenient it was for the authorities that the anti-Semitic imagination blamed the Jews for the rapid expansion in land speculation in the 19th century.

17: Israel, 1989, p68; Rose, 2004, p52.

18: Sorkin, 1992, p180.

produced a stunning two-volume history, *The House of Rothschild*. As Marx was writing his essay, the Rothschild family already dominated European banking. James Rothschild, in particular, was financing the introduction of the railway system in France, the most potent symbol of the new industrial economic order that was emerging. Ferguson points us to a passage in Marx which follows that stark phrase "the emancipation of mankind from Judaism". Marx cites Bauer's claim that "the Jew who is only tolerated in Vienna...determines the fate of the whole empire through his financial power",[19] and Ferguson demonstrates that Bauer is referring to Rothschild.

Marx is responding to this personification of "Jewish economic power". For Bauer this justified denial of political rights to Jews, while Marx argued that Jews were entitled to political rights. He put this particularly succinctly in a later polemic with Bauer in *The Holy Family*:

> The Jew has all the more right to the recognition of his free humanity [ie the rights of man], as free society is thoroughly commercial and Jewish and the Jew is a necessary link in it.[20]

Heinrich Heine, poetic genius and another of the young Marx's Hegelian comrades of Jewish origin, has left us an intriguing insight into James Rothschild, the "Robespierre of finance":

> Money is the new religion. It possesses the moral force or power which religion has lost... I see in Rothschild one of the great revolutionaries who have founded modern democracy... Robespierre and Rothschild...signify the gradual annihilation of the old aristocracy... [They are] Europe's...fearful levellers... Rothschild destroyed the predominance of land, by raising the system of state bonds to supreme power, thereby mobilising property and income and at the same time endowing money with the previous privileges of the land. He thereby created a new aristocracy, it is true, but this, resting as it does on the most unreliable elements, on money, can never play as enduringly regressive a role as the former aristocracy, which was rooted in the land... For money is more fluid than water, more elusive than air... In the twinkling of an eye, it will dissolve and evaporate.[21]

19: Marx, 1967, p244.

20: Carlebach, 1978, p180.

21: Ferguson, 1998, pp213-214. Heine's last line here makes an interesting comparison with one of the most famous lines in Marx and Engels' *The Communist Manifesto*: "All that is solid melts into air, all that is holy is profaned, and man is at last compelled to face, with sober senses, his real conditions of life, and his relations with his kind."

Abram Leon's The Jewish Question

The great innovation in Leon's argument is that he provides the detailed historical analysis of the Jewish trading communities using the concept of a "people-class", which is no more than hinted at in Marx's original essay. His analysis anticipated the findings of academic Jewish studies by over half a century and overturns the Zionist "lachrymose" view of Jewish history in Europe which sees only "centuries of Jewish suffering" as a result of endemic anti-Semitism.[22] Finally, it answers the question of how the Jewish people became an urban people.

An interesting starting point is the so-called "exile of the Jews" at the time of the fall of the Second Temple in AD 70. In fact "exile" is a myth because a flourishing diaspora already existed in the Roman Empire and beyond: "A majority of Jews lived outside the Roman province of Judaea".[23] Leon argues that this diaspora was already urbanising, characterised by "commercial prosperity"[24] and beginning to be led by merchants. The Jewish community was a third of the half million strong population of Alexandria, the greatest commercial port city of the ancient world prior to Rome,[25] and Jewish trading communities were sometimes so successful that they triggered conversions to Judaism in the urban areas, while the Jewish peasantry was assimilating into the "pagan", then increasingly Christian and, later, Islamic countryside. Large numbers of Phoenicians and Carthaginians became Jewish, bringing with them "their commercial skills".[26] Islamic expansion throughout the Mediterranean arena and beyond enhanced the Jewish trading role: "Jewish traders served as important mediators in a world divided by Islam and Christianity... By the 9th century Hebrew had become a leading international language".[27]

Ibn-Hurdadbih, the head of the Caliph of Baghdad's postal and intelligence service in the middle of the 9th century, described a group of international Jewish merchants known as the "Radanite Jews" who traded over vast distances from the "Frankish" lands (roughly today's France) to the Caspian Sea (on the northern coast of today's Iran). Scattered along this well travelled trading zone were Jewish colonies, which organised the exchange of forest products, horses and hides, swords and slaves of both

22: Rose, 2004, p43.
23: Barclay, 1996, p4, footnote 1. Where possible here I use a Leon theoretical framework but with updated sources.
24: Leon, 1970, chapter 2.
25: Modrzejewski, 1995, p73.
26: Baron et al, 1975, p21.
27: Baron et al, 1975, pp28-29.

sexes from the west for luxury goods from the east.[28] Jewish prosperity and political influence at this time impacted on the empire of the Khazars. Its elite actually converted to Judaism late in the 9th century as a way of maintaining its political independence and integrating itself into the Jewish trading network.[29]

The early feudal period in European history "was also the period of greatest prosperity for the Jews. Commercial and usurious 'capital' found great possibilities for expansion in feudal society. The Jews were protected by the kings and princes, and their relations with other classes were in general good".[30] An unlikely source, Abba Eban, the former senior Israeli cabinet minister and classics scholar, writes in his bestselling coffee table book, *Heritage, Civilisation and the Jews*, that Charlemagne, the commanding European figure of the early Middle Ages, "protected the Jews because of their services to trade and finance". At this time "the original Ashkenazim Jews" arose as "economic pioneers, men of great mercantile enterprise. They were also deeply devoted to learning".[31]

Leon's conclusion was that "the Jews have been preserved, not despite their dispersal, but because of it".[32]

However, as mercantile capitalism[33] began to develop, and European nation states began to form, the Jewish trade role was threatened by the emergence of local traders. In the 11th century:

> Western Europe entered a period of intense economic development. The first stage was...the creation of a corporative industry and a native merchant bourgeoisie... The growth of cities and of a native merchant class brought with it the complete elimination of the Jews from commerce. They became usurers...[but] the relative abundance of money enabled the nobility to throw off the yoke of the usurer. The Jews were driven from one country to another... In certain cities...the Jews became loan makers to the popular masses...in this role...they were often the victims of bloody uprisings.[34]

A good example was the expulsion of the Jews from England in

28: Abramsky, 1986, pp15-18.
29: Abramsky, 1986, p16.
30: Leon, 1970, p82.
31: Eban, 1984, p119; Rose, 2004, pp46-50.
32: Leon, 1970, p122.
33: It is expedient here to assume the concept "mercantile capitalism" is complementary to that of "market feudalism" used by Chris Harman in *A People's History of the World*.
34: Leon, 1970, pp82-83.

1290 where they had been official bankers to the king.[35] But the most spectacular example was the expulsion from Spain and the official "hunt" for them by the Spanish Inquisition. Spain, with its "discovery" of the Americas, dominated the new mercantilism. Leon goes on to argue that Jews leaving Spain, and the rest of western Europe, were absorbed into eastern Europe, especially Poland and parts of Russia, since feudal structures survived and strengthened there. The Jewish "economic situation [became] very good" in Poland. They enjoyed "a special internal autonomy",[36] even managing estates on behalf of absentee landlords.

This is the single most important factor explaining why the majority of the world's Jews were living in Poland, Russia and other parts of eastern Europe at the dawn of the modern period. Leon's innovative approach is particularly important as it locates the roots of Zionism in the very specific history of these eastern European Jewish communities in the latter part of the 19th century.[37]

However, one result of the Leon perspective was to disconnect Jews from the rise of capitalism. Jewish merchants could not possibly be the "bearers of the new mode of production".[38] Leon takes a particular sentence in Marx's *Capital*, a rare reference to Jews, which suggests that Jews, far from being central to the new economic system in western Europe, were economically marginalised in the backwaters of eastern Europe. Marx is discussing a point that trading peoples, historically, were outside society: "The trading peoples of old existed like the gods of Epicurus in the intermundia, or like the Jews in the pores of Polish society".[39]

But Leon never tackled the paradox of the young Marx's preoccupation with Judentum and the mature Marx's abandonment of it.

How early modern Europe readmitted the Jewish traders

Leon writes, correctly, that "it is...inaccurate to regard the Jews as founders of modern capitalism. The Jews certainly contributed to the development of the exchange economy in Europe but their specific function ends precisely

35: Rose, 2004, p50.
36: Leon, 1970, pp190-193. See also later in this article.
37: Rose, 2004, pp53-55, chapter 6.
38: Leon, 1970, p76.
39: Marx, 1991, p447. The footnote about Epicurus in this edition of *Capital* reads: "According to the Greek philosopher Epicurus (341 BC–270 BC), the gods existed only in the intermundia, or spaces between different worlds, and had no influence on the course of human affairs. Marx had studied Epicurus's conception for his doctoral dissertation." See also Leon, 1970, p77.

where modern capitalism begins".[40] But he also insists, without qualification, that the "mercantile economy...expelled the Jews. The Jew 'banker to the nobility' was already completely unknown in western Europe at the end of the Middle Ages... The collapse was a total one".[41]

This is very misleading because the mercantile economy revived the specific mercantile function of the Jews in the period preceding modern (ie industrial) capitalism. Jonathan Israel's path-breaking study, *European Jewry in the Age of Mercantilism 1550-1750*, provides a similar picture to Leon's for the late Middle Ages but also points to the re-entry of the Jewish traders. Famously Oliver Cromwell favoured re-entry of the Jews into revolutionary England in 1655 "to trade and trafficke".[42] This led to the revival of their trading function in the context of the massive expansion of the European and Atlantic market:

> No less important than army contracting, and perhaps more so, was the increasing role of Jews in state finance and international payments generally. This rested essentially on Amsterdam's role as Europe's chief bullion and money market combined with Jewish dominance of the gold, silver and other metal trades in central Europe. It arose also from the Jews' particular need of government favours and concessions as well as their exceptional vulnerability to government pressure... But most crucial of all was the wide, not to say pervasive, reach of the closely knit...financial network and its ability to raise large sums with great speed, often on mere trust, and to remit money swiftly from one part of Europe to another.[43]

At the centre of this system were the "Court Jews"—bankers to the nobility:

> In the course of time, the Court Jews not only accumulated riches and honours but evolved a lifestyle to match. Gradually they were exempted from many, but by no means all, of the irksome restrictions and curtailments which the Christian state imposed on the Jew.[44]

Earlier we saw Draper and Carlebach describing Jews as a single economic unit. Jonathan Israel analysed the unit's tight social and

40: Leon, 1970, p182.
41: Leon, 1970, p153.
42: Israel, 1989, p159.
43: Israel, 1989, p132.
44: Israel, 1989, p142.

economic structure, based on the metallic money economy that reflected the mercantile period:

> The vertical ties...lent Jewish society its inner cohesion—[the] commercial collaboration and the patronage network implicit in Jewry's institutions, charities, and welfare system—were of much greater significance than any occasional friction between rich and poor. First, at the apex of the pyramid, stood the elite of financiers, Court Jews, and princely agents; next came the much more numerous body of substantial merchants, manufacturers..., thirdly, and probably most numerous of all, was the mass of pedlars, hawkers, old clothes men, and other petty tradesmen; fourthly and less numerous but, nevertheless, a substantial proportion of Jewish bread-winners, were the craftsmen and artisans; finally, at the base of the pyramid was a depressed mass of vagrants, beggars, and other unemployed.[45]

Jewish emancipation reformers such as Moses Mendelssohn were determined to break up this medieval caste-like structure.[46] That "break up" was inevitable under the twin pressures of the French Revolution, which lifted religious and occupational restrictions on the Jews, and the intense pressure for internal reform to bust the authority of the orthodox rabbis within the Jewish community.

Clearly we require a more complex and nuanced analysis of the Jewish contribution to capitalism and its economy. This will now be attempted with the some case studies.

The Spanish Inquisition

This momentous and horrific process for Jews and, indeed, the rest of Europe has been neglected by those historians grappling with the issues discussed here. Yet it formed the strategic springboard for the Catholic monarchs as they united Spain, and as Spain emerged as Europe's principal maritime and mercantilist power at the end of the 15th century.[47] The Inquisition was about obliterating all opposition, including Spain's Muslims in the south. But it was preoccupied to the point of blind obsession with Jews, including converted Jews, known as *conversos*.

The Jews, who had formed about 2 percent of Spain's population,

45: Israel, 1989, p171.

46: Israel, 1989, p132; Rose, 2004, p57.

47: On the significance of Spanish mercantile capitalism, see Harman, 1999, p174, footnote 36.

were expelled in 1492. But tens of thousands preferred baptism to expulsion. This troubled the new Spanish authorities. Were the conversions genuine? The "new Christians" were rapidly integrating into the religious, political and economic hierarchy of the new Spain. In Castile conversos were very influential, both in many of the municipal councils of the growing towns and cities, and in local Catholic hierarchies. In Aragon converso officials dominated the crown administration of Ferdinand. The converso royal treasurer helped fund Columbus's first voyage to America; conversos, and even practising Jews, were part of the crew. The scholarly consensus today is that most of the conversions were genuine. But there are two provisos. First, the very ferocity of the Inquisition revived pro-Jewish religious sentiment among some converso families. Second, a converso "consciousness" developed as a kind of hybrid combining Christianity and Judaism.[48]

We have here a lightning conductor for Judaism's crisis at the dawn of capitalism. Judaism certainly did not pilot Spain's maritime and imperial prowess. But neither was Judaism successfully excluded as the Spanish Empire tried to roll over the rest of Europe as well as dominate the high seas.

Shakespeare's Jew: Shylock

In a standard left wing interpretation, Shylock, the Jewish moneylender, has been reduced to this grim restricted status by traditional Christian anti-Semitism, reinforced by the new mercantile period that witnessed the rise of proud and adventurous Christian merchants such as Shakespeare's Antonio, the Merchant of Venice. However, the genius of Shakespeare allows for competing interpretations to coexist, each with a plausible claim on our attention, by virtue of the highly complex, contradictory and fast changing world at the beginning of the 17th century.

A simple linguistic manoeuvre helps. What happens if you replace the word "moneylender" with the word "banker"? Can Shylock, as banker, become the equal of Antonio, the merchant? Is this, also, what the play is about? After all "merchants" and "bankers" are often interchangeable economic roles. This cannot properly be explored within the confines of the play itself. But a legitimate exercise can test the claim against the real historical circumstances within which the play was written.

Tawney's classic, *Religion and the Rise of Capitalism*, illustrated how English Protestantism was changing the meaning of usury or moneylending

48: The evidence on the Spanish Inquisition presented here is taken from Kamen, 1997. I would like to thank Sebastian Balfour, professor of Spanish history at the London School of Economics, for drawing my attention to this book.

and modernising it at the very moment that Shakespeare was writing. A recent study, Norman Jones's *God and the Moneylenders*, has fleshed out the empirical detail to support this argument:

> The great money-men of early Jacobean London combined speculation and moneylending to make enormous profits, playing the role of bankers in a nation that had no banks.[49]

These men, like the wonderfully named Sir Baptist Hicks, were already becoming the new aristocrats and government officials in Shakespeare's England. Even more intriguing is what was happening to the real merchants of Venice at this time. Christian merchants were losing the battle to control Jewish merchants. The special Venice Charter of 1589, though it fell short of granting formal legal equality, decisively expanded Jewish merchant rights. Later in the 17th century a sixth of Venice's largest ships in its shipping fleet would be owned by Jews.[50]

Spinoza

Any serious investigation of the Jewish contribution to the rise of capitalism must include the intellectual or ideological contribution. Spinoza is one of leading philosophers of the Enlightenment. "No one, during 1650-1750, remotely rivalled Spinoza's notoriety as the chief challenger of the fundamentals of revealed religion, received ideas, tradition, morality".[51] He helped clear the ideological and superstitious baggage and obstacles for the rise of science, an essential component of capitalism, especially its later industrial phase. He also helped lay the foundations for a scientific study of society itself.

Spinoza's life history actually symbolises those menacing lines in Marx's *Jewish Question* where Marx equates Judaism with money making and sees emancipation from both as the precondition for the quest for universal freedom.

One of the best studies of Spinoza provides a fascinating description of the philosopher's break with both the Judaism and the merchant activities of his family in Amsterdam. They were originally Portuguese conversos

49: Jones, 1989, p200.
50: Davis and Ravid, 2001, pp18-19, 88-94, 95. I would like to thank Michael Rosen for our discussion on the extraordinary claim that Shylock's demand of a "pound of flesh" from Antonio can be considered as a circumcision threat: Shapiro, 1996, pp126-130. Shylock could be saying, "Treat me as an equal or I'll make you into a Jew."
51: Israel, 2002, p160.

who had re-established Jewish ties after emigrating to the city and were highly successful international traders. Spinoza was already struggling to make his break, but it was finally precipitated when the family business went into crisis after England blockaded Dutch trade.[52]

Isaac Deutscher has insisted, though, that Spinoza, and even Marx, should still be relocated in a Jewish tradition. They are:

> "non-Jewish Jews"...Jewish heretics who transcend Jewry but who belong to a Jewish tradition, who...were exceptional in that as Jews they dwelt on the borders of various civilisations... Their minds matured where the most diverse cultural influences crossed and fertilised each other... It was this that enabled them to rise above their...times...and strike out mentally into wide new horizons and far into the future.[53]

The point is made in a different way by the principal architect of the Jewish Enlightenment, Moses Mendelssohn, in his effort to "reinstate" Spinoza. He saw it as a condition for Judaism's survival in the modern world.[54] It doesn't really matter whether Mendelssohn was successful. The fact that the modern spokesman for the old religion had to pay such homage to the philosopher of atheism speaks for itself.

The Jews of Poland

Abram Leon was absolutely right about the way Poland appeared as a magnet attracting Jews fleeing expulsion from across Europe as new national markets emerged. However, Poland did not simply integrate the Jews into its feudal structure. The paradox here is that, while the internal feudal structure entrenched itself ever more deeply and oppressively amongst the peasantry, externally, Polish agriculture temporarily became a vital component of the western European mercantile economy.[55] The Jews gravitated to the east of the country, which was much less developed and where the landed magnates wielded undisputed control. Western Europe wanted cheap Polish grain, which could be transported by eastern Poland's river network. Jewish migrants began to settle in the numerous small towns and villages belonging to these great landlords.[56] An *Arenda*

52: Israel, 2002, pp166-167.
53: Deutscher, 1968, pp26-27.
54: Israel, 2002, pp658-659.
55: Another example of "market feudalism": Harman, 1999, pp155-158. See also Kula, 1976.
56: Israel, 1989, pp27-29.

system developed in which Polish nobles leased their estates to Jewish management.

> Jews were thus the main agents...of a vast traffic encompassing the whole of Europe...for just as they sold the produce of the land for shipment to Holland and beyond, it was they who distributed the western cloth, salt, wine and luxuries, such as spices and jewellery.[57]

In 1648 Ukraine exploded. Over half the landed estates in Ukraine were managed by the Jewish Arenda on behalf of absentee Polish landlords. Led by a minor noble, Chmielnicki, the Ukrainian peasantry, aided and abetted by Cossacks and Crimean Tartars, rose up in rebellion against Polish rule and its Jewish agents. The targets were the Polish nobility, the Catholic clergy and the Jews who, as they were more numerous than the others, took the brunt of the losses. The Arenda system would eventually stagnate. Polish feudalism would sink into atrophy, paving the way for the partitioning of Poland by Russia, Prussia and Austria at the end of the 18th century. The mass of now impoverished Jews ended up in the Tsarist Russian Empire.[58]

The rise of the House of Rothschild

The roots of the Rothschild banking family lie deep in the Frankfurt Jewish ghetto:

> A traveller arriving in 18th century Frankfurt, as he crossed the main... bridge could hardly miss the *Judensau*—the Jews' Pig, on the wall of the Jewish ghetto. Obscene graffiti depicted a group of Jews debasing themselves before—or rather beneath and behind—a fierce sow.[59]

As Goethe, the city's most celebrated literary son, wrote, this was not just "private hostility, but erected as a public monument". Yet:

> There was to this persistent discrimination more than ancestral prejudice. An important factor was that the Gentile business community genuinely feared the economic challenge which they believed would be posed by an emancipated Jewish population. The fact that a slum like the *Judengasse* could

57: Israel, 1989, p30; Rose, 2004 pp53-55.
58: Rose, 2004, chapter 6.
59: Ferguson, 1998, p35.

produce mathematics teachers and doctors tells us something important about its culture; it was not as closed as it seemed. Despite—perhaps partly because of—the grim conditions in which they lived, the Frankfurt Jews were anything but an underclass in cultural terms.[60]

Out of this world stepped Mayer Amschel Rothschild in 1790, an antique dealer with a growing circle of suppliers and customers, to whom credit was extended from time to time. By 1797 he was one of the richest Jews in Frankfurt, and a central part of his business was banking, doing business with cities not only across Germany, but also with Vienna, Amsterdam, Paris and London.[61] His big break came with an introduction to Prince William of Hesse-Kassel, who was one of the wealthiest German princes because he sold the "services" of his Hessian army to the highest bidder, "usually Britain".[62] Rothschild became, in effect, William's "Court Jew".[63]

When Napoleon's French revolutionary armies chased William and his "court" out of the Frankfurt region, Rothschild helped the now exiled William conceal some of his fortune.

One financial move helped transform the Rothschild family into a vital component of the new industrial-economic complex now emerging in Europe, with England at the helm. Rothschild had sent one of his sons, Nathan, to the very heartland of the new industrial machine, Manchester, England. Nathan bought textiles, which were produced cheaper there than anywhere else, and sent them back to his father and brothers for sale in Europe.[64] Now Nathan would handle the exiled William's English investments.[65] This "helped Nathan make the transition from Manchester merchant to London banker"[66]—in fact, Britain's leading banker. He was quite simply "the richest man in Britain and therefore, given Britain's economic lead at this time, almost certainly the richest man in the world".[67]

Ferguson credits him with laying the foundations of the international bond market.[68] In effect, the Rothschild brothers "were establishing that system of international monetary cooperation which would later be

60: Ferguson, 1998, p39.
61: Ferguson, 1998, p45.
62: Ferguson, 1998, p61.
63: Ferguson, 1998, p63.
64: Ferguson, 1998, p49.
65: Ferguson, 1998, p66.
66: Ferguson, 1998, p71.
67: Ferguson 1998, pp300, 304.
68: Ferguson, 1998, p125.

performed routinely by central banks, and on which the gold standard came to depend".[69]

The Dual Revolution

Ferguson is particularly perceptive about the source of the Rothschilds' power: "The most outstanding personal qualities may sometimes require exceptional circumstances and world-shattering events to come to fruition".[70] This had nothing to do with "Jewish power". On the contrary, Ferguson argues, it derived from emancipation afforded to the Rothschilds by the Dual Revolution. This was made up of the economic changes driving England's Industrial Revolution, which was about to spread across continental Europe, and the far-reaching struggle for mass democracy triggered by the 1789 French Revolution, which included, crucially, full Jewish democratic rights, and which would also spread across Europe and certainly cross the English Channel.

A new historical epoch was erupting with the possibility of an expansion and transformation of human resources and human potential, unimaginable to all previous generations.

Abram Leon once wrote of the "commercial and artisan heritage of Judaism, heritage of a long historical past".[71] There can be no question that this heritage is a factor helping shape the new epoch, though we should add the concept "intellectual" alongside the phrase "commercial and artisan". The Dual Revolution produced the international Rothschild banking family as a part of an already existing and developing banking process. The poet Byron celebrated and satirised England's most famous banker "twin-set" in *Don Juan* in 1823:

> Who keep the world, both old and new, in pain
> Or pleasure? Who makes politics run glibber all?
> The shade of Bonaparte's noble daring?—
> Jew Rothschild, and his fellow Christian Baring.[72]

The Dual Revolution not only guaranteed modern freedoms for Jews, but also stimulated an outstanding Jewish contribution to European civilisation, of which "wealth creation" was just one part.

69: Ferguson, 1998, p137.
70: Ferguson, 1998, p47.
71: Leon, 1970, p236.
72: Ferguson, 1998, pp111-112.

As Hobsbawm puts it, while most Jews were trapped in increasing poverty in eastern Europe:

> The smaller communities of the west seized the new opportunities...even when the price they had to pay was a nominal baptism, as in semi-emancipated countries it often still was, at any rate for official posts... More striking than Jewish wealth was the flowering of Jewish talent in the secular arts, sciences and professions... By 1848 the greatest Jewish mind of the 19th century and the most successful Jewish politician had both reached maturity: Karl Marx (1818-1883) and Benjamin Disraeli (1804-1881)... The Dual Revolution had given the Jews the nearest thing to equality they had ever enjoyed under Christianity. Those who seized the opportunity wished for nothing better than to "assimilate" to the new society.[73]

Marx and capitalism

Capitalism has no religion. And it certainly has no roots in any particular religious or ethnic group. However, Marx noted that capitalism, as it developed, precipitated the Reformation, which tore Christianity into two parts, Catholicism and Protestantism. In his mature writings he associates the cult of money with Protestantism rather than Judaism. He writes:

> The cult of money has for its corollary asceticism, abstinence, sacrifice, saving and frugality, contempt for the pleasures of the world, temporal and transitory, the eternal hunt for wealth. From whence the relation of English Puritanism and Dutch Protestantism with the action of making money.[74]

In *Capital* he shows how industrial production and the exploitation of labour power displace trade as the dynamo of the new economic system:

> Now it is not trade that revolutionises industry, but rather industry that constantly revolutionises trade...commercial supremacy is now linked with the greater or less prevalence of the conditions for large-scale industry. Compare England and Holland for example. The history of Holland's

73: Hobsbawm, 1962, pp234-235.

74: Traverso, 1994, p21. Marx here anticipates, by half a century, the conservative sociologist Max Weber, who, in his book *The Protestant Ethic and the Spirit of Capitalism* drew a link between the new Christianity and the 'entrepreneurial spirit'. Tawney's book, referred to earlier, is a response to Weber.

decline as the dominant trading nation is the history of the subordination of commercial capital to industrial capital.[75]

Marx, likewise, shows the banking system to be absolutely decisive in the development of modern capitalism, since credit "accelerates the material development of the productive forces and the creation of the world market", but "at the same time credit accelerates the violent outbreaks of...crises... It is this dual character that gives the principal spokesmen for credit...their nicely mixed character of swindler and prophet".[76]

Jews and capitalism

A theory cannot be summarised properly in one or two sentences. Still, Leon Trotsky came closest to capturing the Jewish dilemma regarding capitalism on the eve of the Nazi Holocaust:

> In the epoch of its rise, capitalism took the Jewish people out of the ghetto and utilised them as an instrument in its commercial expansion. Today decaying capitalist society is striving to squeeze the Jewish people from all its pores.[77]

Of course, we now know that capitalism not only recovered after the war, but also witnessed the most sustained period of expansion in its history. Assimilation of Jews in the countries where the vast majority lived, in America and Western Europe, proceeded at an unprecedented rate. Jews became the most successful ethnic minority on any measure of equal opportunities and social mobility. However, Trotsky's remarks remind us of the Holocaust's genocidal legacy. It left such a deep scar that it contributed to the forced imposition of a Jewish state on Palestine.

Here we can legitimately adapt Trotsky's remarks: capitalism now took the Jewish people and utilised them as an instrument for its oil imperialist expansion on Arab lands. In fact Trotsky had expressly warned against this with a prescience bordering on prophecy: "The future development of military events may well transform Palestine into a bloody trap for...the Jews".[78]

75: Marx, 1991, p451.
76: Marx, 1991, pp572-573.
77: Traverso, 1994, p204; Trotsky, 1940a.
78: Trotsky, 1945, p379.

References

Abramsky, Chimen, et al (eds), 1986, *The Jews in Poland* (Blackwell).

Barclay, John M G, 1996, *Jews in the Mediterranean Diaspora* (T & T Clark).

Baron, Salo, Arcadius Kahan, et al (eds), 1975, *Economic History of the Jews* (Keter).

Berlin, Isaiah, 1959, *The Life and Opinions of Moses Hess* (Cambridge).

Carlebach, Julius, 1978, *Karl Marx and the Radical Critique of Judaism* (Littman Library of Jewish Civilisation).

Davis, Robert, and Benjamin Ravid, 2001, *The Jews of Early Modern Venice* (John Hopkins University).

Deutscher, Isaac, 1968, *The Non-Jewish Jew* (Oxford University).

Draper, Hal, 1977, *Karl Marx's Theory of Revolution, volume one: State and Bureaucracy* (Monthly Review). The section "Marx and the Economic Jew Stereotype" is available online at www.marxists.de/religion/draper/marxjewq.htm

Eban, Abba, 1984, *Heritage, Civilisation and the Jews* (Weidenfeld & Nicolson).

Ferguson, Niall, 1998, *The House of Rothschild, volume one: Money's Prophets 1798-1848* (Penguin).

Fischer, Lars, 2007, *The Socialist Response to Antisemitism in Imperial Germany* (Cambridge University).

Frankel, Jonathan, 1981, *Prophecy and Politics: Socialism, Nationalism and the Russian Jews 1862-1917*, (Cambridge University).

Harman, Chris, 1999, *A People's History of the World* (Bookmarks).

Hobsbawm, Eric J, 1962, *The Age of Revolution 1789-1848* (Weidenfeld & Nicolson).

Israel, Jonathan, 1989, *European Jewry in the Age of Mercantilism 1550-1750* (Oxford University).

Israel, Jonathan, 2002, *Radical Enlightenment: Philosophy and the Making of Modernity 1650-1750* (Oxford University).

Jones, Norman, 1985, *God and the Moneylenders* (Blackwell).

Kamen, Henry, 1997, *The Spanish Inquisition: A Historical Revision* (Weidenfeld & Nicolson).

Leon, Abram, 1970 [1946], *The Jewish Question* (Pathfinder), www.marxists.de/religion/leon/

Leopold, David, 2007, *The Young Karl Marx: German Philosophy, Modern Politics and Human Flourishing* (Cambridge University).

Marx, Karl, 1967 [1843], *On the Jewish Question*, in Loyd David Easton and Kurt Guddat (eds), *Writings of the Young Marx on Philosophy and Society* (Anchor). A different translation is available online at www.marxists.org/archive/marx/works/1844/jewish-question/

Marx, Karl, 1991 [1894], *Capital*, volume three (Penguin). A different translation is available online at www.marxists.org/archive/marx/works/1894-c3/

Mendes-Flohr, Paul, and Jehuda Reinharz (eds), 1995, *The Jew in the Modern World* (Oxford University).

Modrzejewski, Josef, 1995, *The Jews of Egypt* (Princeton University).

Penslar, Derek, 2001, *Shylock's Children: Economics and Jewish Identity in Modern Europe* (University of California).

Rose, John, 2004, *The Myths of Zionism* (Pluto).

Rose, John, 2005, "Eleanor Marx", *Socialist Worker*, 10 December 2005, www.socialistworker.co.uk/art.php?id=7941

Shapiro, James, 1996, *Shakespeare and The Jews* (Columbia University).

Sorkin, David, 1992, "The Impact of Emancipation on German Jewry: A Reconsideration", in Jonathan Frankel and Steven Zipperstein, *Assimilation and Community: The Jews in Nineteenth Century Europe* (Cambridge University).

Tawney, Richard H, 1966 [1926], *Religion and the Rise of Capitalism* (Penguin).

Traverso, Enzo, 1994, *The Marxists and the Jewish Question* (Humanities).

Trotsky, Leon, 1940, "Imperialist War and the Proletarian World Revolution", resolution adopted by the emergency conference of the Fourth International, 19-26 May 1940, www.marxists.org/history/etol/document/fi/1938-1949/emergconf/fi-emerg02.htm
Trotsky, Leon, 1945 [1940], "On the Jewish Problem", *Fourth International*, volume 6, number 12 (December 1945), www.marxists.org/archive/trotsky/1940/xx/jewish.htm

The world economy—a critical comment

Jim Kincaid

Chris Harman has invited comment on his recent writings on the world economy and there is certainly much for Marxists to debate. We are confronted with a global economy, hugely uneven to be sure, but which in 2007 was producing no less than 25 percent more goods and services than just six years earlier. However, many on the left, including Harman, believe that the basic tendency of the system remains inflected towards stagnation in profits and rates of accumulation—as it has been, in their view, since the 1970s. They argue that the recent period of growth since 2001 has been based mainly on the build-up of unsustainable levels of debt, notably, rapid increases in household borrowing in the US based on soaring house prices. It is certainly true that this increase in debt helped finance a series of huge US trade deficits, which in turn boosted fast export-led growth in China and other emerging economies.

At the start of 2008 a wider international downturn is in progress. US house prices have collapsed, recession looms in the US, the dollar has fallen further and a crisis of unpayable debt has repercussed through the international banking system. What is now crucial is whether the new centres of accumulation in China and the rest of the industrialising world are large enough and dynamic enough to limit the recessionary impact of credit crisis. No one can be sure, but what happens over the next two to three years will be a decisive test.

In this article I will argue three main points against Harman's vision of the world economy, and the theory on which it is based.[1]

(1) Despite repeated phases of temporary downturn—and massive increases in economic inequality—the basic story of the world economy over the past 25 years has been one of rising profits, and growth in output and levels of capital accumulation. Advances in productivity have not undermined profitability as would be expected according to Harman's analysis.

(2) Harman uses Marx's declining rate of profit analysis in rather an abstract way, focusing too much on the average rate and giving insufficient weight to the countertendencies that Marx saw as limiting or reversing the tendency for overall profitability to fall.

(3) He also makes too little use of Marx's account of the forces that drive accumulation in the capitalist system: its tendency to expand geographically; to increase the mass and rate of profits by opening up new branches of production; to build up the quantities of capital deployed in the world's productive system and multiply the numbers of workers employed by capital. Harman's attention is not focused on the astonishing growth in recent decades in the size of the system, and the momentum given to it by enormous rises in the mass of extra profit generated. His analysis is limited rather narrowly to rates of profit. He neglects trends in the rate of formation of new capitals, relying here on an incorrect view that concentration of capital in most strategic sectors of the world economy blocks the entry of new firms into successful competition. He therefore seriously underestimates the role that the current dynamism of China and other large emerging countries can play in limiting the recession now starting and in sustaining a further phase of overall growth.

I do agree, however, with Harman's broader perspective that political crises, rooted in the economics of the system, will continue and are likely to deepen. We face tightening ecological constraints, intensifying contrasts between extremes of wealth and poverty, and the social destructiveness inflicted by the never-ending competitive restructuring of capital.

An important measure of the performance of the world economy is the annual percentage rate of growth in production. Figure 1 gives the IMF's estimate of world GDP growth (in real terms) since 1970. It shows that in about half of the 36 years covered growth was above or close to 4 percent a year. There were five periods of downturn, but in only two years did annual

1: For a re-education in critical Marxism during a period as one of the editors of *Historical Materialism* I am grateful to Sebastian Budgen and his cosmopolitan array of colleagues. My thanks also to Pete Green for discussion and productive disagreement, and to Sally Kincaid for computer assistance.

growth fall to less than 2 percent. The last two recessions followed each other quite quickly but were shallower and briefer than the more serious contractions at the start of the 1980s, and that of 1990-1. In the five years ending in 2006 growth averaged over 4 percent, and the same has turned out to be the case in 2007, despite the serious financial crisis which started last August. World population has increased since 1970, but the average rate of economic growth per person has still averaged nearly 3 percent per year.

Figure 1: World real GDP growth
Source: IMF World Economic Outlook, April 2007

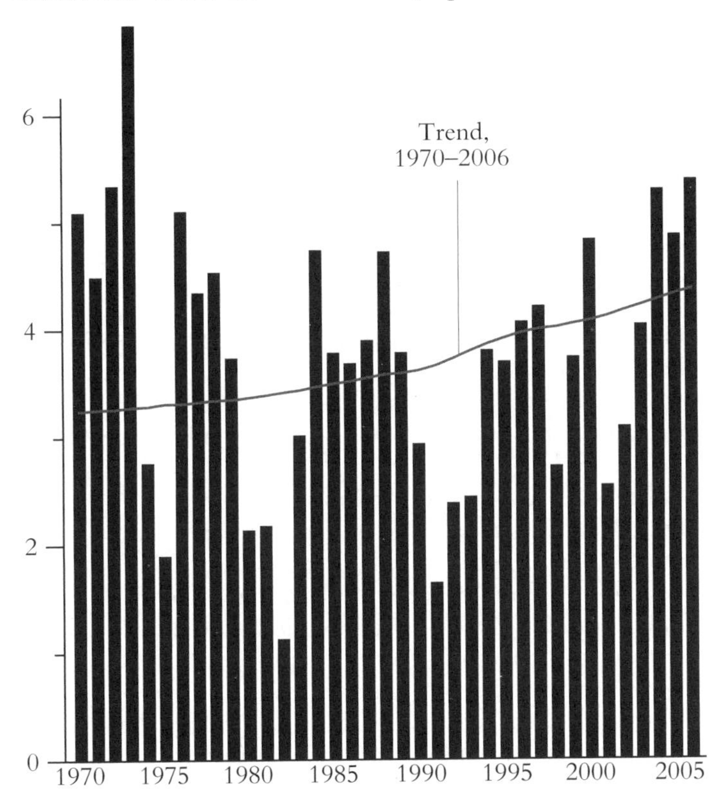

The distribution of all this extra output has, of course, been extremely unequal. In most countries the upper income groups have secured enormous increases in their wealth and income. About one billion people live on the equivalent of $1 a day or less, which represents a fall since 1980 from 27.9

percent to 21.1 percent of the world's population. But total world population has grown over this period, which means that there are large numbers of people who have seen a serious degree of improvement in their material situation. In some areas, especially in Africa and the Middle East, there has been a horrifying deterioration in living standards.[2] But in China, for example, over the period since 1980, the numbers of people living on less than $1 a day fell from 600 million to below 180 million, a decrease of more than 400 million people just in one generation.[3] The bitter knowledge that an even greater rate of poverty reduction might have been attained in China (and elsewhere in the world) should not blind us to the magnitude of this achievement by the toil of working people and advances in productive techniques.

Thus the picture is one of a capitalist system that is highly dynamic and growing in terms of overall output, though with huge social unevenness over geographical space, and a marked cyclical pattern over time. But how sustainable are current levels of growth? Will the present financial crisis and associated downturn in the giant US economy be just another brief and shallow recession like the five previous ones? Or are we starting to see the impact of more serious limits in the recent pattern of growth?

To assess the situation more deeply we have to consider levels of capital accumulation and profitability in the system. In an article published last year Chris Harman reviews the literature about trends in profits and comes to the following conclusions:

(1) In the industrial countries high profits in the 1960s were followed by a period of sharp decline which ended in the early 1980s with the profit rate about one third of the level of 20 years before.

(2) "Profit rates did recover from about 1982 onwards, but they only made up about half of the decline that had taken place in the previous period".[4]

Harman does not comment on the overwhelming evidence that an enormous rise in profitability in much of the world economy took place in the period between 2001 and 2007. On the global situation, one authoritative source, the Bank for International Settlements, reports that "corporate profits in 2005 appear to have reached historical highs as a proportion of global GDP".[5] Harman's discussion is confined to Britain, the US and China. He concedes that "for the moment profits in Britain appear to be high...15.6 percent for all non-financial corporations in the fourth quarter

---

2: See Bush, 2007, for a valuable recent account.

3: World Bank, 2006, p9; "Somewhere Over The Rainbow", *Economist*, 24 January 2008, www.economist.com/world/international/displaystory.cfm?story_id=10564141

4: Harman, 2007a, p150.

5: Bank for International Settlements, 2006, p24.

of 2006—the highest figure since 1969".[6] On the US Harman gives no figures, but suggests only that the upsurge of financial speculation and private equity takeovers will mean that "corporate profits will be being puffed up until they lose touch with reality". In actual fact the ratio of share prices to company profits has not been exceptionally high in recent years in the US. Fred Moseley concludes in a recent article:

> The rate of profit in the US is now approaching the previous peaks achieved in the 1960s. The last several years especially, since the recession of 2001, have seen a very strong recovery of profits, as real wages have not increased at all, and productivity has increased very rapidly (4-5 percent a year). And these estimates do not include the profits of US companies from their production abroad, but include only profits from domestic US production.[7]

In the period 2004-6 annual profits in the US continued to rise, from $1,331 billion in 2004 to an estimated $1,653 billion in 2006.[8]

On profits in China, reliable sources report massive recent increases. For example, Lardy estimates that "from 1998 through to the first half of 2006, profits of industrial enterprises in China soared from 2 percent of GNP to over 10 percent".[9] The World Bank's Beijing office, in its latest report, says that "profit margins in China's industry have continued their trend increase, supported by continued rapid growth of efficiency and labour productivity".[10]

Of course, overall profit levels in the system vary cyclically and there is no reason to suppose that the recent upsurge in profitability in China and much of the rest of the world economy will not be followed at some point by a downturn. It may well have started in 2008. But what would prevent a further recovery in the near future, as happened after the temporary downturn around 2001? In October 1987 world equity markets crashed by 30 percent in a few days, but interest rate cuts implemented by

6: Harman, 2007a, p158.

7: Moseley, 2007. The *New York Times*, 4 August 2007, reported, "In the 1960s, about 7 percent of US corporate profits came from overseas. By the first quarter of 2007, that share was up to 29 percent...government figures also indicate that the profits of American companies from their overseas operations have been growing at a much faster pace of 13.7 percent in the current decade, nearly twice the 7 percent rate of growth of profits in the domestic US economy."

8: "Economic Report of the President", 2008, table B 90, p331, www.gpoaccess.gov/eop/

9: Lardy, 2007, p14.

10: World Bank, 2007, p7.

the US Federal Reserve and other central banks contained the crisis. The effect on jobs and profitability in the industrial sector was very limited. The same was true of the massive downward lurch in US bond markets in February 1994. As Marx noted, because the financial markets are often dominated by speculative pressures they can, for long periods, become relatively autonomous from the productive economy. Thus, he concluded, financial crises, even if severe, need not necessarily have a corresponding impact on commodity production.

There are many problems of measurement and definition of profits, but really no clear evidence of any pattern of long term fall in the rate of profit.[11] Profit rates in the industrial countries were high in the late 1950s and early 1960s. They fell sharply in the late 1960s and 1970s, rose again after 1982, hit a short but steep downturn around 2001 and recovered strongly through to 2007.

Harman reads Marx as arguing that the deepest and most persistent tendency in the system is for average profit rates to decline, though the decline can be slowed, or even in some periods reversed, by the operation of countervailing processes. Harman explains:

> Marx's basic line of argument was simple enough. Each individual capitalist can increase his (occasionally her) competitiveness through increasing the productivity of his workers. The way to do this is by using a greater quantity of the "means of production"—tools, machinery and so on—for each worker.[12]

What matters here is not physical productivity—a few workers operating a semi-automated production line, or a huge container ship—compared with how it was done 50 years ago. Productivity is seen as increasing the organic composition of capital, ie a rise in the ratio of machinery to the labour that operates it. Thus the hours of socially necessary labour-time incorporated in the means of production tend to increase, but the number of hours of unpaid labour—the surplus value extracted—does not rise correspondingly, and may even fall. Thus Harman's argument is that the increase in productivity, which results from the pressures of

11: Harman 2007a, p150, reprints Brenner's data showing a fall in the rate of profit of manufacturing capital in the US, Germany and Japan from 1970 to 2000. See also Harman, 2008, p33. Neither Brenner nor Harman take account of the shift in manufacturing to the Global South, the massive rise of a labour intensive service sector in the advanced economies, and the post-2000 rise in profit rates in G3 manufacturing (even in Japan).
12: Harman, 2007a, p142.

competition, must inevitably put "a downward pressure on the ratio of profit to investment—the rate of profit".[13]

This is certainly a correct summary of Marx's argument that productivity advance undermines the rate of profit by raising the organic composition of capital. However, this is stated by Marx as a tendency only, and one which operates at a very high level of abstraction. Marx modifies his abstract argument in a number of crucial ways as he develops, stage by stage, an analysis of how capitalism operates as a concrete historical system. Harman gives insufficient weight to the ways in which Marx qualifies his abstract argument.

For example, Harman writes that "each capitalist has to push for greater productivity in order to stay ahead of the competition. But what seems beneficial to an individual capitalist is disastrous for the capitalist class as a whole".[14] This is a surely gross overstatement. Over the neoliberal epoch since 1980 productivity has been rising persistently throughout most of the world. Yet all the evidence shows that the global capitalist class is doing fine. It has been growing rapidly in numbers, and in the amount and proportion of the world's wealth which it owns and controls.

Harman emphasises that aggressive price-cutting competition by innovative firms undermines profits and devalues the capital of rivals who are stuck with old technology, and may put them out of business altogether. These losses have to be deducted from the overall rate of profit in each branch of production. But such capital devaluation and lowering of profits are directly damaging only for those companies with outmoded means of production.

As Marx points out, the firms which find new ways of increasing productivity are able, as they cut costs, to secure an above average rate of profit—surplus profit in Marx's phrase—and the average overall rate of profit in a capitalist economy is raised by such surplus profits. It is true that as their competitors adopt the new methods there is a general fall in prices of the commodities being produced, and rates of profit for all producers in a given branch of production fall towards an average. But Harman says nothing about the effect of the surplus rates of profit secured during the time period before the competition catches up. In his scheme, innovations seem to be adopted

13: Harman, 2007a, p142. For an account of value as the monetary expression of socially necessary labour-time, see Marx, 1981, chapters 13-15. Harman's account of the organic composition of capital, and its presumed effect on profit rates, is similar to that presented in Fine and Saad-Filho, 2004. For a detailed critique of this approach, see Kincaid, 2007. Ben Fine and Alfredo Saad-Filho have responded with a detailed defence of their position. This will be published by *Historical Materialism* in 2008, together with a critical reply by myself which also proposes an alternative reading of Marx's account of profits and value theory.

14: Harman, 2007a, p142.

virtually instantaneously by all producers. Nor does he discuss the possibility of continuous ongoing waves of innovation which would allow some firms to enjoy surplus profits for much longer periods. Yet repeated advances in productivity have been the pattern in recent years in many industries.[15]

And, in any case, as Harman does accept, in the longer run phases of devaluation and elimination of less efficient capital clear the way for higher profits among survivors. However, he discusses the law of value and competition only in terms of bankruptcies, conceding that the level of bankruptcy in the 1980s and 1990s had been higher than he expected, because the role of the state in propping up failing firms had been more limited than he had anticipated. But here he underestimates the depth and savagery of the restructuring process in Europe and the US in this period, and he says nothing about the major form which it took. This was the internal restructuring as large firms eliminated their less profitable operations, closed loss-making factories, and imposed huge redundancies.

The current credit crisis

In the previous issue of *International Socialism* Harman argues that the present credit crisis arose because in recent years the US and European banking system has been awash with cash for which the banks needed to find borrowers. Where did this extra money came from? Harman notes that corporate investment in the industrial countries, which fell during the recession of 2000-1, did not recover to previous levels in the years following. Instead industrial companies fed unused money capital into the banking system. The banks recycled the money to the household sector especially in the form of vast increases in mortgage lending. This pushed up house prices and allowed more borrowing which drove up house prices even more—in a self-feeding bubble, which, while it lasted, sustained consumer demand.

Harman accepts Riccardo Bellofiore's characterisation of this as "privatised Keynesianism". The mechanism went into reverse when the subprime crisis imploded in August 2007. The effect will be, Harman argues, that the fundamental long-term forces of low profits and industrial overcapacity will now reassert themselves strongly: "Only the financial

---

15: Harman considers that computers have increased productivity in only limited sectors of the economy. "The US has experienced high levels of productivity due to computerisation, but they have mainly been centred in the computer industry itself and in the retail trade"—Harman, 2007b, p13. This is a very questionable statement. The effects of computerisation, and associated internet technology in the US and in the rest of the world have been enormous and they have been experienced right across the spectrum of commodity production, including cultural products as well as goods and services.

bubble stopped recession occurring earlier." As over-indebted households cut back on their spending the fall in consumer demand will spread recession in the G7 countries. The emergent countries will be hit as their exports to the industrial countries collapse.

Certainly Harman is quite correct to stress that in the four-year period up to 2005 industrial investment lagged behind profits in the G7 countries and that this was one source of the housing-based bubble in bank lending. The IMF reports that in the G7 countries "on average over 2002-4 the excess saving of the corporate sector—defined as the difference between undistributed profits and capital spending—was at a historical high of 2.5 percent of GDP".[16] Harman sees this lag in corporate investment as decisive evidence of a basic stagnation since the 1970s in profits and rates of accumulation in the industrial countries. Here there are two objections to be made. First, the IMF report emphasises that "one factor behind the increase in corporate saving has been the strong rise in profitability that has underpinned higher corporate saving despite an increase in dividends paid".[17] Thus the lag in investment was not due to a lack of profitability, as Harman's basic line of argument tends to suggest.

Second, there is evidence that industrial investment in G7 has recovered considerably since 2004. For example, the latest OECD report on the question shows that net saving by the corporate sector was down to only 1 percent of GDP by 2006.[18] However, even if based on high profits, this is still a sizeable surplus of savings over investment. Nevertheless, before invoking the lack of demand which Keynes saw as underlying the slump of the early 1930s, we have to consider how far the present very high rate of investment and consumption in the emergent world will plug the gap left by a diminished rate of growth in consumer demand in the industrial countries. I will return to this, but first it is essential to consider the recent investment lag in a Marxist framework, rather than rest content with a Keynesian underconsumptionist perspective.

Productivity and the organic composition of capital

If the rate of productivity growth in department one of the economy, which produces means of production, is faster than in department two, which produces wage goods, then the effect will be a fall in the ration of constant capital

16: IMF, 2006, p136.
17: IMF, 2006, p140.
18: OECD, 2007, chapter 3. Such data, based on national accounts, underestimate the levels of investment by G7 companies in emergent countries such as China.

to labour. There is strong evidence that this has recently been the case. A carefully researched report by the Bank for International Settlements concludes:

> Record profits, high cash levels [in companies] and low interest rates have not prompted record corporate investment... Indeed corporate investment as a share of GDP remained low in the G3 economies by past standards... That nominal investment/GDP ratios are still low in most advanced economies remains a puzzle. Part of the explanation could be the fall in the relative price of business fixed investment. For instance, in Japan and the US the price of capital goods has declined by between 25 and 40 percent since 1980.[19]

The IMF notes that the lag in investment levels in 2002-4 was temporary and in part due to "a short-term reaction to the high corporate debt levels of the early 2000s". But the IMF adds that "there has also been a longer-term downward trend in the relative price of capital goods" and this explains "about one half of the decline in the nominal investment ratio". Looking to the future, the IMF suggests that "technological progress will likely continue to lower the prices of capital goods, especially in information technology".[20]

This is the process that Marx calls the cheapening of the elements of constant capital. Because productivity has soared in the sectors of the economy which make means of production, industrial companies have been able to meet production targets with relatively limited investment in money terms. The result—in combination with rapidly rising profits—has been to leave many companies with ample cash reserves and debt-to-profits gearing ratios at unprecedented low levels. If the recession that has now started in the US does turn out to be limited in its effect on production and jobs in the world economy, a major reason will be the current financial robustness of the industrial sector.

Profitability and the cheapening of constant capital

The cheapening of the elements of constant capital has contradictory effects on the underlying profitability of the system. It slows down, or maybe even

---

19: Bank for International Settlements, 2006, p24. Incidentally Japan shows investment at 14 percent of GDP, ie higher than the US at about 9 percent and Europe at 10 percent. Lack of growth in Japanese GDP has not precluded a fairly high rate of profitability and investment in Japan.

20: IMF, 2006, pp141-142, 145. Harman, 1984, pp20-23, discusses capital-saving investment. He argues that "we can expect that there will always be more innovations calling for increased capital than those calling for less". I remain unconvinced. Why should capital-saving investment be less prevalent than investment to economise on labour?

reverses, the build-up in the organic composition of capital—which, as I noted earlier, is central in Harman's "declining rate of profit" reading of Marx. But, as Marx noted in a little studied section of *Capital*, lower prices for means of production have the effect of releasing capital from the circuits of productive capital.[21] It is this loss of demand that allows Harman to argue that a lower rate of growth in investment in money terms could spread recession. But to clarify which of these conflicting tendencies predominates in the global economy today requires attention to some crucial and often overlooked elements in Marx's analysis of accumulation and profitability.

Some other countertendencies

Chris Harman's attention tends to be focused on only one form of accumulation—the intensive accumulation of mechanisation and rising productivity. But in Marx we also find discussion of a second sort—extensive accumulation in which investment takes place in new branches of production and in geographically new centres of production, and in which extra workers are drawn into capitalist production. The extra workers come from two main sources. They are either workers whose jobs have vanished because of the increase in productivity or workers new to capitalism, for example, those drawn from the peasant sectors of the economy.

The scale of extensive accumulation in the Global South in the recent period has been extraordinary, much of it based on local accumulations of capital derived from surplus extraction from peasants, rather than capital exported from the industrialised North. Chinese wages, for example, are 1/20th of US wages, so rates of exploitation can be very high, though limited of course by the intensity of competition in the markets for commodities produced. Richard Freeman estimates that the labour force available to global capital, either for direct employment or as a reserve army of labour, tripled from just under one billion workers in 1980 to over three billion in 2000. He calculates that 1.5 billion workers were added to the actual or potential capitalist labour force by its incorporation of China, India and the former Soviet bloc. The other half billion workers came from population growth in countries that were already part of the capitalist system in 1980, especially in Africa and Latin America. Freeman also notes the sizeable effect of extensive accumulation in cutting what Marxists call the organic composition of capital: "The entry of China, India and the former Soviet bloc into the global economy cut the global capital/output ratio by

21: Marx comments on what he calls the release of capital which can result from productivity advance in either the means of production or wage goods sectors. Marx, 1981, pp206-209.

55 percent, to just 60 percent of what it otherwise would have been".[22] As I noted earlier, Marx argued that this fall in the average organic composition of capital will tend to boost rates of profit.

China as a driver of world accumulation

I have already shown that profits in China have soared in recent years. Harman argues that the Chinese economy is still far from being large enough to act as locomotive in pulling the advanced countries out of the impending recession. He notes that, at dollar exchange rates, China produces only 6 percent of world GDP. But this takes no account of the large undervaluation of the renminbi against the dollar. In any case the capacity of China, a driver of the world economy, is much greater than GDP figures would suggest. GDP in China has been expanding at the rate of 10 percent a year, and nearly half of this is reinvested. The increase in investment in China accounts for a large part, maybe 20 percent, of the annual increase in world capital formation in recent years. Profits from production in China are not just retained domestically, or transformed into the vast Chinese holding of US financial assets, but are widely shared among the foreign capitals that invest in China, thus helping to offset the damage done to profits elsewhere in the world by Chinese competitive success.

Foreign capital, American and Japanese especially, is sharing lavishly in the profitability of the "Chinese" economy. For example, the US government reported that "in the first six months of 2006, US corporate profits in China passed $2 billion, up more than 50 percent from the first half of last year. US companies were on pace to earn more in China in 2006 than they earned there during the entire 1990s".[23] Demand from China for raw materials and other imports is fuelling growth throughout much of East Asia, Latin America and the Middle East. Nor is the Chinese productive system as dependent on exports to the US as is claimed by many, Harman included. The World Bank reports that only one quarter of growth in 2006-7 was accounted for by an increase in exports.[24]

22: Freeman, 2005, p1. Marx also discusses the boost to the average rate of profit provided by the opening up of new branches of production, especially for what he calls luxury consumption—often the production of services which combine a low organic composition of capital with low wages and thus a high rate of profit—"both the rate and mass of surplus value in these branches of production are unusually high"—Marx, 1981, p344. Enormous numbers of such new branches of production have been opened up in the modern capitalist economies.

23: *USA Today*, 25 October 2006.

24: The World Bank, 2007, p2, says that in China net external trade contributed 3 percent of total GDP growth of 12 percent in 2006-7. See also Lardy, 2007, p4.

In a useful recent article the *Economist* points out two developments that will limit the global impact of recession in the US and Europe, and act to offset the G7 investment lag that Harman stresses. First, half of China's exports now go to other emerging economies. Indeed, Chinese exports to Brazil, India and Russia were up by 60 percent in 2007, and to oil exporters by 45 percent. Exports to the US account for only 8 percent of Chinese GDP, 4 percent of India's, 3 percent of Brazil's and 1 percent of Russia's. Second, in 2007 consumer spending in the emergent countries rose almost three times as fast as in the developed world:

> Investment seems to be holding up even better: according to HSBC real capital spending rose by a staggering 17 percent in emerging economies last year compared with only 1.2 percent in rich countries.[25]

Exports from the US and Europe to the emerging world are rising rapidly, especially in the capital goods sector. It is certainly not possible to predict how severe and extensive the crisis unfolding in 2008 will turn out to be. But so far it does not look like a classic Marx-Minsky type of crisis of corporate indebtedness. In general industrial companies in most major countries have responded to the high profitability since 2001 by running down debt, buying back shares and building up cash reserves—all of which will help them ride out a downturn. In addition it is important to register that in recent years China has been only the most dramatic example of a more general pattern of more rapid growth in the industrialising than in the industrial economies. In 2006 GDP in India increased by 9.2 percent, in Russia by 6.7 percent, in Latin America by 5.5 percent. In the emerging and developing world as a whole it grew by 7.9 percent, compared to 3.2 percent for the advanced economies. This growth is deeply uneven between different regions and social classes. Moreover, such figures do not measure what is essential for a Marxist analysis, namely rates of accumulation. But they unmistakeably indicate that very high levels of capital accumulation are taking place.

My fundamental difference with Harman is that his discussion of the rate of profit fails crucially to register the political and economic consequences of the sheer momentum of current accumulation. He is preoccupied with average rates of profit, but what drives the system is the quest for surplus profit. We often use Trotsky's phrase "combined and uneven development", but in practice discussion focuses mainly on the

25: "Decoupling Debate", *Economist*, 6 March 2008, www.economist.com/finance/displaystory.cfm?story_id=10809267

unevenness of capitalist development. My plea is for much more attention to current patterns of combined development. Large numbers of new workers each year are entering direct capitalist employment. Many of them move into jobs in technically advanced factories and offices at high levels of productivity, and are thus subject to high rates of exploitation. In the present period what we urgently need to be thinking and debating about are the economic and political consequences for the system—and for our politics—when accumulation is both rapid and global, and when extensive and intensive accumulation are closely combined.

References

Bank for International Settlements, 2006, "76th Annual Report" (June 2006), www.bis.org/publ/arpdf/ar2006e.htm

Bush, Ray, 2007, *Poverty and Neoliberalism: Persistence and Reproduction in the Global South* (Pluto).

Freeman, Richard, 2005, "What Really Ails Europe", http://theglobalist.com/StoryId.aspx?StoryId=4542

Fine, Ben, and Alfredo Saad-Filho, 2004, *Marx's Capital* (Pluto).

Harman, Chris, 1984, *Explaining the Crisis: A Marxist Reappraisal* (Bookmarks).

Harman, Chris, 2007a, "The Rate of Profit and the World Today", *International Socialism 115* (summer 2007), www.isj.org.uk/index.php4?id=340

Harman, Chris, 2007b, "Rate of Profit Warning", *Socialist Review*, November 2007, www.socialistreview.org.uk/article.php?articlenumber=10144

Harman, Chris, 2008, "From the Credit Crunch to the Spectre of Global Crisis", *International Socialism 118* (spring 2008), www.isj.org.uk/index.php4?id=421

IMF, 2006, *World Economic Outlook* (April 2006), www.imf.org/external/pubs/ft/weo.htm

Kincaid, Jim, 2007, "Production versus Realisation: A Critique of Fine and Saad-Filho on Value Theory", *Historical Materialism*, volume 15, number 4.

Lardy, Nicholas R, 2007, "China: Rebalancing Economic Growth", www.petersoninstitute.org/publications/papers/lardy0507.pdf

Marx, Karl, 1981, *Capital*, volume three (Penguin). An alternative version is available online: www.marxists.org/archive/marx/works/1894-c3/

Moseley, Fred, 2007, "Is the US Economy Heading for a Hard Landing?", www.mtholyoke.edu/courses/fmoseley/HARDLANDING.doc

OECD, 2007, *Economic Outlook 82* (December 2007), www.oecd.org/dataoecd/60/0/39727868.pdf

World Bank, 2006, "Global Economic Prospects 2006", http://go.worldbank.org/CGW1GG3AV1

World Bank, 2007, "China Quarterly Update" (September 2007), http://siteresources.worldbank.org/cqu_09_07.pdf

Misreadings and misconceptions

Chris Harman

We cannot understand the system we live in or how to fight it simply by the repetition of slogans. We need serious analysis and debate. For that reason, I welcome Jim Kincaid's rejoinder to my articles in recent issues of *International Socialism*. But I think it is wrong in some important respects. He misreads what has been happening to the system, does not fully grasp theoretically the dynamic of capitalism and misconstrues some of the things I have said.

Western profitability

Let's start with the facts. Kincaid lays a lot of emphasis on the growth in the output of "goods and services" between 2001 and 2007. No one has denied that growth.[1] I wrote an article on the impact of Chinese growth on Europe two years ago that emphasised it.[2] But if you look closely at the IMF graph he provides, it is clear that the growth rate in 2002-6 was no greater than that in 1970-3 (with a much lower peak than in 1973)—that is, in the years that ended in the crisis which spelled the end of the long post-war boom. Furthermore, the graph starts in 1970. If it started five years earlier, in the mid-1960s, it would show an overall downward trend, not an upward one. In other words, worldwide growth is still substantially lower than in the 1960s (see figure 1).[3]

1: Although measurement of growth of services is a very contentious area, since non-traded services are measured in terms of the incomes of those paid to those who provide them. See, for instance, Kumar, 2006, pp43-44, which questions growth figures for India.

2: Harman, 2006.

3: This graph recently appeared in Li, 2008.

World GDP growth rate, 1961-2006
Source: World Bank, World Development Indicators

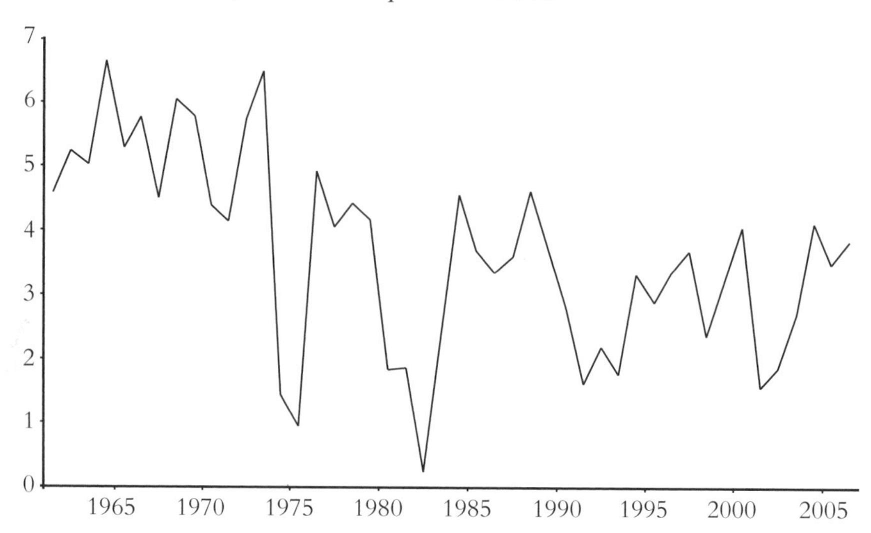

Kincaid then goes on to assert that "the past 25 years have seen rising profits, growth in output and levels of accumulation". Interesting here is the time span he gives. Twenty five years takes us back to 1982-3. A key point in my argument has been that profit rates started recovering in the early 1980s from the very low point they reached in the late 1970s. But in 2000 they were far from the level of the late 1940s, the 1950s and early 1960s that had sustained the long boom. Differing calculations by, for instance, Gerard Duménil, Robert Brenner and Fred Moseley all come to this conclusion, as I showed in my article on the rate of profit last year.[4]

What is the picture over the past six or seven years, since recovery from the recession at the beginning of the present decade? Kincaid says profitability has recovered. He quotes the Bank for International Settlements' assertion that profits as a share of global GDP "reached historical highs" in 2004. But profit as a share of GDP is not the same as the rate of profit. The profit share has grown because of increasing rates of exploitation worldwide (something I have repeatedly referred to). But that does not mean the ratio of profits to investment (ie the rate of profit) is at a record high. Kincaid also quotes Moseley. But Moseley's own figures[5] show profit rates in the long boom as

4: Harman, 2007.

5: www.mtholyoke.edu/courses/fmoseley/fig1.xls

hovering between 18 percent and 22 percent (between 1947 and 1968); they then fall through the 1970s to between 11 percent and 12 percent; from there they rise to about 14 percent to 15 percent in the late 1980s; and then in the mid-1990s to between 16 percent and 18 percent. From this level they fall back to between 14 percent and 15 percent in the early 2000s before rising to 19 percent in 2004. In other words, in one year in the past 25—2004—they reached the lowest figure in the long boom. Reports on the balance sheets of major US companies suggest that profits began to decline in 2005-6.

Brenner has also calculated recent profit rates. His overall trends are not that different to Moseley's, except that he shows the rise in profit rates from 2001 onwards as not exceeding the peak of 1997, and then beginning to decline in 2005-7—that is just before the onset of the financial crisis. The pattern shown by David Kotz's calculations is even starker. He shows the profit rate in 2005 as 4.6 percent, compared with 6.9 percent in 1997.[6]

Reported profits and profit rates are always open to question, since they depend to some degree on what firms can get auditing accountants to agree to. According to Susan Dev, professor of accounting at the London School of Economics, quoted in a study of the Murdoch empire:

> Profits are not facts; they are just opinions... This is one of the great truths of accounting—privately admitted but frequently denied in public by accountants... When a company draws up its accounts it needs to make a lot of assumptions. This is mainly because at the end of the year there is a lot of unfinished business, which creates uncertainties. For example, there are unpaid debts, and a judgment has to be made about whether these will be paid. There are lots of assets and a judgment has to be made about how long these will last. All these are subjective judgments: one company may decide that all the debts will be paid; another that none will be. The second company will then write off the debt and declare less profit that year. Profit then is a matter of opinion.[7]

I pointed in my last article to some of the indications profit rates have been inflated using such methods, noting increased investment by non-financial companies in US finance and real estate. Now we know what has been happening inside one of the US's manufacturing giants, General Electric, whose boss, Jeffrey Immelt, revealed in April that profits were

6: Kotz, 2008.

7: Belfield and Hird, 1991, pp232-233. For a longer discussion of how British firms inflated their figures in the late 1980s, and how the Bank of England accepted the inflated figures, see Harman, 1993, pp20-21. For the same phenomenon in the 1990s, see Harman, 2001, p101.

going to be half those he promised in December. As the *Economist* reported, in the past:

> GE's profits grew with the sort of predictable consistency that was made possible by laxer accounting standards and a talent for making good any unexpected shortfall with last-minute sales of assets held by the firm's notoriously opaque finance arm, GE Capital.

But the credit crunch exposed how inflated the supposed values of its assets were:

> It seems that the main cause of GE's last-minute failure to hit its latest target was that the seizure of the capital markets prevented several asset sales from being completed in time. Despite its image as an industrial company—making wind turbines, lightbulbs and so on—40 percent of GE's revenue now comes from GE Capital, so these incomplete deals made a big difference to the overall results.[8]

Accumulation

Just as misleading as Kincaid's claims over profitability are those over accumulation. Here we are fortunate. The IMF published an empirical study on global saving and investment in 2005. It concluded:

> Global saving and investment (as a percent of GDP) fell sharply in the decade following the first oil price shock in the early 1970s, but were then relatively stable until the late 1990s. More recently, however, they again declined, hitting historic lows in 2002 before modestly recovering over the past two years.

It recognised that "these global trends mainly reflect developments in the industrial countries, where both saving and investment have been trending downward since the 1970s", and these were compensated for to some extent until the late 1990s by investment "increasing substantially" in "the emerging market and oil producing economies". Nevertheless it points out that in these parts of the world too "investment" has "fallen" since "the time of the Asian financial crisis" and "remains below the levels of the mid-1990s".[9]

8: "Immeltdown", *Economist*, 17 April 2008, www.economist.com/business/displaystory.cfm?story_id=11058445

9: IMF, 2005.

In other words, far from global accumulation showing a rising long-term trend, as Kincaid contends, it has been falling (see figure 2).

That does not mean that there are not ups as well as downs in the system as a whole, or that there has been no long-term growth. As we have long argued in this journal, conditions so far have been quite different to those of the slump of the 1930s, just as they have been quite different to those of the post-war long boom. The system has, so far, been able to avoid a long drawn-out slump. But it has not been able to avoid repeated crises and long periods of stagnation in one part of the world system or another—witness the very deep recession of the former Eastern bloc countries in the 1990s and the near stagnation of the Japanese economy since 1992.

Figure 2: Global saving and investment
Source: IMF World Economic Outlook, April 2005

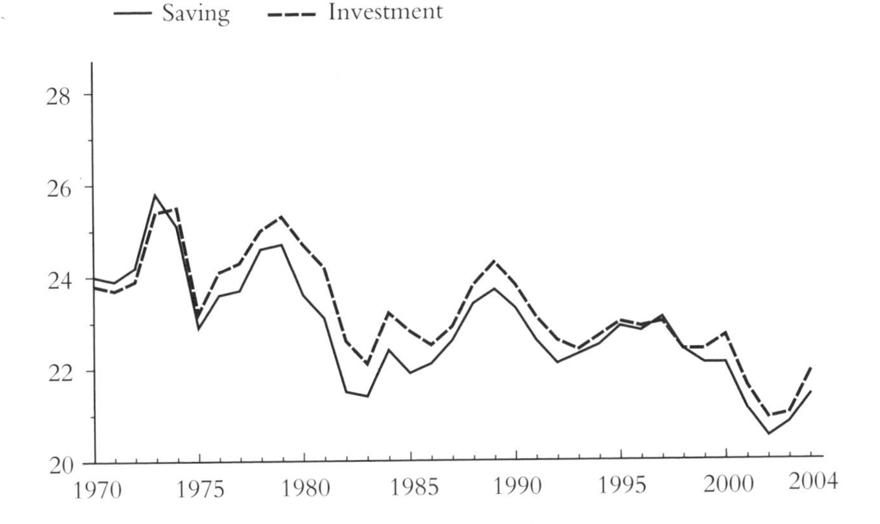

Theoretical slips

Kincaid objects to my theoretical arguments over the rate of profit. I do not want to rehearse them all here. I did this many years ago in a book that is still in print, and repeated the points last year in this journal.[10] So I will restrict myself to his claim that somehow increased productivity and "ongoing waves of innovation" have created "surplus profits", which by

10: Harman, 1984, chapter one; Harman, 2007.

allowing some firms to get greater profitability increase the profitability of the system as a whole. I am sorry, but he makes an elementary mistake here if he is attempting to proceed on the basis of Marx's analysis.

Increased productivity anywhere in the system has the effect of reducing the socially necessary time required to produce its output, and therefore the value of that output. This counter-intuitive insight was one of Marx's great advances on the classical economists who preceded him. Unfortunately, it is one that some Marxists have abandoned.[11] Now, it is true that the first capitalists to innovate get excess profits. But as the innovation is generalised across a sector, profit rates fall. The temporary surplus profits might conceivably lead to a more rapid rate of investment, as Kincaid claims. But here the evidence on investment seems to indicate otherwise. Kincaid admits with his quote from the Bank for International Settlements that "corporate investment as a share of GDP remains low in G3 economies". Here he wants to have his cake and eat it.

He claims that "internal restructuring" by firms through shutting down "less profitable operations" can increase the rate of profit, not merely by increasing the pressure on the remaining workforce, but by "devaluing capital". He forgets that shutting down plants that firms have spent money on does not miraculously do away with that spending. If they have borrowed to pay for that plant, they still have to repay what they owe. If they have shareholders, they still expect a return on their original investment, not its reduced current value. That is why bankruptcies have historically been a major way of the system recuperating from crisis: some capitals can recover their profitability by cannibalising the value contained in others. You cannot cannibalise yourself.[12]

Similarly, the fact that rapid innovation can cut the cost of new investment does not in any way help the capitalists who have already invested. Their existing investments suffer from more rapid obsolescence, adding to their depreciation costs. This was one very important feature of the mid to late 1990s, when "the standard measure of the average depreciation rate" became "invalid".[13]

In fact, despite Kincaid's claims, the burst of investment in the mid to late 1990s did not produce any miraculous growth in productivity. Productivity growth was higher in the US than in the 1980s, but lower than

---

11: It is the great flaw, for instance, of Duménil and Lévy, 2004.

12: The exception is when a crisis or a war does such damage that those who have invested or lent are so desperate that they are prepared to write off vast sums, as in Germany or Japan after the Second World War.

13: Tevlin and Whelan, 2000.

in the 1960s. Productivity increases that did take place were concentrated in certain sectors of the economy—in the sector making the new technology itself, in retail (especially one giant firm, WalMart) and in finance.[14] Robert Gordon found that computerisation produced little increase in labour productivity growth in 88 percent of the economy.[15] This "productivity paradox" has bewildered many mainstream economists.[16]

Finally, Kincaid tries to save his case by claiming, on the basis of the Bank for International Settlements report, that less is spent on investment in means of production because they have got cheaper. But it does not matter why less is spent on investment. If it falls while real wages are still being held back (indeed, cut in the US and Germany), then a gap opens up between supply and demand in the economy as a whole due to what Keynesians call a "liquidity trap", an excess of saving over investment[17]—unless something else fills the gap. That is what the credit and housing bubbles did, and that is why without them there would not have been a recovery from the 2001 recession on anything like the scale that actually occurred.

The China option

Kincaid's final set of arguments claims that economies outside the advanced industrial world, especially China, "will plug the gap left by a diminished rate of growth of consumer demand in the industrial countries".

Accumulation in China, and to a lesser extent in India, has been rising. But the Chinese economy today is not big enough to be a locomotive that can pull the rest of the world economy behind it. At current exchange rates the IMF gave its GDP in 2006 as $2,600 billion—just behind Germany, just ahead of the UK and less than a fifth of the size of either the US or the EU. GDP can also be measured in "purchasing power parity", which is based on domestic buying power. A revised World Bank estimate recently cut this down from 60 percent of US GDP to about 50 percent,[18] but, in any case, a country cannot trade according to purchasing power

14: See, for instance, Ark, Inklaar and McGuckin, 2003.

15: Gordon, 2000.

16: See the discussion in Hutchinson, 2008.

17: Despite Kincaid referring to "Keynesian underconsumptionism", the first to have this insight was Marx with his refutation of Say's Law. The problem Keynesians have is that they cannot deduce it from the dynamic of accumulation because they are stuck with a slightly amended version of neoclassical value theory and its assumptions about equilibrium.

18: Selim Elekdag and Subir Lall, "Global Growth Estimates Trimmed After PPP Revisions", *IMF Survey Magazine*, 8 January 2008.

parities. A country that only accounts for 4 or 5 percent of global buying power cannot compensate for the effect of a major economic crisis in a country that accounts for over 20 percent.

But the faults with Kincaid's argument go deeper than that. He claims that "extensive investment", using relatively more labour than fixed capital, is taking place in China and the other "Brics" (that is, Brazil, Russia, India and South Africa as well as China), so providing a massive boost to profitability and investment worldwide. He justifies his claim by referring to a much hyped article by Richard Freeman, which claims:

> The labour force available to global capital tripled from just over one billion workers in 1980s to over three billion in 2000—either for direct employment or as reserve army of labour... 1.5 billon workers were added to the actual or potential labour force by its importation of China, India and the former Soviet bloc.

According to Freeman this has supposedly "cut the global capital/labour ratio by just 55 to 60 percent of what it otherwise would have been". The claim is astonishing in its disregard for facts. First, the idea that India was until quite recently not part of the world system should immediately be seen as wrong by any intelligent person: its share of world exports was three times higher in 1950 than it is today.[19] I would also argue that Russia and China were part of the world system, and in looking at its dynamics you have to take into account their fairly high rates of growth and accumulation 30 or 40 years ago.

Perhaps more importantly, the figures for worldwide employment are fanciful. In 2001 the non-agricultural workforce of the developing and transition economies was 1,135 million.[20] Self-employment accounted for a high proportion of these: 32 percent in Asia, 44 percent in Latin America and 48 percent in Africa.[21] Those proportions have grown everywhere with urbanisation. That reduces Freeman and Kincaid's 1.5 billion potential new workers to about 700 million. Just as importantly, only a proportion of those who seek work gain employment in the formal sector in modern industry. Most are in very low productivity jobs, often working for firms with only a couple of workers.

Industrial employment in China actually fell from 78 million in 1997

19: The growth of world trade means a smaller share counts for more today.
20: "Summary of Food and Agricultural Statistics 2003", available from www.fao.org
21: ILO, 2002.

to 54.4 million in 2001.[22] The fall was due to large-scale redundancies in the old industrial sectors, which was not compensated for by increased employment in newer sectors, despite the massive investment taking place there. In India in the early 1990s only 42.1 percent of the urban workforce were "regular employees", while 41.7 percent were categorised as "self-employed", and 16.2 percent as "casual employees".[23] Things have not changed since. Employment growth is only a little over 1 percent a year and even less in manufacturing.[24]

These figures show something Marx would have recognised. International competition is leading the less industrialised countries to pursue patterns of accumulation based on high ratios of fixed capital to workers. The "organic composition of capital" has been rising at a rapid rate in China. "During 1980-2003 China's capital stock grew by 11.3 percent per annum on average, far outpacing the annual average of 1.6 percent growth in labour".[25] This is intensive accumulation, not the extensive version that, according to Kincaid, means a new wave of expansion for the global system.

His and Freeman's mistake is to fall for a fallacy of neoclassical economics, the notion that means of production and labour are interchangeable according to their supposed marginal productivities. Marx had to deal with an earlier version of this argument, according to which accumulation depended on the growth of population. He, by contrast, recognised that the growth or otherwise of the employed population depended on the pattern of accumulation. This could lead to the apparent paradox of accumulation being accompanied by a massively expanding surplus population:

> The additional capital formed in the course of accumulation attracts fewer and fewer labourers in proportion to its magnitude. The old capital...repels more and more of the labourers formerly employed by it.[26]

This phenomenon in China leads to another very important process that Kincaid does not recognise. The rate of profit has been undergoing a long-term fall.

Kincaid writes of profit margins and the share of profits in GDP rising. This may be true, although it is very difficult for anyone to calculate

22: Brooks, 2004.
23: Figures in Unni, 2001, p2367.
24: Dasgupta and Singh, 2006.
25: Lavina and Fan, 2008, p762.
26: Marx, 1961, p628.

the real profits of Chinese industrial enterprises because of the degree to which they are able to borrow from the banks, which have taken on a very large amount of bad debt.[27] In any case, profit margins and profit share are not the same as profitability. Recent studies of profitability in China show it as falling. Phillip O'Hara calculates it as declining from 47 percent in 1978 to 32 percent in 2000.[28] Another study by Editha Lavina and Emma Xiaoqin Fan points to the same trend, but different figures, showing a fall from 13.5 percent in 1980s to 8.5 percent in 2003.[29]

This can account for something else that Kincaid fails to recognise the significance of. The excess of "saving" (accumulated past surplus value) over investment is as much a feature of the Chinese economy as of those in the advanced industrial world. The sums that have been flowing across the Pacific to finance US indebtedness could have instead been invested productively in China. Chinese capital, whose current savings amount to a massive 50 percent of national output, has not felt profit rates are high enough to sustain investing more than 90 percent of these savings. There was already recognition five years ago that "investment in many sectors—including property, cement, steel, cars and aluminum—is being overdone".[30] There are now even greater fears as the massive upsurge in investment forces up world raw material and food prices in a way that has not happened since the boom in the advanced industrial economies of 1971-3 that precipitated the first serious post-war crisis in 1974-6.

Chinese premier Wen Jiabao was rather more worried about the contradictions of Chinese capitalism than Jim Kincaid when he told the National People's Congress in March 2007 that "the biggest problem with China's economy is that the growth is unstable, unbalanced, uncoordinated, and unsustainable".[31]

That is what capitalist booms are like. That does not necessarily mean the boom is going to collapse tomorrow. It does mean it is not going to go on forever, as so many people who write on China seem to think. It also

---

27: According to official estimates 20 percent of all loans were "non-performing" in 2003—an unofficial estimate suggests that non-performing loans reached 45 percent of GDP—*Financial Times*, 23 September 2003.

28: O'Hara, 2006. But, as with all calculations of profit rates, there can be doubts as to the accuracy of the statistics O'Hara bases his calculations on—particularly since his figures indicate a decline in the rate of exploitation, which hardly fits in with the declining proportion of wages and consumption in GNP shown in Aziz and Li, 2007.

29: Lavina and Fan, 2008, p748.

30: *Financial Times*, 18 November 2003.

31: Cited in Aziz and Dunaway, 2007.

means that China, far from alleviating the problems of global capitalism in 2008, is adding to the double crisis that besets it, with the credit crunch originating in the US on the one hand, and the surge in energy and food prices on the other.

On one thing Kincaid is right. It is still just possible that those running the bits of the world system will be lucky enough to find some way out of their immediate problems. Maybe pumping money into the US economy will somehow reflate the bubble—as happened in 1987 and 1998. Maybe that will happen without further fuelling the pressures on energy and food prices. But the chances are not very great. And in 1987 and 1998 they only succeeded in postponing a wider crisis for a couple of years.

Meanwhile, their failure to achieve a sustained recovery of profit rates forces them to try to push down hard on the rest of us. That is what the counter-reforms wrapped up in the ideology of neoliberalism are about. Precisely because capitalism has not recovered from the problems it first discovered in the 1970s, the attacks are going to continue hitting people and breeding further resistance. This is the most important point.

References

Ark, Bart van, Robert Inklaar and Robert H McGuckin, 2003, "ICT and Productivity in Europe and the United States, Where do the Differences Come From?", *CESifo Economic Studies*, volume 49, number 3, www.ggdc.net/~inklaar/papers/ictdecompositionrev2.pdf

Aziz, Jahangir, and Li Cui, 2007, "Explaining China's low Consumption", IMF, working paper, www.imf.org/external/pubs/ft/wp/2007/wp07181.pdf

Aziz, Jahangir, and Steven Dunaway, 2007, "China's Rebalancing Act", *Finance and Development*, volume 44, number 3, IMF, www.imf.org/external/pubs/ft/fandd/2007/09/aziz.htm

Belfield, Richard, and Christopher Hird, 1991, *Murdoch, the Decline of an Empire* (Little, Brown).

Brooks, Ray, 2004, "Labour Market Performances", in Eswar Prasad (ed), *China's Growth and Integration into the World Economy*, IMF, www.imf.org/external/pubs/ft/op/232/op232.pdf

Dasgupta, Sukti, and Ajit Singh, 2006, "Manufacturing, Services And Premature De-industrialisation in Developing Countries", Centre for Business Research, University of Cambridge, working paper, www.cbr.cam.ac.uk/pdf/WP327.pdf

Duménil, Gerard, and Dominique Lévy, 2004, *Capital Resurgent* (Harvard).

Gordon, Robert, 2000, "Does the 'New Economy' Measure up to the Great Inventions of the Past?", *Journal of Economic Perspectives*, volume 14, number 4, http://econ161.berkeley.edu/Teaching_Folder/Econ_210c_spring_2002/Readings/Gordon_Inventions.pdf

Harman, Chris, 1984, *Explaining the Crisis: A Marxist Reappraisal* (Bookmarks).

Harman, Chris, 1993, "Where is Capitalism Going", *International Socialism 58* (spring 1993).

Harman, Chris, 2001, "The New World Recession", *International Socialism 93* (winter 2001), http://pubs.socialistreviewindex.org.uk/isj93/harman.htm

Harman, Chris, 2006, "China's Economy and Europe's Crisis", *International Socialism 109* (winter 2007), www.isj.org.uk/index.php4?id=160

Harman, Chris, 2007, "The Rate of Profit and the World Today", *International Socialism 115* (summer 2007), www.isj.org.uk/index.php4?id=340

Hutchinson, Robert, 2008, "Knowledge and Control: A Marxian Perspective on the Productivity Paradox of new Technology", *Rethinking Marxism*, volume 20, number 2.

ILO, 2002, "Women and Men in the Informal Economy", www.ilo.org/public/english/employment/infeco/download/menwomen.pdf

IMF, 2005, *World Economic Outlook* (April 2005), www.imf.org/external/pubs/ft/weo/2005/01/

Kotz, David, 2008, "Contradictions of Economic Growth in the Neoliberal Era", *Review of Radical Political Economy*, volume 40, number 2.

Kumar, Arun, 2006, "Flawed Macro Statistics", in *Alternative Economic Survey, India 2005-2006* (Daanish).

Li, Minqi, 2008, "An Age of Transition", *Monthly Review* (April 2008), www.monthlyreview.org/080401li.php

Lavina, Editha, and Emma Xiaoqin Fan, 2008, "Diverging Patterns of Profitability, Investment and Growth in China and India during 1980-2003", *World Development*, volume 36, number 5.

Marx, Karl, 1961 [1867], *Capital*, volume one (Moscow), www.marxists.org/archive/marx/works/1867-c1/

O'Hara, Phillip Anthony, 2006, "A Chinese Social Structure of Accumulation for Capitalist Long-Wave Upswing?", *Review of Radical Political Economics*, volume 38, number 3.

Tevlin, Stacey, and Karl Whelan, 2000, "Explaining the Investment Boom of the 1990s", US Federal Reserve (March 2000), www.federalreserve.gov/Pubs/feds/2000/200011/200011pap.pdf

Unni, J, 2001, "Gender and Informality in Labour Markets in South Asia", *Economic and Political Weekly* (Bombay), 30 June 2001.

Some notes on the crunch and the crisis[1]

Fred Moseley

I agreed with much of Chris Harman's latest article. My comments below will focus on the disagreements in order to further the discussion.

(1) As I have argued before, I think there has been a substantial recovery of the rate of profit in the US economy, although not a complete one. Even if we accept Robert Brenner's estimates, the rate of profit has recovered more than half of the prior decline. And if we add in foreign profits and also top executive salaries (both of which are excluded from the Nipa estimates of profit, which Brenner uses),[2] the recovery would be even greater. The criticisms Harman makes of the profit data do not apply to the Nipa estimates, because these estimates are based on actual profits and exclude capital gains and losses.

Three decades of stagnant real wages and increasing exploitation have substantially restored the rate of profit, at the expense of workers. This important fact should be acknowledged.

(2) Plus, debt-to-profit levels for non-financial corporate business have come down (except for the leveraged buyouts), so most of these companies are in decent financial shape, and are not at great risk of bankruptcy.

1: This feedback article is a response to Chris Harman's "From the Credit Crunch to the Spectre of Global Crisis", *International Socialism 118*, www.isj.org.uk/index.php4?id=421

2: Nipa: the national income and product accounts tables produced by the US Bureau for Economic Analysis (www.bea.gov).

(3) However (as Harman discusses), the higher profit levels have not resulted in a significant increase of investment. Instead capitalists have chosen to pay themselves more dividends and higher salaries—as is clearly evidenced by higher dividend/profit ratios and the large stock buy-backs (which increases the incomes of executives who have stock options).

This means that there is even less of a "trickledown effect" of higher profits (if there ever is one).

(4) The main problem in the current crisis is the financial sector. Harman says that the crisis is not due mainly to the bankers' greed and shortsightedness. I agree with that, but I would say the problem is more fundamental—the nature of the capitalist financial system, which is inherently speculative.

The best theorist of the capitalist financial system is Hyman Minsky, not Karl Marx. The current crisis is more of a Minsky crisis than a Marx crisis. I am not saying that we should throw away Marx (obviously), but rather that we should supplement Marx with Minsky, especially for analysis of the modern capitalist financial system.

Minsky's theory emphasises: (i) Speculative finance—borrowing short term and lending long term. This is exactly what has exploded in recent decades (investment banks, hedge funds, mortgage companies, etc). (ii) A period of prosperity breeds over-confidence; lenders take more and more risks in hopes of higher returns (eg subprime mortgages). (iii) Eventually the bubble bursts, not all the speculative borrowers are able to refinance their short-term loans (they have to dump assets, many go bankrupt, etc) and the crisis spreads.

I think this is a very good framework for understanding and explaining the current credit crisis in the US.

(5) The government will try to solve this crisis by bailing out the speculative lenders in one way or another. Such proposals are already being made and frantically developed. I think that is where we should focus our class struggle energies in the months ahead—"no bail-outs for the bankers" and "stop foreclosures now". This raises fundamental questions about the nature of the capitalist financial system.

Reviews

What's wrong, and what can be done

Paul McGarr

Jonathan Neale, ***Stop Global Warming: Change the World*** *(Bookmarks, 2008), £11.99*

There are already many excellent books on climate change. Indeed, it seems like almost every environmental commentator either has written or is in the process of producing one. Not all are good, but many are excellent. A few of my personal recommendations would be George Monbiot's *Heat*, Fred Pearce's *When the Rivers Run Dry* and *The Rough Guide to Climate Change*.

What distinguishes the different books is not usually the description of potential catastrophe facing the world, which they all contain, though the way they present this certainly varies according to the author's particular approach. Instead different perspectives usually emerge more sharply when the discussion turns to the key questions—what can be done to stop climate disaster, and, once that has been settled, how can this be achieved?

And so it is with Jonathan Neale's new book. What informs it, gives it its strength and sharply distinguishes it from all others is Jonathan's politics. The heartbeat of the book is the perspective of a committed socialist activist and revolutionary, who sees the world as fundamentally shaped by class. Jonathan has played an active role in the great global resistance movements of the past decade and is at the heart of efforts to build a new global movement to resist climate change. He writes, "This book is part of that movement", and, in my opinion, anyone who is or wants to be part of that movement should read it.

Before getting to the politics of climate change Jonathan does give a very useful overview of the scale of the problem. He points out that the climate has constantly changed during the Earth's history and will do so in the future. The debate today, however, is about abrupt or relatively sudden climate change provoked by human actions, principally, of course, the rising atmospheric concentration of greenhouse gases such as carbon dioxide and methane.

The global climate is a complex, dynamic system (what scientists call non-linear) in which an accumulation of small quantitative changes can, at some "tipping point", produce sudden and large qualitative changes. Jonathan, following most serious commentators, argues that the rising concentration of carbon dioxide in the atmosphere will very likely bring the global climate to such a tipping point in a few decades unless something drastic is done.

In a useful section of the book Jonathan quickly deals with two key arguments. First, he nails the lie that poor people are the problem—that it is the hope of people in countries such as India and China to escape from poverty and improve their living standards that is the issue. As Jonathan points out, there is no reason why economic growth that benefits people in these countries has to follow the blueprint of the US. It is possible to have economic development and rising living standards without this necessarily leading to the runaway growth of fossil fuel dependent energy use.

Second, Jonathan deals head-on with the mistaken argument that sacrifice by ordinary people is a necessary part of the answer to the threat of abrupt climate change. He explicitly rejects analyses that lump together all those in the richer industrialised countries as a homogenous "we". Instead he argues that an analysis of how we can build a movement on climate change must start from class, and this applies with just as much force to countries such as China and India, which are equally riven by class division.

His central argument is that "climate justice must also mean a global movement to lift all the world out of poverty... this cannot be done by a movement that focuses of what people will lose to save the climate". He insists, "At root climate activists face a choice. We can look to the ruling class, the wealthy and the powerful for the solution. Or we can look to ordinary people."

The core of Jonathan's book is, then, a detailed examination of the key factors driving greenhouse gas emissions and of possible solutions. He sets himself one important limitation to his argument: "I restrict myself to what we can do now with the technology we already have, rather than a discussion of solutions that may one day be possible with enough research and investment." He rightly insists, "We already have the technology to stop global warming." The real issue is the political will to implement the solutions that are within our grasp.

Jonathan sums up at the outset some central elements in his argument. One is that putting the required technologies into effect will require massive government intervention. Another is that market solutions of the kind favoured by most of the political establishment simply will not work. And he also argues that the fashionable avenue of personal consumer choices cannot solve the problem.

For those who say governments cannot act in the way he argues is needed, Jonathan gives a very useful and informative account of what happened in the US economy during the Second World War. Back then the state's War Industries Board directed what was produced and how in great detail, and imposed petrol rationing across the country. In a few months the whole US economy was transformed. If that could be done to win a war, why can't something similar be done to save the planet?

In separate chapters Jonathan goes on to look at key areas such as electricity generation, buildings, transport and industry. In each he argues for solutions based on what is currently possible but also with an eye on how to build real struggles. He concludes each section by suggesting immediate concrete demands to fight for.

Much of the detail in these sections shares common ground with the best literature on climate change. For instance, Jonathan explains why renewable and clean energy such as wind, solar and (perhaps most promising of all) concentrated solar power transported from sunny areas of the world on high voltage direct current cables, could easily meet the world's electricity needs. The section on transport clearly puts forward the argument for car-free cities and a massive expansion of public transport.

Jonathan is right about the aim of car-free

cities and that we should fight for this. But I think he could have perhaps spent more time dealing with measures that could immediately begin to reduce car use. He could have developed a little more the point that dealing with transport problems in cities also demands action on how housing, workplaces, education and much else are structured.

For instance, in most major cities in Britain government policy on schools directly contributes to car use. The myth of "parental choice" leads to the madness of school "league tables" and the transportation of children away from their local area to schools all over the city. Scrapping so-called "parental choice" and requiring young people to attend their local school, and organising safe walking and cycling routes to get there, would end the chaos of the "school run" and have a major impact on car use in our cities.

The expansion of giant out of town supermarkets makes shopping essentially dependent on cars. Action on this—either banning such developments and insisting shops be located where people can walk or banning cars from such developments and instead ensuring free, reliable means of getting people there and back—could also have an immediate and major impact on car use. Similar arguments apply to workplaces and commuting.

Winning the war on climate change does not just mean the choice of cars or public transport (though it does mean that). It is also about battling to reshape cities to both improve people's quality of life and reduce carbon emissions.

Having looked at solutions that will work, Jonathan turns to those that will not, but which are advocated by many. In a convincing summary he shows why biofuels, favoured by some of the world's most powerful governments, are not part of the solution. He also explains why proposed solutions relying on hydrogen fuelled transport are unproven, as are solutions based on removing carbon from power station emissions and storing it, for example, in depleted oil reservoirs under the seabed. Jonathan also sketches out why nuclear power is not the answer.

Another major section of the book sets out the reasons why the rich and powerful will not act to deal with the threat of climate disaster. This gives an excellent overview of the rise of neoliberalism as the key ideology of global capitalism and how this is a response to the crisis of profitability that has dogged the system for the past few decades. Jonathan discusses the central role of the carbon corporations—the oil, gas, car and steel corporations—in global capitalism and how they shape government policy. He takes us through a very informative case study of how the US car corporations turned to producing SUVs, which illustrates how reluctant these carbon corporations are to embrace fundamental change.

Jonathan also deals with the road to the 1997 Kyoto climate agreement and explains the limitations of that deal. And he also explains why all the various market-based emissions trading schemes simply will not solve the problem of climate change.

He puts an important argument about the personal lifestyle and consumer choice solutions that are increasingly pushed as the answer to climate change in the media. Such solutions are "at the wrong end of the pipe". Action to deal with climate change must be directed towards the source of the problem, not aimed at mitigating its effects. Changing how society produces energy, organises transport, constructs buildings and develops cities is the key—and this requires large-scale government action and regulation, not individual lifestyle choices.

A key focus in the book is that understanding and dealing with the threat of climate disaster require a wider understanding of capitalism. For those who, having read Jonathan's account, want to go deeper into this debate and to examine the specifically Marxist explanation of these connections, I would suggest the writings of John Bellamy Foster as a good place to begin.

Jonathan goes on to give a very good overview of the rise of global resistance to neoliberalism and then to war over the past decade, an overview that is especially useful as it is written by someone who has played an important role in many of the mobilisations. It is out of these movements that he sees the hope for building a global movement that can simultaneously fight for action on climate change and to lift people across the world out of poverty.

Before drawing his argument to a conclusion Jonathan gives two case studies of how climate change interacts with a host of other social and political factors to shape real human disasters. One of these is a brilliant discussion of the impact of Hurricane Katrina in New Orleans in 2005 and what has happened since. This is the clearest and best account I have read of this disaster, and Jonathan writes with a sure touch as he guides the reader through the story. This section alone would be worth getting the book for.

The second case study Jonathan uses is that of the Darfur region of Sudan. This, at least to me, was less successful. Jonathan admits the story is horrendously complicated, and it is. Though this may betray my own ignorance, I must confess that I found this section over-long, the detail a little bewildering and that I was still left confused at the end.

Jonathan ends his book with a call and a debate. The call is for "a mass climate movement", on the model of the global resistance to neoliberalism and war. I feel there is an important discussion to be had as to whether such a movement, focused specifically around the climate, can be built. Or will such a movement be principally made up of struggles and mobilisations around more specific issues that are linked to climate change but also draw in other social issues? The massive 2007 Mexican tortilla protests, which were linked to the push for biofuels by the US and the impact on food prices, show how such struggles could emerge.

The final discussion Jonathan initiates considers whether the fight to stave off climate disaster is "like the fight for the welfare state", one which can be won, at least to a degree, within the existing system, or if it is more "like the French Revolution", requiring a willingness to pose a revolutionary challenge to the system if we are to win it.

As Jonathan argues, it may be possible to mobilise enough pressure from below to force action by those at the top of the system in time, but we simply do not know if this will work. We do know that, first, a willingness to pose a revolutionary challenge to capitalism is often the best way to mobilise to win changes from the system and, second, that, were such a revolutionary challenge to succeed, it would certainly stave off climate disaster.

Not all farmers were bad...

John Newsinger

Daniel J Leab, ***Orwell Subverted: The CIA and the Filming of Animal Farm*** *(Pennsylvania State University, 2007), £36.50*

The Halas and Batchelor animated film of *Animal Farm* had its world premiere in New York on 29 December 1954. A fortnight later on 13 January 1955 it premiered in London. The film was celebrated as the first British feature length cartoon, as a cartoon that dealt with serious issues and as a breakthrough for the British film industry.

As we now know, the film was only made possible by CIA funding. Daniel Leab's appropriately titled *Orwell Subverted* is an exhaustive study of CIA involvement in the making of the film, a study that furthers the earlier work of Frances Stonor Saunders (*Who Paid the Piper? The CIA and the Cultural Cold War*), throws additional light on the CIA's cultural activities, and reveals, for the first time, the detailed changes that the agency insisted on in the making of the film.

That George Orwell, a socialist who had fought with the Poum militia in the Spanish Civil War, and who remained committed to the socialist cause until the day he died, should have his great anti-Stalinist classic hijacked by the CIA, is, of course, the ultimate indignity. But was there something about the book and its politics that lent itself to the use made of it by both the CIA and the British secret propaganda agency, the Information Research Department (IRD), established by the Labour government in 1948? Both organisations sponsored the publication of *Animal Farm* and *Nineteen Eighty Four* throughout the world. The IRD even produced a comic strip version of *Animal Farm*, which it tried to place in newspapers in as many countries as possible. Orwell's anti-Stalinism was, to be blunt, made use of for counter-revolutionary and pro-imperialist purposes. How much of the responsibility for this was his?

There is no doubt Orwell's anti-Stalinism led him into dangerous waters in the late 1940s. His understanding of the Soviet Union as a bureaucratic or oligarchic collectivist society led him to conclude that capitalism was the lesser evil in a conflict between the two systems. While this was certainly better than proclaiming Russia a socialist paradise, it did involve Orwell, on a number of occasions, making it clear that in the event of a war between the US and the Soviet Union he would support America. It also saw him collaborating with the IRD shortly before his death, although the nature of this collaboration has been widely and wilfully misinterpreted.

Even so, the important point was that, while capitalism might be the lesser evil, he still regarded it very much as an "evil". Indeed, he argued that the only cause actually worth fighting for was a United Socialist States of Europe, something that he recognised both the Americans and the Russians would regard as a threat. He condemned British subordination to the US as an obstacle to socialist advance.

Ironically, the best demonstration of the socialist credentials of Orwell's anti-Stalinism is provided by the changes the CIA required to Joy Batchelor's scripts, which were faithful to the book. Leab identifies three areas where Joseph Bryan, a psychological warfare expert who monitored the film for the agency, wanted modifications. The "investors", he told Batchelor, were very concerned at her sympathetic portrayal of Snowball (based on Leon Trotsky). Her script suggested that Snowball was "intelligent, courageous, dynamic" and that if he had not been assassinated he might have succeeded in "creating a benevolent, successful state… this implication we cannot permit". Instead the CIA wanted Snowball portrayed as a "fanatic intellectual whose plans if carried through would have led to disaster no less complete than under Napoleon".

The second objection was that the script had to make it absolutely clear that not all farmers were bad, that many cared for their animals and that there were farms where the animals were content. The CIA was certainly not going to finance a film attacking capitalism as well as Stalinism.

They also wanted to change the final episode of the novel, in which the farm animals can no longer tell the pigs from the men, so that it involved only pigs. This was, of course, a crucial change because what Orwell was intent on demonstrating in the novel was that the Soviet Union was behaving exactly the same as the imperialist powers—in fact, had become indistinguishable from them. This was clearly not acceptable to the CIA. Finally, they wanted the film to end with an animal uprising against the pigs, something that was very much part of CIA policy at the time.

Clearly, crucial aspects of Orwell's novel were recognised by the agency as being incompatible with their objectives. Why then were they so easily able to hijack it?

The answer is simple. It was the British left's continuing celebration of the Soviet Union as socialist that effectively handed the book over to the enemies of socialism in the 1950s. Orwell's warnings about Stalinism went unheard by the majority of British socialists, and it was their sympathy for Stalinism that enabled the book to be used against them, although even then its meaning had to be violated.

The difficulty of the fight that Orwell was engaged in when trying to convince the left that Stalinism had nothing to do with socialism would, of course, have been familiar to the founders of this journal. The problem was perfectly illustrated by the historian E P Thompson's 1951 response to allegations of tyranny and repression in Eastern Europe. He contemptuously dismissed the claims as "the Big Lie technique of Goebbels over again". Indeed, the lie was "so monstrous that we cannot be troubled with it, we turn our backs on it, and direct the argument on to more practical questions".

Even someone of Thompson's stature was prepared to cast himself in the role of Squealer! Nothing Orwell could write or do could possibly make any impression on this certainty, this refusal to recognise the Stalinist repression that was sweeping across Eastern Europe. It was to take the action of the working class in Poland and Hungary in 1956 to make these people listen. Hope lay with the proles.

The party that never was

Kim Moody

*Robin Archer, **Why Is There No Labor Party in the United States?** (Princeton University, 2007), £19.95*

It's an old question, but despite the academic fashion of dismissing it, it's a question that returns in both academic and political guise with some regularity. Robin Archer of the London School of Economics gives some new answers. Not many have ventured to argue that racism was not a barrier to the formation of a labour-based party in the US. Even fewer dared to say one reason was that there was just too much Marxist influence in the US labour movement. Furthermore, Archer roundly dismisses some of the old chestnuts of the "American exceptionalism" debate. Prosperity, upward mobility, individualism, lack of feudal remnants, early male suffrage, among others, all make the dustbin of history—as they should. In conclusion, he argues, "It is the importance of repression, religion, and socialism that helps explain the failure to establish a labour party."

Archer's method of disposing of so many of the old cliches and arriving at his novel conclusions is to compare American labour with Australian, the "most similar case", rather than the more traditional comparison with Europe. Australian labour, he argues, had all the features of its American counterpart in the 1880s and 1890s, the period Archer designates as crucial. Its working class was more prosperous, had more upward mobility, was at least as individualistic in outlook, had white male suffrage and was deeply racist.

Yet, with all this working against it, the Australian working class, or at least its white members, managed to form a labour party in 1891. It was anti-Chinese, anti-Melanesian and for a "White Australia". So why didn't white US labour, most of which was similarly racist in outlook, form an all-white labour party? Aside from the fact that it would have had disastrous consequences, it would not have been possible in the US for reasons I will return to.

The other aspect of Archer's approach is that he pins the failure to form a labour party on a single event, the 1894 convention of the American Federation of Labour (AFL). It was at this convention that the delegates debated the merits of the "Political Programme", which was in fact the programme of the British Independent Labour Party. It had been proposed by the socialists in the AFL, and its preamble called for independent class political action. Its supporters presumed that its passage meant the AFL would take action to form an alliance with rebelling rural populists to create a farmer-labour party.

Although it appeared that the programme would pass, Samuel Gompers and the other advocates of "pure and simple unionism" manoeuvred to secure its defeat. Archer's argument about the role of socialism and too much Marxism comes down to the bitter factionalism of that debate between two sides whose leaders had learned their strident "either-or" polemical style in the Marxist-dominated socialist movement of the 1870s and 1880s. This is in counterposition to the Australians who, he argues, were utopians or ethical socialists rather than Marxists, and hence presumably less polemical and more willing to compromise.

Ironically, the "pure and simple unionism" that emerged full-blown in the 1890s, had its origins in the debates between the followers of Marx and Engels, who saw trade unions as an important development in working class organisation, and the Lasallians, who, with their "iron law of wages", saw unions as a waste of time. Adolf Strasser and his understudy, Samuel Gompers, both of the Cigar Makers' union and major shapers of "pure and simple unionism", were active in the International Workingmen's Association as supporters of Marx in the US and were founders of the Social Democratic Workingmen's Party in 1874 upon the demise of the International.

This was renamed the Socialist Labour Party in 1877. Within the socialist movement they emphasised the centrality of unions and came to advocate a form of organisation based on the English "new model" craft unions. Their views would be picked up by the other father of "pure and simple unionism", Peter J McGuire, who converted from Lasallianism to Marxism around 1880, after which he helped found the Carpenters' Union. Marx and Engels were, of course, highly critical of the English craft unions precisely for the narrowness that Strasser et al embraced as the recipe for stable organisation. All three would lead the fight against the Political Programme. In the end, Strasser et al chose craft narrowness and bureaucratic stability over socialism of any sort. Nevertheless, Archer sees the rigid factional style they learned as socialists as the reason the programme failed.

In fact, the failure of the AFL to take a first step towards a labour party involved much more than the sectarian styles of former and current Marxists. For one thing, the depression of 1893 had reduced the ranks of the unions dramatically. Despite the birth of the United Mine Workers in 1890 and the radical Western Federation of Miners in 1893, and the huge dockers' strikes in St Louis and New Orleans in 1892, any chance of mass general unionism at that time was set back by the depression. Only the more conservative skilled building workers' craft unions held on as the urban building boom continued and their employment expanded.

They formed Gompers's base and would continue to do so into the new century. It was this base that allowed him to defeat subsequent moves towards an independent class party. It was also the survival of these craft unions that further convinced Gompers and others that exclusive craft unionism was the only model of stable organisation.

Archer treats racism as simply a matter of ideology. Racist Australians formed a labour party, so why couldn't racist Americans do the same? He misses something important about the period he is examining. Racism grew out of and perpetuated a clear structure of oppression in slavery. With its abolition at the end of the civil war not only did that structure disappear, but the former slaves became citizens—unlike Austrialia's Chinese, Melanesians and Aboriginies.

While racism as an ideology remained a powerful force in itself, it took some time for a new institutional basis of racial oppression to be constructed. Institutional segregation and political disenfranchisement were not successfully imposed until the opening of the 20th century. African Americans in most of the South still voted and even held political office until 1900, despite the final withdrawal of federal troops in 1877 and the reign of Ku Klux Klan terror and lynch law that followed.

This was, in fact, the basis of the rise of populism, as black and white tenant farmers in the South united with those in the West to form the People's Party after 1890. The strategy of the socialists in the AFL who put forth the "Political Programme" was to fuse with the populists to create a broader working class base for the People's Party. Whatever the depths of subjective racism in the unions or among farmers, this could only be accomplished on an inter-racial basis—something that was made explicit at the time by both populist farmer and radical trade union leaders.

From its founding in 1886 to the 1894 convention, the AFL not only admitted black workers, but actively sought them as members, particularly in the South. This was less a matter of lofty principles or consistent practice than of economic necessity. African American workers still held important jobs on the waterfronts of the South. The dramatic inter-racial strikes of dockers in St Louis and New Orleans in 1892 underlined this fact. The AFL actually refused to issue charters to two unions that explicitly excluded African Americans in the early 1890s.

Although this was deeply compromised by the practice of separate locals of the same union for blacks and whites in the South, the idea of an all-white labour movement was still inconceivable. All of this would change rapidly after 1894, as the movements for segregation and disenfranchisement gained momentum and swept workers and unions in their wake.

In effect, with the defeat of the "Political Programme" in 1894, the moment was lost. Racial exclusion by AFL unions would increase so that by 1916 W E B DuBois named 16 unions that explicitly excluded blacks and others that did so informally. Populism was absorbed by the Democratic Party in 1896 and then splintered along racial lines. Key organising drives in meat packing (1917-8) and steel (1919) were undermined by craft narrowness and racial division, derailing the next move towards a labour party in the early 1920s.

A more central reason for the failure to form a labour party in the 1890s, in fact, the most important one Archer names, was the absence of the "new unionism" of unskilled and semi-skilled workers that characterised both Australia and Britain. America's equivalent of the new unionism, the Knights of Labour, had all but disappeared by the critical mid-1890s. The base of the AFL, formed in 1886 of mostly craft unions, was too narrow to support the kind of mass based organisation that would have been needed to break through the dominant two-party system, even if the Political Programme had passed and racism had been subordinated to the rudimentary class consciousness that existed at the time. Archer attributes the failure of the Knights of Labour mainly to the brutal repression that followed the 1 May 1886 general strike and the Haymarket "riot" two days later. The depression of 1893 killed any chance of a revival of mass general unionism at that time.

Yet underlying the inability of the working class then in formation in the US to create mass unions on a nationwide basis or a class-based party was something deeper and more powerful—something that is usually overlooked in discussions of "American exceptionalism", including Archer's. That is the exceptional nature of the process of capital accumulation in the US from the 1870s through the early 1900s.

What made this process of accumulation different from that of other industrialising nations of the time was not only the size and speed of industrial growth, but the rapid urbanisation that accompanied it and the geographic and demographic scale of the process, which no other nation experienced. The space of this review precludes a full analysis, but a few figures will give some idea. In the 20-year period from 1880 to 1900 the population grew by 51 percent, itself rather a leap. But the urban population grew by 174 percent. Real GDP grew by half in those 20 years, while real manufacturing output grew by 138 percent. The number of production workers in manufacturing almost doubled, while the value-added they produced more than tripled in real terms.

All of these figures might lead to the conclusion that this rapid industrialisation should have laid the basis for mass industrial or general unionism. But the process was not simple or linear. Every year thousands of businesses went under as the new trusts, cartels and finally corporations deploying new technology destroyed old sites of unionism, creating an accelerating power imbalance in terms of organisation and resources—one that union activists were well aware of.

Between 1870 and 1900 industry and urbanisation swept across 2,500 miles from Pittsburgh in the East to the Rocky Mountains in the West. Not only did urban giants like Chicago, Milwaukee and St Louis explode, but the plains and prairies were dotted with smaller industrial cities, and the Western mountains filled with mines, mills and booming urban centres from Canada to Mexico.

As this vast accummulation of capital created less skilled jobs and burgeoning cities, workers, millions of them immigrants, were constantly on the move. This made both stable unions and political organisations difficult on a national scale. The typical housing of unskilled workers was the boarding house. In this context, the Knights of Labour and newer industrial unions like the United Mine Workers (1890) or the radical Western Federation of Miners (1893), all of which had favoured independent political action and might have provided the mass base for a labour party, faced retreats due to not only repression and depression, but the scale, scope, and spread of industry during the period, all of which made stable organisation on a national scale extremely difficult.

Archer's comparative study is an interesting one and a number of the points he makes are valid. Yet the comparison with Australia ultimately falters on the enormous differences in the economic context. It would be another 40 years before this new industrial working class would launch the mass industrial unions that might have been the base for a labour party. Of course, it didn't happen then either, but that is another story.

Where we came from

Peter Wearden

Colin Renfrew, ***Prehistory: The Making of the Human Mind*** *(Weidenfeld & Nicholson, 2007), £14.99*

After three decades of debates on human evolution dominated by Richard Dawkins's *Selfish Gene*, sociobiology and evolutionary psychology, it is a great relief to read a synthesis of the current knowledge and understanding of human evolution that avoids biological reductionism and recognises the importance of theory in understanding our human past. Colin (Lord) Renfrew, until 2004 Professor of Archaeology at Cambridge, might not appear immediately as someone the left would look to for new insights into human evolution. It is interesting, therefore, that he gives Gordon Childe, Lewis Henry Morgan, and indeed Karl Marx and Frederick Engels, as well as many academics, due credit for their contribution to the theoretical interpretations of the nature of humanity and human evolution.

Renfrew traces the development of the idea of prehistory as reflected by the developing techniques (radio carbon dating, DNA analysis, etc) and knowledge base in archaeology. However, the key issues are developed in chapter five, "The Sapient Paradox". The principal aspect of the paradox is that in the past 60,000 years (and quite possibly 100,000 or 150,000 years) no significant changes have taken place in the human genotype, while massive social and behavioural transformations are obvious.

Renfrew concludes, "The changes in human behaviour and life that have taken place since that time...sedentism (settled villages), cities, writing, warfare—are not in any way determined by the very limited genetic changes." Renfrew does not dismiss genetics but argues that human capacities, for example for language, pre-exist, and are common to all cultural *homo sapiens*. The capacity does not determine the use of that capacity.

Renfrew divides human development into two key phases. In the "speciation" phase the emergence of different branches of the *hominid* family is dominated by processes of natural selection. He does not ignore cultural aspects, and he recognises that aspects of learned behaviour are retained and passed on culturally in many species, particularly, but not exclusively, primates. However, they are secondary processes.

After this phase the relationship is reversed. While the processes of natural selection may still be apparent, human development can only be explained by understanding economic, political and cultural development. This takes place in the "tectonic" phase. In his only reference to Richard Dawkins, he rejects Dawkins's explanation of human cultural development. He argues that the analogy between the cultural "meme" and the "gene" is "a misguided one". He rejects the idea that cultural and biological "evolution" can be seen as essentially comparable as "just not appropriate".

Here he confronts the second part of the paradox. If all currently existing human capacities existed 60,000 years ago, why did it take another 50,000 years for agriculture, cities, states, written language, etc to emerge? Renfrew argues that it is the "material engagement" of human societies with their environment that is the starting point for the understanding of human cultural development (similar to the starting point of Marx and Engels in *The German Ideology*). He then explains the development of "institutional facts", comparable to the "definite social relations" of Marx and Engels. These institutional facts are just as real as material facts but are dependent on "material engagement" of human societies with their world. However, the reason why such relationships are so apparently unchanging across the millennia is not clear.

The second half of the book tracks the various "developmental trajectories" of prehistoric societies and early civilisations. His analysis ranges from stone tool production in *homo erectus* finds to the shift from transient hunter gatherers, to sedentary village societies, to the civilisation of China, Mesoamerica, Sumer, Egypt and the Indus, from egalitarian hunter gatherers to hierarchical class based states. While he considers rise of villages and towns, agriculture and the state (consistent with many of the ideas in Engels's *Origin of the Family, Private Property and the State*), he demonstrates that particular directions of development are neither universal nor inevitable.

In considering the rise of symbolic representation Renfrew develops the idea that the "mind" develops an external dimension in the customs, rules and "religious" ideas that are given concrete form. But there is no single form of such "institutional facts" that is higher or better. Indeed he argues that the "ideographic script" of China has been just as effective in extending theoretical and conceptual capacities as the "alphabetic script" of Western Asia and Europe. Again Renfrew deals with these issues with brevity, clarity and a keen sense of the need for and relationship between theory and evidence.

Renfrew's account is far more satisfying than the contorted attempts to reduce human society to a genetic imperative. He recognises two key contributions of genetics to evolutionary theory: first, in explaining how variations between organisms occur, which can then be worked on by natural selection; second, in how the analysis of Y-chromosomes and mitochondrial DNA can illuminate the genealogy of human (and pre-human) populations.

The key weakness in his analysis is the failure to adequately address the transition from the speciation to the tectonic phase. However, that he sees this as a qualitative transition, resulting from the gradual development of social, cultural and cognitive capacities, allows us a basis for better understanding. Marx's proposition of human production, involving the separation of production from consumption through a process of exchange governed by customary rule, may constitute one of the founding "institutional facts" of human cultural society, and might be of value in unlocking this qualitative transformation.

This book is eminently readable and a genuine synthesis of current knowledge of the transition to cultural humanity and of the key developments of prehistory. It avoids crude empiricism and is a brilliant example of the effective combination of theory and evidence to provide real explanatory power. It is also a good antidote to postmodernist obfuscation.

Renfrew's demands for a more complete "cognitive archaeology" recognise that "new approaches to investigate the concept of mind...lie in the domain of philosophy as much as of neuroscience". Renfrew points us in the direction of a materialist analysis of the origins of humanity and of ancient societies in their economic, institutional/political and ideological content and meaning. This seems vastly more productive than the search for a gene for every aspect of human behaviour.

Politics without enough economics

Adrian Budd

*Bob Jessop, **State Power: A Strategic-Relational Approach** (Polity, 2008), £17.99.*

State Power presents the most recent developments in Bob Jessop's three decades long research project on state theory. Inspired by his reading of Marxism, Jessop has become one of the world's most influential writers on state power and state theory. His prodigious output is illustrated by the inclusion of no less than 69 articles and books either written or co-written by Jessop in the bibliography of *State Power*.

The first two parts of this work explore the evolution of what Jessop calls his "strategic-relational approach" to state theory and the ideas of key influences on his thinking (including Karl Marx, Antonio Gramsci, Nicos Poulantzas and Michel Foucault). Part three applies this approach to important contemporary debates on the state—including the relations between states and neoliberal globalisation, and the emergence of the European Union and its implications for state theory.

Following Poulantzas, and echoing Marx's argument that capital is not a thing but a social relation, Jessop's starting point is that the state is a social relation. Furthermore, while much political theory sees the state as external to the economy and social relations, Jessop argues that it is centrally involved in their constitution and reproduction. Conceiving the state as a social relation, and like all such relations subject to change and development, allows us to recognise the historical fluidity of its forms and functions—captured in Max Weber's argument that "there is no activity that states always perform and none that they have never performed" (p3).

Whether or not this is strictly accurate (for all states have maintained juridical and coercive instruments to defend ruling class interests), Jessop insists that the state is "just one institutional ensemble among others within a social formation" (p7). Nevertheless, the state has particular powers and a specificity that derives from the fact that this ensemble alone has "overall responsibility for maintaining the cohesion of the social formation of which it is merely a part" (p7).

Mobilising these basic principles in the analysis of capitalist states, Jessop is concerned to avoid two opposed problems in Marxist state theory. First, he rejects the structuralism of "capital-logic" theory, whereby the state automatically supports the interests of capital and promotes a single logic of capital accumulation and capitalist reproduction at any particular stage of development.

But if capital-logic theory produces a closed, non-dialectical image of state power, the alternative "class-theoretical" approach is equally problematic. For here the state is purely the result of the contemporary balance of class forces, which can be assessed empirically with no reference to capitalism's overall systemic properties. Put simply, Jessop's alternative approach combines elements of structuralism with a recognition of the significance of the actions and intentions of social actors. As such, it represents a sophisticated engagement with the structure-agency problem of Marxism and wider social theory.

For Jessop, structural constraints are real but not absolute and, rather than simply foreclosing political choices, operate strategically and selectively. Since the state is not a homogeneous structure but can only be activated by actual people and forces acting in specific ways in specific conjunctures, it is strategically selective in the sense of privileging certain actors and identities and encouraging particular strategies among those "seeking to control, resist, reproduce, or transform them" (p46).

Meanwhile, social actors always operate reflexively and pursue strategies that, while operating within the limits of the structural, may nevertheless stretch those limits and thereby give states and state power a certain elasticity. For the strategic-relational approach, then, state structures frame the strategies pursued by political forces, simultaneously enabling and setting limits to their strategic calculations.

One strength of this approach is that, despite Jessop's abstract presentation and apparent disdain for concrete evidence to illustrate his arguments, it provides theoretical resources to explain institutional variations between capitalist states, and policy differences between countries and parties. Only in part three, where Jessop applies his model to contemporary problems, does he seek to make his approach concrete.

The chapter on globalisation is one of the strongest in the book and provides a number of important insights that challenge the thesis that transnational economic integration entails the straightforward transcendence of state power. Jessop insists that there is no easy spatial contradiction between economic globalisation and the nation-state. While in the last few decades states have become increasingly linked to global processes and so "interiorise the interests of foreign capital", they have also engaged in measures to "project the interests of national capital abroad" (p189).

In any case, economic globalisation should not be overstated, for "the specificity of many economic assets and their embedding in extra-economic institutions mean that much economic activity is place and time-bound" (p189). And, with the persistent intertwining of the state with wider social relations, capitals and states remain locked in relations of "reciprocal interdependence".

Even under economic transnationalisation, "bourgeois reproduction is still focused on the nation-state" (p136). This is not only generally true, but is also the case in the region where integration between and across states is most developed, the European Union. The development of the EU, Jessop argues, does not represent the transcendence of the nation-state but "multilevel governance in the shadow of national government(s)" (p200).

Nevertheless, while Jessop has a healthy scepticism towards the thesis of the decline of the nation-state held by many who write on globalisation, there are problems with his analysis of the role of states under globalisation. For, where many Marxists, including those writing in this journal, emphasise the role of the world's major states in the establishment of globalised capitalism, Jessop underplays state power. He regards globalisation as a "multicentric, multiscalar, multitemporal, multiform, and multicausal process" (p178).

What Jessop's approach overlooks is the hierarchical organisation of the capitalist world order and, in particular, the centrality of the power of the US. Certainly, US power is contested in various ways (by other states, by regional groupings of states such as the EU, and by the global anti-war and anti-capitalist movements). But the US remains the world's dominant imperialist power and continues to spearhead the reordering of the global political economy along neoliberal lines. Part of its purpose, as transnationalist international theory one-sidedly emphasises, is to make the world safe for transnational capital. But the US is also driven by another imperative—to retain its position in the international pecking order of capitalist states.

There is, then, an international dimension to state power that is relatively under-explored in Jessop's work. This gap, and in particular the relative neglect of US global power, leads Jessop into a series of unsupported assertions. How, and under what circumstances, might the US's neoliberal political-economy model be so challenged that "the European model in particular may regain ground in the coming decade" (p193)? Similarly, short of a dramatic challenge to capitalist power, and that of the US in particular, how might restraints on international capital, such as the Tobin Tax on international financial transfers or an energy tax on fossil fuels (p195), be brought about?

Other international dimensions of state power are also under-explored by Jessop, including war and nationalism, which are barely mentioned in *State Power*. Jessop is right to note that state theorists who place war at the centre of their analysis generally treat it as separate from, and unexplained by, wider social relations. But the corollary of the rejection of this approach should not be to marginalise war from Marxist approaches to state theory.

A more general criticism of Jessop's strategic-relational approach concerns his treatment of capitalism as a mode of production, a concept that he notes but which is not central to his approach. The capitalist mode of production—defined primarily by the dominant social relations of competition and exploitation, which themselves entail law, politics and other extra-economic relations—does not impose particular state policies in every detail. States, therefore, may have different tax regimes, different welfare structures, and a host of other particularisms. In this respect Jessop is right to argue that strategic actors have some room for manoeuvre.

Nevertheless, capitalism does impose boundary conditions for state policy and the behaviour of social actors, such that, whatever their differences, in a capitalist world order the world's states are all capitalist. It is not clear why Jessop understates the powerful determining constraints of the mode of production, but his emphasis on the strategic choices of political actors is surely a contributory factor.

He argues that "a state could operate principally as a capitalist state, a military power, a theocratic regime, a representative democratic regime answerable to civil society, an apartheid state, or an ethico-political state" (p8). In reality, capitalist states have at various times assumed these particular forms while remaining capitalist. He goes on to argue that there "is no unconditional guarantee that the modern state will always (or ever) be essentially capitalist" (p8). The opposite is true. While there is no guarantee that the state will be apartheid, etc, under capitalism the state must always be a capitalist state, whatever the formal character of its politics.

The under-developed analysis of the mode of production as a structural constraint also leaves Jessop open to the criticisms that his approach results in an over-politicised view of the state and presents a reformist model of political change. Jessop acknowledges both criticisms.

This does not mean that, within the limits implied by my criticisms above, his approach is incapable of providing important insights into the operation of state power, particularly in periods of relative stability. For, given capitalism's underlying dynamism—involving class struggle, conflicts between capitals, the rhythm of booms and slumps, international competition, etc—states do privilege certain accumulation strategies over others and engage in periodic strategic reorientation. In this sense the state can be seen as relatively autonomous in the narrow sense of standing above the interests of individual capitals and concerned with the long-term reproduction of capitalism as such.

Nevertheless, the over-politicisation of the strategic-relational approach and its neglect of the structural constraints on state power imposed by the capitalist mode of production come at the major cost of marginalising those forces whose interests and strategies may clash with the strategic selectivity of the state. As for many academic Marxists in recent decades, class struggle remains a background assumption but discussions of class agency focus almost entirely on capital. As a consequence, the activities of socialist parties, trade unions and radical social movements barely figure in Jessop's book.

More broadly, Marxism's emancipatory vision is glaringly absent from *State Power*. One expression of this is the extremely academic, opaque and convoluted language Jessop uses. This is academic Marxism of the kind that requires a dictionary to navigate its thickets, and a memory sufficient to remember how a sentence started when you reach its conclusion. There is something deeply ironic in Jessop's use of Gramsci to argue that state power is not merely material but also concerns linguistic elements designed to marginalise subordinate dialects and the language of resistance.

The contradiction between an approach to state theory inspired by Marxism and a language designed to alienate all but the hardiest of workers is illustrative of the ultimately enigmatic nature of Jessop's book. It draws upon extensive reading to provide some useful methodological pointers for analysing states and challenges crude approaches to state power, including simplistic oppositions of state and civil society, politics and economics, global and national. In its own way it is an optimistic book, and it highlights the strategic spaces available for the potential elaboration of radical policy programmes. Nevertheless, to develop socialist strategy, to push against the limits of the possible, requires us to take what is useful here and move decisively beyond it.

Under pressure

Sheila Cohen

Gerry Mooney and Alex Law (eds), ***New Labour/Hard Labour? Restructuring and Resistance inside the Welfare Industry*** *(Policy Press, 2007), £22.99*

This book arrives at a time when the bewildering doublespeak beloved of New Labour reaches ever giddier heights. Who knew that teachers would come to be described as "transition managers" and "learning community leaders"—their proposed title when plain old Islington Green comprehensive mutates this year into a "business academy"?

But it is a lot worse than the mangling of the English language favoured by New Labour and its ideological twin, "New Public Management". As the book shows, New Labour is turning welfare into workfare in a very different sense from its own spin. Increasing intensification and extensification of work—harder work for longer hours—make a mockery of the government's much-vaunted "work-life balance" policies. And nowhere is that more true than in the once (comparatively) privileged public sector.

Public sector? What public sector? As *New Labour/Hard Labour* spells out, privatisation, outright and piecemeal, has reduced a once strong and cohesive sector to ribbons of subcontractors, "providers" and, most of all, insecure, disadvantaged workers. Rather than public sector, the editors employ the term "welfare industry" to "capture what is generally referred to as the 'mixed economy' of welfare providers...[and] non-state sectors, such as the voluntary sector and private provision" (p2).

In order to survey this territory, editors Gerry Mooney and Alex Law provide an introductory overview, along with two other broader chapters on restructuring the welfare labour process and industrial relations under New Labour (this co-authored by Peter Bain, the "working class militant and scholar" whose death is mourned in the book's dedication). These introductory pieces, which incorporate the labour process based perspective followed throughout the book, are followed by eight chapters on specific sectors that provide a welcome departure from the usual absence of a worker voice in books about work. Another unusual, and welcome, feature is the emphasis on workers' resistance.

As always, those directly involved at the "ward-face", as one contribution terms it, provide the most vivid expressions of the experiences and contradictions of work. Nursery nurses, mostly very young women, fulminate against "new initatives every year...planning; evaluation; observation; recording", "new responsibilities being given to us—from all sorts of directions", "too many different demands on us... Do this, do that, do the next thing", and in general, "more and more for less and less" (p165).

Such intensification of labour, in an area which might be thought least suitable for it, is accompanied by abysmal levels of pay. As one worker argues eloquently, "All we have heard from management is that we need to get more qualifications, more training, go on more courses...but when we ask will we get more pay for this there is silence." These pressures led to a long dispute which, as the authors point out, marked "Scotland's biggest all-out indefinite strike since the miners" (p183).

Such pressures and resentments recur in the experience of social workers transformed into "care managers" by Blairite social policy. As one complained, "Being a care manager is very different from being a social worker... Care management is all about budgets and paperwork and...financial implications...whereas social work is about people."

The shift from qualitative to quantitative considerations typical of New Public Management is accompanied by extensive monitoring, leading to ever increasing stress: "Much of the stress at work is fear; social workers are scared of their managers, scared of all the monitoring stuff." And the pressure "is always downwards"—from higher to middle to lower management and ultimately to the workforce. These workers' own six-month dispute over the introduction of "one-stop" call centres for potential social work clients in Liverpool is described as "about the clash of two different value systems. On the one hand the business drives and values of management and Blairism...and on the other the values of humanity and social justice" of the striking social workers (p200).

NHS ancillary workers cited in a study of hospital private finance initiative schemes express parallel disgust and indignation at attacks not only on themselves but on the once-cherished principle of free health care: "For us, the question is how the hell a company can just want to make money out of ill people" (p83). A Unison union survey of such contracted-out workers "revealed a mass of inequalities" around pay and hours, with problems ranging from zero-hours contracts to "onerous workloads", and extensive and burdensome worker-monitoring. Yet workers' nostalgia was not for NHS terms and conditions alone but for "the old hospital [where] it was like a family and everybody pulled together". It was a mixture of all these factors—destruction of hospitals and loss of beds alongside job losses and pay cuts—that provided the final straw flaming into a ten-month dispute by ancillary workers at Dudley hospitals in 2000-1.

Staying with the health service, an examination of the labour process in nursing cites a mixture of deskilling and multi-skilling, increasingly supplemented by the same "target" and "audit" driven approach documented in so many of these accounts. As one nurse puts it, "Most attempts I make to adopt a patient-centered approach...are stopped in their tracks by management's continued demands for audit controls on the ward... my time is taken up filling in standardised care plans and audit forms" (p98).

This emerges among other contradictions between Taylor-style management imperatives and the so far irreducible reality that patients are human beings in distress. As one nurse recounts, "[We] were expected to sit at the desk doing paperwork—sitting with patients who were often concerned about procedures was actively discouraged...I suppose doing so was not getting the job done" (p101).

Some of these contradictions, the authors suggest, are dealt with through the management of "emotional labour", an increasingly central weapon in the management arsenal. Yet, while a number of quotes from nurses emphasise that, for example, "it's not a job you would do for the money... It's something that you should care deeply about", the relentless intensification of labour in nursing often works to undermine such dedication. A 2006 Unison report quoted here shows, among other things, that 55 percent of all nursing staff worked unpaid additional hours, leading to as many as eight out of ten nurses considering leaving the profession altogether (p108).

The same pressures and contradictions are documented throughout this book, in the work of teachers, social workers, civil servants and even academics. The relentless emphasis on "performance outcomes" characteristic of New Labour and New Public Management has taken the shape of what would once have been seen as absurd and impossible—performance-related pay for teachers. Despite resistance, the ever increasing workload, "initiative overload" and target-driven culture now endemic have driven almost a third of teachers to consider leaving the profession.

Models of management within schools "come from the world of mass production in the car industry" (p125) where "constant improvement" or—as its critics term this bundle of techniques, management by stress—have long been managerial mantras. This pattern is repeated in, of all places, the lofty echelons of the university, where the "modularisation" of teaching is one example of "how the principles of an industrial process based on cellular production, flexible working...and teamwork...have been imported into academia" (p144).

One of the most interesting accounts in the book is of the labour process in the once decidedly non-commodified "voluntary" or charitable sector. Over the past few years this has gravitated to a hard-nosed "third sector" model in which charities' chief executives are paid fat-cat salaries, while workers, as the current Shelter dispute demonstrates, face traditional capitalist attacks on their pay and jobs. The emergence of the "contract culture" is the prime culprit in the increasing use of such organisations as "providers" of the kind of services once routinely offered by the public sector. As this writer comments, non-profit organisations "are effectively being shaped in the image of the statutory organisations they are there to displace" (p237).

The interest of such developments is that they have, perhaps predictably, proletarianised the idealistic, often highly educated employees of these organisations. Workers for the voluntary sector provider of criminal justice services described here felt increasingly that their project was becoming "less client and more target focused" (sound familiar?). The ensuing culture of bullying and victimisation became a source of solidarity and militancy between formerly individualised workers: "I think they were all for driving us out... and I think that they didn't realise that we wouldn't just go, and wouldn't just bow our heads...as soon as they threw some false accusations at us" (p257).

These workers had turned to their union to help them with a dismissal case but had been turned down on the basis that "compensation...was likely to be minimal" now that the worker concerned had found another job (p257). This highlights another theme running throughout the book—the weakness and apparent unwillingness of many union leaderships to support a fighting approach against the relentless attacks on their members.

Unison, in particular, offers "high quality empirical research" as its principal weapon against the private finance initiative and other privatisation-related horrors, a strategy which has had negligible impact on those in power: "The lessons of this evidence-based approach to politics require serious contemplation", as the author of the chapter on PFI points out (p86). The insistence of Unison and other healthcare sector unions on maintaining support for Labour is also questioned here: "As one Labour conference participant put it in 2001: 'Why feed the hand that bites you?'"

If I have a criticism of this book, it is that its best intentions—the emphasis on worker experience and worker resistance—promise more than they deliver. While workers' voices enliven the pages, there are solid sections of theoretical analysis unleavened by such testimony.

The relatively abstract and academic arguments frequently undertaken here can have little relevance for workplace activists whose struggles might benefit more from a fusion of Marxist theory with everyday practice. The emphasis on a labour process approach might be said to undercut this criticism; but longstanding debates such as the alternation between "direct control" and "responsible autonomy" can add little to the understanding of a world of work in which few have the opportunity to be wooed by a mirage of craft-based independence. If anything, *New Labour/Hard Labour* illustrates how the predictions of deskilling in the book which launched the "labour process debate" appear to have been only too grimly fulfilled.

Organic intellectual

Alan Kenny

George Paizis, ***Marcel Martinet: Poet of the Revolution*** *(Francis Boutle, 2007), £12.50*

Marcel Martinet was born in France in 1887. In his lifetime he would witness huge social transformations. Technological advances around the turn of the century meant big changes to workers' daily lives as they found themselves in much larger workplaces. The year 1914 would see the beginning of the horrors of the First World War, and with 1917 came revolution in Russia and the hopes of its flames spreading across Europe. In this, the first book on Martinet in English, George Paizis introduces us to a man who consistently used his passion for poetry and literature to intervene in this tumultuous political period.

The changes in the structure of the working class and the subsequent upturn in struggle (a tenfold increase in the number of strikes between 1900 and 1910) saw the development of larger trade unions and the birth of new workers' parties. 1895 saw the creation of the Confédération Générale du Travail (CGT) and 1905 the foundation of the French Socialist Party (SFIO). In 1914 these organisations would face the test of how to respond to the First World War. The subject of this book found himself at the centre of these debates.

One measure of his impact is that, as Leon Trotsky was writing a preface to Martinet's 1922 novel *La Nuit*, Lenin asked him why he was spending so much time in doing so. Trotsky explained the role that Martinet and his group had played during the war. Lenin's response was to ask whether these people could lead the newly formed French Communist Party, instead of the existing leadership, which Lenin did not trust.

As a young man, Martinet had thrown away a prestigious education in L'Ecole Normale Supérieure to devote himself to his writing. This was a deliberate rejection of everything that an elite education could furnish upon his life—it was a "refusal to climb". What interested the young poet in his studies and writing was the relationship between art and the people. He wanted to go beyond the novels of authors such as Emile Zola, who he saw as being an outside observer of the world of the poor.

Martinet orientated on workers. He wanted "to study the world with them and embark on the conquest of the world, which is to say the liberation of mankind and social equality". Martinet would begin to argue in his essays for what he described as a "proletarian art". He believed that a class that was beginning to fight for itself needed an art that could express this.

This orientation led him to a group of revolutionary syndicalists within the CGT, who produced the fortnightly publication *La Vie Ouvrière* (the worker's life). As the SFIO and CGT dropped their initial opposition to the First World War, and the Second International collapsed, Martinet's group would continue its steadfast agitation against the war in the face of a deep pessimism in the movement.

The middle section of the book is devoted to Martinet's anti-war poems which scream with rage not only at the horrors of war but also at the failure of workers' organisations to oppose it. In "Cadavres" ("Corpses") the poet describes the potential of the working class to change the world and its betrayal by its leaders in sending armies of men to fight. It cries out to us that the men that could have been fighting against the ruling class are being sent to die in their war. Hope returns in the poem as he imagines armies of corpses rising up from the dead in revolution.

Hostilities were declared at an end in November 1918. Russia's October Revolution of 1917 now served as a beacon of hope as a revolutionary wave washed across Europe. The group around *La Vie Ouvrière* joyfully relaunched their paper in 1919 and circulation reached 20,000. But the post-war militancy would dramatically lose momentum as the CGT called off a general strike in July 1920 after discussions with President Clemenceau. A major split would now take place in the SFIO, with Martinet joining the newly formed Communist Party. His group would be a part of the left in the new party, fighting to turn it into a mass revolutionary organisation.

Martinet would spend three years as literary editor of the party's daily, *L'Humanité*. In this time he would write over 50 articles on his ideas on the relationship between art and struggle. Martinet was determined to open up the pages of *L'Humanité* to the working class. He felt that previously it had been a publication for those "in the know". Martinet would leave *L'Humanité* as Stalinism began to take hold in Russia. Twelve years later he would describe his departure as "motivated not by defection but by fidelity to the proletarian revolution". This commitment to socialism from below shines through in Paizis' book.

The book is useful in two ways. First, it introduces us to an important writer many of us will not have read before. In many ways Martinet prefigured Bertolt Brecht in his dramatic style. He marks a turning point in the literature of the period, one which Paizis identifies as the "individual on his way back to the collective". Second, the book helps us to understand the contours of the class struggle at a critical period for the European left. The book is able to do this because, as Trotsky said, what is most admirable in Martinet is the "organic combination of poet and revolutionary".

Valuable but flawed

Mike Haynes

Boris Kagarlitsky, ***Empire of the Periphery: Russia and the World System****, (Pluto, 2008), £40*

Boris Kagarlitsky is a prolific commentator on modern Russia who has also written on wider trends in global capitalism. It is impossible to read a book by him without learning something, and *Empire of the Periphery: Russia and the World System* is one of his best. It is also impossible to read a book by him and not be frustrated and irritated as wrong turns are taken, loose ends left untied and contradictions unresolved.

First, the good stuff. Kagarlitsky's aim is to write an economic and social history of Russia locating it as a peripheral state in the global system. Starting with the formation of an embryonic Russian state in the early Middle Ages, Kagarlitsky takes us through a thousand years of history drawing on some of the best historians of Russian history. His aim—more challenging to a Russian reader than to one in the West—is to undermine both positive and negative claims about Russia's unique position or nature. Anyone unfamiliar with the history of Russia will find this a readable account; anyone familiar with it will still learn some valuable things.

More, Kagarlitsky tries to situate this history in a theoretical framework formed out of three major elements. The first is the sense of capitalism as a global system where he links "world system analysis" and more traditional Marxist accounts.

The second, for which we should also be grateful, is a long overdue recognition of the achievements of the great Marxist historian of Russia, Mikhail Pokrovsky (1868-1932). Pokrovsky died before the final consolidation of Stalinism but was an early posthumous victim as his body of work was systematically rubbished and then ignored by a lesser generation of propagandists. In fact Pokrovksy was a brilliant historian and a committed revolutionary, a researcher, writer and Bolshevik. Some readers will know of his work through an appendix in Leon Trotsky's *History of the Russian Revolution* where Trotsky takes issue with some of his ideas. But the depth of Pokrovsky's achievement needs to be judged from his whole work, parts of which are available in translation (if often neglected and indifferent ones).

The third element of the Kagarlitsky schema is the idea of the Russian economist and Stalin victim Nikolai Kondratiev (1892-1938) that capitalist development takes the form of long cycles of growth and stagnation, overlaid on which is the normal economic cycle.

It is at this point, however, that the frustrations set in. Kondratiev, after the introduction, barely gets a look in. This is perhaps just as well, given the uncertain state of discussion of his long-wave theory. No one doubts that capitalist growth is uneven. But no one has convincingly shown that this unevenness is cyclical or that a mechanism could exist that adequately explains it.

But this is a lesser issue. The big one is how Kagarlitsky deals with capitalism as a global system and Russia's role within it. Kagarlitsky writes too often from the inside, Russia, looking out at the whole world system. The evolution of the global economy and the way that it generates a range of semi-peripheral and peripheral forms (and for that matter advanced ones too), of which Russia's experience was and is one, is less convincingly sketched.

Kagalitsky does not have a strong enough theoretical framework to understand the global economy and its forms. This includes, most importantly, the role of the state. He correctly argues that writers such as Immanuel Wallerstein and Andre Gunder Frank often stumble because of the lack of a strong political economy focused on production. But Kagarlitsky too does not address the real debates about how far we should take the argument that capitalism can only exist as a global economy. Indeed towards the end of the book he retreats from some of the important arguments made by Wallerstein.

Wallerstein's work, despite its problems, is attractive because of the power of his claim that capitalism is a global system with a global history. What it needs is a clearer account of the way that different forms of competition—market, state, military, etc—operate and a better grasp of the role of the state including a proper confrontation with the idea of state capitalism. This is not only a question of Russia (and its satellites in the Soviet era) but also of what happens in, for example, capitalist war economies.

Yet in a long footnote Kagarlitsky notes that Wallerstein argues that the USSR remained part of the capitalist world system throughout and that, whatever its internal form, it has to be considered as part of the capitalist world. This is not the place to evaluate this argument (or Kagalitsky's summary of it), save to say that readers of this journal would argue that it is a fudge to imply "of capitalism, in capitalism but not capitalism". The problem is that Kagarlitsky will not even go as far as this. He argues that Russia effectively dropped out of the capitalist global economy in the 20th century. The reason is that it ceased to be part of the international division of labour (at least until the 1970s).

Reducing global capitalism to a market imposed division of labour unthinkingly replays older claims about capitalism. Do capitalism and the determining force of the capitalist global system really cease to exist when the international division of labour is suspended (as in war) or countries try to control and limit it (state-led development)?

In a book of history we should not expect too much theory but when Kagarlitsky retreats in this way from a central claim of the book he surely needs to tell us rather more positively what Russia was between the revolution and the 1980s? No less, if it was separated from the capitalist world system did it, as was claimed in Stalin's time, exist as part of an alternative non-capitalist world system?

In Kagarlitsky's analysis we cannot even be sure when capitalism was abolished because he notes that in the 1920s Russian development was more heavily affected by the international division of labour than in the 1930s. Then, from the 1970s, it was on the way to becoming part of it again, a position that was finally realised in the 1990s. But even then his account remains uncertain, for he claims that in the late

1990s parts of the system were still not integrated (are they now?).

This is not only theoretically confusing but politically confusing too. What is the enemy in Russia, and where do we look for it? The lack of clarity in the argument means that politically it can be made to appeal as all things to all people, which is no doubt part of its attraction.

Kagarlitsky also puts too much emphasis on Russia's peripheral status. The account would have been more nuanced had he dealt more with the idea of Russia as part of the semi-periphery, and taken seriously the question of uneven and combined development. Russia is not a peripheral Third World state. It was not a Third World state in 1914. It is not one today, despite an offhand comparison with Zimbabwe.

Overall Kagarlitsky's historical journey is fascinating. The long period under discussion allows him to draw some striking comparisons in his final chapter on modern Russia: the restoration of agricultural exports in a country of malnutrition, once a characteristic of the 19th century; the collapse of the livestock herd in the countryside, once seen as unique consequence of collectivisation; surviving on a sea of oil, once seen as the Achilles heel of the Brezhnev regime; and so on. However his implicit theoretical journey is much less convincing. But the fact that, like the Russian economy, this mixture is uneven should not prevent readers interested in Russia and arguments about its past, present and future learning from this book.

A Marxist look at the legions

*Steve Roskams**

*Neil Faulkner, **Rome: Empire of the Eagles** (Pearson, 2008), £21.99*

Neil Faulkner's new book should be a welcome publication for the readers of this journal, following his earlier articles here (for example in *International Socialism 109* and *116*). Its attempt at a "grand narrative", its explicit anti-imperialist stance, and its willingness to relate our understanding of past societies to the modern world, in particular the anti-war movement, are all important and delivered here in a refreshing, accessible style.

Tackling the Roman Empire in this way is particularly valuable, the ancient world having been used by apologists for more recent empires. The classically educated generals and bureaucrats, who ran these modern systems, sought to justify their excesses in terms of bringing "civilisation" to uncivilised regions, just as Roman legions had previously claimed. Indeed, such sentiments are retained by many who study antiquity today. Thus the author's target is an important one.

Yet this is no conventional historical materialist account of Rome. Following Marx's own, admittedly slim, writings about the ancient economy most Marxists have depicted the slave mode of production as its pivotal, underlying dynamic (see, most obviously, the great work by Geoffrey de Ste Croix, *The Class Struggle in the Ancient*

* This review has benefited considerably from conversations with Chris Fuller, long term comrade and committed anti-war activist. I am, of course, responsible for the outcome of those discussions in this text.

Greek World). Faulkner takes the opposite stance. Slavery did not produce the majority of surplus then and, anyway, was not different in kind from other forms of exploitation. Thus the concept has little explanatory value in analysing Roman imperialism (even when discussing the latifundia at Settefinestre—for many, a site central to debates over the role of slave labour in Italy—he makes no mention of the proposed slave quarters there).

What, then, does Faulkner want to put in the place of Marxist "conventional wisdom"? His core thesis is that foreign conquest was the central mechanism for surplus extraction under Rome: a "system of robbery with violence" (as stated on the front and on the back of the dust cover of the book, and twice in its introduction).

The reviewer faces two problems in trying to assess the validity of such a major change in our thinking about the classical world. First, there is no proper referencing system within the text, so it is difficult to see where some sources of evidence derive from. Second, the account is avowedly empiricist: interpretations are "not imposed on the evidence" but portrayed as "growing out of it".

Despite these obstacles, Faulkner's central point comes over quite clearly in the text. Its five chapters are organised along chronological lines, running from the founding of the Roman city state, around 750 BC, to AD 476, when the Germanic leader Odoacer established the Kingdom of Italy, seen here as marking the fall of the Western Empire. At almost every stage the account concentrates on military and political dynamics and bases itself on documentary evidence.

This focus and the lack of archaeological evidence mean that the material conditions that faced producers in the landscape are rarely discussed in any detail. The realities of such exploitation are something which written sources would always struggle to elucidate, being produced in the main by elites with their own interests and perspectives.

It would be wrong to suggest that Faulkner considers solely military and political dynamics. The opening chapter provides a convincing account of the material conditions of early Rome, of how social cohesion grew and elites emerged from this melting pot.

The underpinning explanatory framework here owes much to standard Marxist thinking: a classic case of the increasingly prevalent mechanisms of production, structured around the basic households, coming into contradiction with an increasingly outmoded framework for surplus extraction, the higher-order clan. (See Chris Harman's article on "Engels and the Origins of Human Society" in *International Socialism 65* for the core argument.) Faulkner also describes how fragmentation within the ruling class generated the specific context in which such contradictions came to a head, another historical materialist perspective—so far, so good.

However, in what follows, such ideas are deployed merely to hint at ways to take analysis forward, rather than as a developed argument—and a critical part of that argument is missing. In the context of an increasing tension between clan and household, warriors are said to arise to provide protection against insecurity. Yet the sources of such uncertainty are not discussed. Later Faulkner charts the move from war carried out by equal citizens to war involving specialist military personnel. This development must have required considerable extra outlay but, once again, it is unclear why these investments are made.

The professionalised army is based, initially, on heavy infantry (hoplites) supplied by citizens of a "middling sort", but the latter category is not explained in relation to economic processes. Understandably, the author appears wary of simple equations with later social formations, most obviously those at the core of the move from feudalism to capitalism (see Christopher Hill's *God's Englishman: Oliver Cromwell and the English Revolution*). Yet subtly acknowledging this with quotation marks is no substitute for real definitions. By the same token, urbanisation arises out of nowhere: cities were founded for "political reasons", with no reference to underlying forces.

The failure to link the process of militarisation and the development of central places to the contradictions embodied in the mesh of means and relations of production is significant. Yet such developments are pivotal to all subsequent arguments on the army as the driving force behind Roman society.

Later chapters chart the army's rise to political power. Maintaining military forces was clearly a burden on the poor, but a price the latter were generally willing to bear because the fruits of conquest replaced surplus from farmland. Peasants, bought off by expansion, are thereafter portrayed as bit part players in history.

Many words are devoted to discussing which emperor was put into power by whom, with explanations of the processes residing in the upper echelons of society, whether machinations among military blocks in various parts of the empire, or between these and senatorial factions back in Rome. For example, when Elagabalus, the choice of the "Syrian" faction, was to be replaced by Severus Alexander in AD 222, the mob had simply to be "courted in the usual way" to guarantee success. Once bribed, it became peripheral. The real danger to the process of succession was not what happened at home with the mob, but abroad with the army. This focus, on external factors rather than internal tensions, brings us very far from any notion of a history from below.

Because of this approach, Faulkner is driven to some very non-Marxist ideas when interpreting the context in which the empire expanded, consolidated its position and fell into decline. One of the most unconvincing is the tendency to explain the past in terms of the character of individual emperors (Augustus: "a truly disgusting man"). It may be useful to remind us that members of a ruling class were not, as often portrayed, paradigms of virtue and civilisation. Yet knowing this hardly takes historical understanding forward much.

A more fundamental problem concerns the process of military expansion and its impact. The Samnites were attacked repeatedly until defeated, a victory pivotal to the development of the Roman state. A century later Phyrrus was tracked across Italy rather than engaged in direct battle, and eventually ground into defeat. Where did Rome's resources come from to sustain such long-term efforts?

On the other side of the equation, what of the fruits of empire? The author states that "glory, booty, slaves and indemnity payments were the principal ways by which surplus was pumped out of producers". Yet no emperor, however astute or devious, can produce more surplus by applying "glory" to the economic process. "Booty" and "indemnity payments" pose questions concerning how one converts gold talents into increasing production at the core and the impact on economic relations in the newly-acquired territories. None of these are examined in any detail.

Finally, "slaves" can only be profitable if one considers their role in domestic service, extraction of minerals and rural production—but Faulkner has already denied himself this form of explanation.

Insofar as the author acknowledges any economic impact of militarisation, he gets perilously close to another favourite technique of bourgeois scholarship when discussing past empires—finding capitalist enterprise in pre-capitalist societies. The maintenance of the empire required larger drafts of soldiers from the land, decimating peasant holdings and allowing elites to expand their property. For Faulkner, these land-grabbing aristocrats become wealthy by producing "cash crops or stock for a burgeoning market".

This idea owes much to Rostovtsev, an ancient historian who fled from the Russian Revolution to the US. His *Social and Economic History of the Roman Empire* (1957) characterised early Rome as a golden age of marketing on the Adam Smith model (in contrast to a late Roman period of intervention by an oppressive state, which caused the empire's downfall—something else which appears in Faulkner's discussion of Late Antiquity).

Beyond these vague notions of "the market", economics gets little mention. Rome is roundly condemned for being voracious and violent, and rightly so. However its competitor, Carthage, is portrayed as being more humane since "powered by merchant profit". Faulkner ignores Rome's engagement in long-distance trade entirely, even when such exchange networks are clearly relevant to his political processes.

For example, Roman emperors came from Italy up to the 1st century AD, from Spain at the start of the 2nd, and from North Africa by the end of that century. This correlates significantly with changes in the regions producing the vast surpluses of olive oil which flowed around the Western Mediterranean over that period of time. Yet this important trade system is not even mentioned in the text.

Finally, environment becomes the main determinant in tackling the ending of expansion. When the empire reached the limits of the plough, and thus the limits of sufficiently complex, relatively sedentary societies that were worthwhile/feasible for Rome to incorporate within its boundaries, it stopped growing. (Hadrian is credited with recognising this and turning it into imperial policy.) Such a suggestion is not only empirically questionable (for example, the position of Hadrian's Wall in northern Britain does not divide an agricultural south from a pastoral north), it is also theoretically weak: there is far more to explaining complex social and economic processes than the "givens" of environmental contexts.

Subsequent crises and ultimate decline are explained by corresponding "conventional wisdom". Rome was now on the back foot, subject to the increasing dominance of an intrusive state (see Rostovtsev, above) and an easy prey for disease. This new situation required a more mobile army but this merely served to plug one gap, only to see another appear elsewhere. The centralism of the state in contrast to the regionalism of the army, plus the growth of tribal proto-states on the margins of empire, has some real power to explain late Roman tensions—but linking them to rampant plague and an interfering state is a real limitation.

Faulkner is driven to these sorts of interpretations because he sees the key explanatory concept as "a state-army nexus built on a tax-pay cycle". This may be a fair description of what happened in one important sphere, but it is not an analysis of the whole process, still less a Marxist analysis. Taxation is a way of organising surplus at a gross level, but is not itself a mode of production. So the state which seeks to organise this taxation will take different forms depending on an underlying dominant mode of production (hence talk of "state capitalism" in the modern world). In ignoring this, Faulkner fails to deploy the fundamental Marxist tools of historical materialism. This is why class conflict makes so few appearances and why, when it does, its roots are unclear. It is very telling here that he believes we can have social revolutions "without a revolutionary class". And when a "peasants' revolt" occurs late in the 2nd century, it becomes significant only when army deserters join with "endemic social banditry"—it is the military component that is, for him, fundamental to the crisis.

Separating the military machine from the rest of society in this way limits the coherence of his critique. It also limits its relevance to the present day, especially in relation to the current anti-war movement. Among non-Marxists protesting against Western intervention in the Middle East there is a common notion that, if only military dynamics in particular, and imperialism in general, could be separated from the more benign aspects of capitalism, we could have a better world by reforming, rather than overthrowing, the present system. This is not possible today—and it is no more feasible when trying to understand imperialisms in the past.

Socialist Worker
Meeting US-backed regimes in the Middle East
Why Gordon Brown wants to cut your pay
DEMOCRACY?
Bush doesn't know the meaning of the word

Socialist Worker
Brown humiliated at polls
Millions bitter over war and inequality
Backlash hits New Labour

Global protests against war
Socialist Worker
Israel unleashes terror on Gaza

Socialist Worker
Brown's government
LIARS CHEATS AND THIEVES
STOP CLIMATE CHANGE
● Labour mired in sleaze
● Policies benefit the rich

MARCH AGAINST THE NAZI BNP
Socialist Worker
Revealed Bush to impose 'security accords'
Iraq: New US plan for total control
BRITAIN'S 'WAR ON TERROR'
● 400 permanent military bases
● US personel get green light to kill

Pick of the quarter

Apologists for imperialism have virtually stopped mentioning Iraq. They keep as quiet as they can about the horrors produced by the US and British sponsored Ethiopian invasion of Somalia. But they crow openly about the *Boy's Own* derring-do of our Nato troops in Afghanistan. This, supposedly, is the good war, not merely for George Bush and Gordon Brown, but also for Barack Obama and Hillary Clinton on the other side of the Atlantic and *Guardian* liberals on this side. This gives particular importance to Tariq Ali's piece "Afghanistan: Mirage of Good War" in the most recent *New Left Review* to arrive (the March-April issue). The article is online at www.newleftreview.org/?page=article&view=2713

In the same issue is a piece by Robin Blackburn on the absurdities and obscenities of finance that culminated in the sub-prime mortage crisis: www.newleftreview.org/?page=article&view=2715

A wider view of what is happening to world capitalism is provided by an interesting article by Minqi Li in April's *Monthly Review*. It looks at the development of the world system over the past eight years, bringing together rates of growth, the imbalances in the relation between the US and China, the potential impact if oil production has passed its peak, and climate change.

It ends by suggesting that the world-systems theorist Immanuel Wallerstein is right to predict "the end of neoliberalism" as a way of managing capitalism. "In the coming years", it predicts, "we are likely to witness a major realignment of global political and economic forces with an upsurge in the global class struggle over the direction of the global social transformation... We probably will observe a return to the dominance of Keynesian or state capitalist policies and institutions throughout the world. However, too much damage has been done." It is an interesting, accessible and thought provoking analysis. www.monthlyreview.org/080401li.php

Another stimulating analysis, of "Accumulation and Crisis in the Contemporary US Economy" is provided by David M Kotz in the spring issue of the *Review of Radical Political Economy*. This is available here: http://people.umass.edu/dmkotz/Contradictions_07_05.pdf

The rapid development of capitalism's other crisis, the food crisis, is forcing Marxists to go through the same learning curve over agriculture as over finance last autumn. Fred Magdoff provides some insights in May's *Monthly Review*: www.monthlyreview.org/080501magdoff.php

This quarter's antidote to miserabilism must come from the strike at the American Axle & Manufacturing plants in Michigan and New York, which supply General Motors (GM). It has stopped production at GM plants employing 40,000 workers and cost GM some $2 billion (yes, $2 billion—that's not a misprint). There is an excellent account of the strike in the May-June issue of *Against the Current*, including a description of how the workers called the management's bluff on its threat to move production to Mexico, knowing that the costs of transporting the axles across the US would outweigh any gains the firm made by paying lower wages on the other side of the border. Read it online: www.solidarity-us.org/node/1472

Two rather different pieces in *Historial Materialism*'s first issue for the year are of interest. Two German Marxists, who share many of the analyses of this journal, Oliver Nachtwey and Tobias Ten Brink, excavate a nearly forgotten but important debate on the role of the world market and the state that took place in the 1970s. And Chris Harman uses a review article on two books on the transition from feudalism to capitalism to develop further some of the arguments over this.

Finally, Tom Behan, who has written on Italian history and politics for *International Socialism*, has a piece entitled "Gillo Pontecorvo: Partisan Film-Maker" in the current issue of *Film International*. It deals with Pontecorvo's experiences as an anti-fascist partisan in the Second World War, and how this fed into his films.

CH and *JC*

What's Happening?? The Truth About Work… & the Myth of "Work-Life Balance"

In 1998 a trade union activists provided contributions to a much appreciated pamphlet, "What's Happening?? The Truth About Work…& the Myth of Partnership".

WHAT'S HAPPENING ??

The Truth About Work...
& The Myth of "Work-Life Balance"

A TU Publications Pamphlet

It focused on strategies and tactics that worked—both in winning victories against management and getting members involved.

Now, ten years later, this new pamphlet gives answers to that frequently ignored question—exactly what goes on at work?

This time, even more contributions—from workers on the buses, in building, telecommunications and more—show us, once again, that, whatever the area of work, management objectives and working pressures are eerily similar.

With an editorial linking these common themes together, What's Happening?? continues to be essential reading for anyone wanting to make a difference who is faced with the problems of falling pay, intensifying workloads and management bullying.

Order copies of What's Happening?? from TU Publications, PO Box 58262, London N1 1ET. Cheques to TU Publications.

Price per copy:
£5 for five or fewer ● £3.50 for more than five copies ● £3 per for more than ten ● £2.50 for more than 20 ● £2 for more than 30.

Back issues

The following issues of *International Socialism* are still available. Each costs £5 plus postage and packaging. To order copies phone +44 (0) 20 7819 1177 or e-mail isj@swp.org.uk

To subscribe, or for more details, visit our website: www.isj.org.uk

International Socialism 118

Spring 2008

Special collection on 1968: Matt Perry on France, Brian Kelly on Martin Luther King and Ian Birchall on Tony Cliff | Egypt's strike wave | China's growth pains | Capitalism goes from crunch to crisis

International Socialism 117

Winter 2008

Costas Lapvitsas interviewed on the credit crunch | Chris Harman on theorising neoliberalism | Iain Ferguson on capitalism and happiness | The African working class and the planet of slums

International Socialism 116

Autumn 2007

The Dutch Revolt | Corporations and climate change | Neil Faulkner or Gordon Childe and Marxist archaeology | Mike Wayne on Realism and film | Putin's Russia | Kevin Murphy on 1917

International Socialism 115

Summer 2007

Robin Blackburn interviewed on the end of slavery | The anti-capitalist movement | Kim Moody on the US working class | Chris Harman on the falling rate of profit | John Molyneux on a century of *Les Demoiselles*

International Socialism 114

Spring 2007

Special collection on Gramsci: the Turin years, hegemony and revolutionary strategy, the *Prison Notebooks* and international relations | LGBT politics | The Big Brother phenomenon

International Socialism 113

Winter 2007

In the name of decency: the contortions of the pro-war left | Britain in focus: Martin Smith on the shape of the working class, Charlie Kimber on Ken Livingstone's reformism, Hassan Mahamdallie on Muslim working class struggles | Daniel Bensaid on strategy

International Socialism 112

Autumn 2006

Hizbollah and the war Israel lost | Choi Il-bung and Kim Ha-yong on South Korea | Collection on 1956: Mike Haynes on the Hungarian Revolution, Anne Alexander on Suez, Paul Blackledge on the New Left, Stan Newens's memories of a seminal year | C L R James: the revolutionary as artist

International Socialism 111

Summer 2006

France's extraordinary movement, Italy's centre-left government, Germany's strategic debate | Megan Trudell on the hidden history of US radicalism | Chris Harman and Robert Brenner debate the origins of capitalism | Andy Durgan on the Spanish Civil War